Bayesian Networks

Bayesian Networks

An Introduction

Timo Koski

Institutionen för Matematik,
Kungliga Tekniska Högskolan, Stockholm, Sweden

John M. Noble

Matematiska Institutionen,
Linköpings Tekniska Högskola,
Linköpings universitet, Linköping, Sweden

A John Wiley and Sons, Ltd., Publication

This edition first published 2009
© 2009, John Wiley & Sons, Ltd

Registered office
John Wiley & Sons Ltd, The Atrium, Southern Gate, Chichester, West Sussex, PO19 8SQ, United Kingdom

For details of our global editorial offices, for customer services and for information about how to apply for permission to reuse the copyright material in this book please see our website at www.wiley.com.

Library of Congress Cataloging-in-Publication Data

Koski, Timo.
  Bayesian networks : an introduction / Timo Koski, John M. Noble.
      p. cm. – (Wiley series in probability and statistics)
  Includes bibliographical references and index.
  ISBN 978-0-470-74304-1 (cloth)
  1.   Bayesian statistical decision theory. 2.   Neural networks (Computer science)   I. Noble, John M. II. Title.
  QA279.5.K68 2009
  519.5'42 – dc22

                                                                                                    2009031404

A catalogue record for this book is available from the British Library.

ISBN: 978-0-470-74304-1

TypeSet in 10/12pt Times by Laserwords Private Limited, Chennai, India.
Printed and bound in Great Britain by TJ International Ltd, Padstow, Cornwall.

Contents

Preface

This book evolved from courses developed at Linköping Institute of Technology and KTH, given by the authors, starting with a graduate course given by Timo Koski in 2002, who was the Professor of Mathematical Statistics at LiTH at the time and subsequently developed by both authors. The book has been aimed at senior undergraduate, masters and beginning Ph.D. students in computer engineering. The students are expected to have a first course in probability and statistics, a first course in discrete mathematics and a first course in algorithmics. The book provides an introduction to the theory of graphical models.

A substantial list of references has been provided, which include the key works for the reader who wants to advance further in the topic.

We have benefited over the years from discussions on Bayesian networks and Bayesian statistics with Elja Arjas, Stefan Arnborg, and Jukka Corander. We would like to thank colleagues from KTH, Jockum Aniansson, Gunnar Englund, Lars Holst and Bo Wahlberg for participating in (or suffering through) a series of lectures during the third academic quarter of 2007/2008 based on a preliminary version of the text and suggesting improvements as well as raising issues that needed clarification. We would also like to thank Mikael Skoglund for including the course in the ACCESS graduate school program at KTH. We thank doctoral and undergraduate students Luca Furrer, Maksym Girnyk, Ali Hamdi, Majid N. Khormuji, Mårten Marcus and Emil Rehnberg for pointing out several errors, misprints and bad formulations in the text and in the exercises. We thank Anna Talarczyk for invaluable help with the figures. All remaining errors and deficiencies are, of course, wholly our responsibility.

1

Graphical models and probabilistic reasoning

1.1 Introduction

This text considers the subject of *graphical models*, which is an interaction between probability theory and graph theory. The topic provides a natural tool for dealing with a large class of problems containing uncertainty and complexity. These features occur throughout applied mathematics and engineering and therefore the material has diverse applications in the engineering sciences. A complex model is built by combining simpler parts, an idea known as *modularity*. The *uncertainty* in the system is modelled using probability theory; the graph helps to indicate independence structures that enable the probability distribution to be decomposed into smaller pieces.

Bayesian networks represent joint probability models among given variables. Each variable is represented by a node in a graph. The direct dependencies between the variables are represented by *directed edges* between the corresponding nodes and the conditional probabilities for each variable (that is the probabilities conditioned on the various possible combinations of values for the immediate predecessors in the network) are stored in potentials (or tables) attached to the dependent nodes. Information about the observed value of a variable is propagated through the network to *update the probability distributions* over other variables that are not observed directly. Using *Bayes' rule*, these influences may also be identified in a 'backwards' direction, from dependent variables to their predecessors.

The *Bayesian* approach to uncertainty ensures that the system as a whole remains consistent and provides a way to apply the model to data. Graph theory helps to illustrate

Bayesian Networks: An Introduction   T. Koski, J. Noble

and utilize independence structures within interacting sets of variables, hence facilitating the design of efficient algorithms.

In many situations, the directed edges between variables in a Bayesian network can have a simple and natural interpretation as graphical representations of *causal* relationships. This occurs when a graph is used to model a situation where the values of the immediate *predecessors* of a variable in a network are to be interpreted as the immediate *causes* of the values taken by that variable. This representation of causal relationships is probabilistic; the relation between the value taken by a variable and the values taken by its predecessors is specified by a conditional probability distribution. When a graph structure is given and the modelling assumptions permit a causal interpretation, then the estimates of the conditional probability tables obtained from data may be used to infer a system of causation from a set of conditional probability distributions. A Bayesian network is essentially a directed acyclic graph, together with the associated conditional probability distributions. When a Bayesian network represents a causal structure between the variables, it may be used to assess the effects of an intervention, where the manipulation of a cause will influence the effect.

The ability to infer causal relationships forms the basis for learning and acting in an intelligent manner in the external world. Statistical and probabilistic techniques may be used to assess direct associations between variables; some additional common sense and modelling assumptions, when these are appropriate, enable these direct associations to be understood as direct *causal* relations. It is the knowledge of causal relations, rather than simply statistical associations, that gives a sense of genuine understanding, together with a sense of potential control resulting from the ability to predict the consequences of actions that have not been performed, as J. Pearl writes in [1].

K. Pearson, an early, pre-eminent statistician, argued that the only proper goal of scientific investigation was to provide descriptions of experience in a mathematical form (see [2] written by Pearson in 1892); for example, a coefficient of correlation. Any effort to advance beyond a description of associations and to deduce *causal* relations meant, according to this view, to evoke hidden or metaphysical ideas such as causes; he did not consider such modelling assumptions to be scientific.

R.A. Fisher, possibly the most influential statistician, considered that causation could be inferred from experimental data only when controlled, or randomized experiments were employed. A majority of statistical studies follow the approach of Fisher and only infer 'correlation' or 'association' unless randomized experimental trials have been performed.

It is not within the scope of this treatment of Bayesian networks to review the sophisticated attempts at characterizing causality and the ensuing controversies (starting with David Hume [3]) amongst scholars; the reader is referred to the enlightening treatment by J. Williamson [4]. This text attempts to take what seems to be a 'common sense' point of view. The human mind is capable of detecting and approving causes of events in an intuitive manner. For example, the nineteenth century physician Ignaz Semmelweis in Vienna investigated, without knowing about germs, the causes of child bed fever. He instituted a policy that doctors should use a solution of chlorinated lime to wash their hands between autopsy work and the examination of patients, with the effect that the mortality rate at the maternity wards of hospitals dropped substantially. Such reasoning about causes and effects is not as straightforward for computers and it is not clear that it is valid in terms of philosophical analysis.

Causality statements are often expressed in terms of large and often complicated objects or populations: for example, 'smoking causes lung cancer'. Causal connections at finer levels of detail will have a different context: for example, the processes at the cell level that cause lung cancer. Causal connections are also contingent on many other conditions and causal laws than those explicitly under consideration. This introduces a level of uncertainty and the causal connections are therefore probabilistic.

Correlations or statistical associations between two variables often imply causation, even if there is not a direct causal relation between the two variables in question; one need not be the cause of the other. A correlation may indicate the presence of hidden variables, that are common causes for both the observed variables, so that the two observed variables are statistically associated.

When there is a possibility that there may be such unknown hidden variables, it is necessary to separate the 'cause' from the extraneous factors that may influence it in order to conclude that there is a causal relationship and a randomized (or controlled) experiment achieves this. For a randomized experiment, the groups for each level of treatment, and a control group to which no treatment is applied, are chosen at random so that the allocation of members to treatment groups is not affected by any hidden variables. Unfortunately, there are situations where it may be unethical to carry out a randomized experiment. For example, to prove that smoking causes lung cancer, it is perhaps inappropriate to force a non-smoker to start smoking.

In the example of smoking and lung cancer, the model is unknown and has to be inferred from data. The statistical analyst would like to establish whether smoking causes lung cancer, or whether there are additional hidden variables that are causes for both variables. Note that common sense plays a role here; the possibility that lung cancer may cause smoking is not considered. In terms of graphical models, where the direction of the pointed arrow indicates cause to effect, the analyst wants to determine which of the models in Figure 1.1 are appropriate.

When he carries out a controlled experiment, randomly assigning people to 'smokers' and 'non-smokers' respectively, the association between the hidden variables and smoking for this experiment are broken and the causal diagram is therefore given by Figure 1.2.

By carrying out a 'controlled experiment', an intervention is made whereby the causal path from the hidden variable is removed, thus ensuring that only the causal connection of interest is responsible for an observed association.

In many situations, a controlled experiment seems the only satisfactory way to demonstrate conclusively that an association between variables is due to a causal link. Without a controlled experiment, there may remain some doubt. For example, levels of smoking dropped when the first announcements were made that smoking caused lung cancer

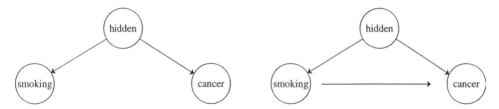

Figure 1.1   Smoking and lung cancer.

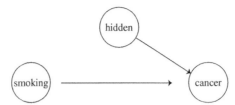

Figure 1.2   Controlled experiment: the control has removed the hidden causes of smoking from consideration.

and levels of lung cancer also dropped substantially. A controlled experiment would have demonstrated conclusively that the drop in lung cancer was not due to other environmental factors, such as a decline in heavy polluting industry that occurred at the same time.

Bayesian networks provide a straightforward mathematical language to express relations between variables in a clear way. In many engineering examples, the variables that should be present in the model are well defined. From an appropriate model that contains the hidden (or non-observable) variables and the observable variables, and where it is clear which variables may be intervened on, it will be possible to verify whether certain 'identifiability' conditions hold and hence to conclude whether or not there is a causal relation from the data, without a controlled experiment.

Many of the classical probabilistic systems studied in fields such as systems engineering, information theory, pattern recognition and statistical mechanics are problems that may be expressed as graphical models. Hidden Markov models may be considered as graphical models. Engineers are also accustomed to using circuit diagrams, signal flow graphs, and trellises, which may be treated using the framework of graphical models.

Examples of applied fields where Bayesian networks have recently been found to provide a natural mathematical framework are reliability engineering [5], software testing [6], cellular networks [7], and intrusion detection in computer systems [8]. In 1996, William H. Gates III, a co-founder of Microsoft, stated that expertise in Bayesian networks had enhanced the competitive advantage of Microsoft. Bayesian networks are used in software for electrical, financial and medical engineering, artificial intelligence and many other applications.

1.2   Axioms of probability and basic notations

The basic set notations that will be used will now be established.

Definition 1.1 (Notations) The following notations will be used throughout:

- The *universal set* will be denoted by Ω. This is the *context* of the experiment. In Bayesian analysis, the *unknown parameters* are considered to be random. Therefore, the context Ω consists of the set of all possible outcomes of an experiment, *together with all possible values of the unknown parameters.*

- The notation $\mathcal{X}$ will be used to denote the sample space; the set of all possible outcomes of the experiment.

- $\tilde{\Theta}$ will be used to denote the parameter space, so that $\Omega = \mathcal{X} \times \tilde{\Theta}$.

The following notations will be used when considering sets:

- If A and B are two sets, then $A \cup B$ or $A \vee B$ denotes their union. If $A_1, \ldots, A_n$ are a finite collection of sets, then $\cup_{j=1}^n A_j$ or $\vee_{j=1}^n A_j$ denotes their union. Also, $A_1 \cup \ldots \cup A_n$ or $A_1 \vee \ldots \vee A_n$ may be used to denote their union.

- If A and B are two sets, then $A \cap B$ or AB or $A \wedge B$ may be used to denote their intersection. If $A_1, \ldots, A_n$ are a finite collection of sets, then $A_1 \ldots A_n$, $A_1 \cap \ldots \cap A_n$, $\cap_{j=1}^n A_j$ or $A_1 \wedge \ldots \wedge A_n$ all denote their intersection.

- $A \subset B$ denotes that A is a strict subset of B. $A \subseteq B$ denotes that A is a subset of B, possibly equal to B.

- The empty set will be denoted by ϕ.

- A^c denotes the complement of A; namely, $\Omega \backslash A$, where Ω denotes the universal set. The symbol $\backslash$ denotes exclusion.

- Together with the *universal set* Ω, an *event space* $\mathcal{F}$ is required. This is a collection of subsets of Ω. The *event space* $\mathcal{F}$ is an *algebra of constructable sets*. That is, $\mathcal{F}$ satisfies the following:

  1. $\phi \in \mathcal{F}$ and $\Omega \in \mathcal{F}$.

  2. Each $A \in \mathcal{F}$ may be constructed. That is, for each $A \in \mathcal{F}$, let $i_A : \Omega \to \{0, 1\}$ denote the mapping such that $i_A(\omega) = 1$ if $\omega \in A$ and $i_A(\omega) = 0$ if $\omega \in A^c$. Then there is a procedure to determine the value of $i_A(\omega)$ for each $\omega \in \Omega$.[1]

  3. If for a finite collection of events $(A_j)_{j=1}^n$ each A_j satisfies $A_j \in \mathcal{F}$, then $\cup_{j=1}^n A_j \in \mathcal{F}$.

  4. If $A \in \mathcal{F}$, then $A^c \in \mathcal{F}$, where $A^c = \Omega \backslash A$ denotes the complement of A.

  5. Each $A \in \mathcal{F}$ satisfies $A \subseteq \Omega$.

For Ω, $\mathcal{F}$ satisfying the above conditions, a *probability distribution* p over $(\Omega, \mathcal{F})$ (or, in short, a probability distribution over Ω, when $\mathcal{F}$ is understood) is a function $p : \mathcal{F} \to [0, 1]$ satisfying the following version of the *Kolmogorov axioms*:

  1. $p(\phi) = 0$ and $p(\Omega) = 1$.

  2. If $(A_j)_{j=1}^n$ is a finite collection such that each $A_j \in \mathcal{F}$ and the events satisfy $A_j \cap A_k = \phi$ for all $j \neq k$, then

$$p\left(\cup_{j=1}^n A_j\right) = \sum_{j=1}^n p(A_j).$$

  3. $0 \leq p(A) \leq 1$ for all $A \in \mathcal{F}$.

---

[1] **A non-constructable set** The following example illustrates what is intended by the term *constructable* set. Let $A_{n,k} = (\frac{k}{2^n} - \frac{1}{2^{3(n+1)}}, \frac{k}{2^n} + \frac{1}{2^{3(n+1)}})$ and let $A = \cup_{n=1}^\infty \cup_{k=0}^{2^n+1} A_{n,k}$. Then A is not constructable in the sense given above. For any number $x \in A \cap [0, 1]$, there is a well defined algorithm that will show that it is in $A \cap [0, 1]$ within a finite number of steps. Consider a number $x \in [0, 1]$ and take its dyadic expansion. Then, for each $A_{n,k}$ it is clear whether or not $x \in A_{n,k}$. Therefore, if $x \in A$, this may be determined within a finite number of steps. But if $x \in A^c$, there is no algorithm to determine this within a finite number of steps.

This is a reduced version of the Kolmogorov axioms. The Kolmogorov axioms require countable additivity rather than finite additivity. They invoke axiomatic set theory and do not therefore require the constructive hypothesis. For the Kolmogorov axioms, the event space $\mathcal{F}$ is taken to be a *sigma*-algebra and the second axiom requires *countable* additivity.

- Let $(\theta, x) \in \Omega = \tilde{\Theta} \times \mathcal{X}$. For each $A \in \mathcal{F}$, let $A_{\Theta} = \{\theta \in \tilde{\Theta} | (\theta, x) \in A\}$ and let $A_{\mathcal{X}} = \{\mathbf{x} \in \mathcal{X} | (\theta, x) \in A\}$. Then $\mathcal{F}_{\Theta}$ will be used to denote the algebra $\{A_{\Theta} | A \in \mathcal{F}\}$ and $\mathcal{F}_{\mathcal{X}} = \{A_{\mathcal{X}} | A \in \mathcal{F}\}$.

Definition 1.2 (Probability Distribution over $\mathcal{X}$) *If $\mathcal{X}$ contains a finite number of elements, then $\mathcal{F}_{\mathcal{X}}$ contains a finite number of elements. In this setting, a probability distribution over $\mathcal{X}$ satisfies:*

- $p(\{x\}) \geq 0$ *for all $x \in \mathcal{X}$,*

- *For any $A \in \mathcal{F}_{\mathcal{X}}$, $\sum_{x \in A} p(\{x\}) = p(A)$.*

- *In particular, $\sum_{x \in \mathcal{X}} p(\{x\}) = 1$*

Definition 1.3 (Notation) *Let $\mathcal{X}$ be a finite state space and let A_X denote the set of all subsets of $\mathcal{X}$ (including ϕ the empty set and $\mathcal{X}$). For probability distribution $p : A_X \rightarrow [0, 1]$ defined above (Definition 1.2), p will also be used to denote the function $p : \mathcal{X} \rightarrow [0, 1]$ such that $p(x) = p(\{x\})$. The meaning of p will be clear from the context.*

The definitions and notations will now be established for random variables.

Definition 1.4 (Random Variables and Random Vectors) *Discrete random variables, continuous random variables and random vectors satisfy the following properties:*

- For this text, a *discrete random variable* Y is a function $Y : \Omega \rightarrow C$ where C is a countable space, that satisfies the following conditions: for each $x \in C$, $\{\omega | Y(\omega) = x\} \in \mathcal{F}$, and there is a function $p_Y : C \rightarrow \mathbf{R}_+$, known as the probability function of Y, such that for each $x \in C$,

$$p_Y(x) = p(\{\omega | Y(\omega) = x\}).$$

This is said to be the 'probability that Y is *instantiated* at x'. Therefore, for any subset $C \subseteq \mathcal{C}$, p_Y satisfies

$$\sum_{x \in C} p_Y(x) = p(\{\omega | Y(\omega) \in C\}).$$

In particular, taking $C = \mathcal{C}$,

$$\sum_{x \in \mathcal{C}} p_Y(x) = 1.$$

- A *continuous random variable* Ξ is defined as a function $\Xi : \Omega \to \mathbf{R}$ such that for any set $A \subseteq \mathbf{R}$ such that the function $\mathbf{1}_{A \cap [-N,N]}$ is Riemann integrable for all $N \in \mathbf{Z}_+$, the set $\{\omega | \Xi(\omega) \in A\} \in \mathcal{F}$, and for which there is a function π_Ξ, known as the probability density function, or simply density function, such that for any $A \subset \mathbf{R}$ with Riemann integrable indicator function,

$$p(\{\omega | \Xi(\omega) \in A\}) = \int_A \pi_\Xi(x) dx.$$

In particular,

$$p(\{\omega | \Xi(\omega) \in \mathbf{R}\}) = p(\Omega) = \int_{\mathbf{R}} \pi_\Xi(x) dx = 1.$$

- A *random vector* $\underline{Y}$ is a vector of random variables. It will be taken as a *row* vector if it represents different characteristics of a single observation; it will be taken as a *column* vector if it represents a collection of independent identically distributed random variables.

  This convention is motivated by the way that data is presented in a data matrix. Each column of a data matrix represents a different attribute, each row represents a different observation.

  A random row vector $\underline{Y}$ is a collection of random variables that satisfies the following requirements: suppose $\underline{Y} = (Y_1, \ldots, Y_m)$ and for each $j = 1, \ldots, m$, Y_j is a discrete random variable that takes values in a countable space C_j and let $C = C_1 \times \ldots \times C_m$, then $\underline{Y}$ is a random vector if for each $(y_1, \ldots, y_m) \in C$, $\{\omega | (Y_1(\omega), \ldots, Y_m(\omega))\} \in \mathcal{F}$, and there is a joint probability function $p_{Y_1, \ldots, Y_m} : C \to [0, 1]$ such that for any set $C \subseteq \mathcal{C}$,

$$\sum_{(y_1, \ldots, y_m) \in C} p_{Y_1, \ldots, Y_m}(y_1, \ldots, y_m) = p(\{\omega | (Y_1(\omega), \ldots, Y_m(\omega)) \in C\}).$$

In particular,

$$\sum_{(y_1, \ldots, y_m) \in C} p_{Y_1, \ldots, Y_m}(y_1, \ldots, y_m) = 1.$$

If $\underline{\Xi} := (\Xi_1, \ldots, \Xi_n)$ is a collection of n random variables where for each $j = 1, \ldots, n$ Ξ_j is a continuous random variable, then $\underline{\Xi}$ is a random vector if for each set $A \subset \mathbf{R}^n$ such that $\mathbf{1}_{A \cap [N,N]^n}$ is Riemann integrable for each $N \in \mathbf{Z}_+$,

$$\{\omega | (\Xi_1(\omega), \ldots, \Xi_n(\omega)) \in A\} \in \mathcal{F},$$

and there is a Riemann integrable function $\pi_{\Xi_1, \ldots, \Xi_n} : \mathbf{R}^n \to \mathbf{R}_+$, where $\mathbf{R}_+$ denotes the non-negative real numbers such that for each set $A \subset \mathbf{R}^n$ such that $\mathbf{1}_{A \cap [N,N]^n}$ is Riemann integrable for each $N \in \mathbf{Z}_+$,

$$p(\{\omega | (\Xi_1(\omega), \ldots, \Xi_n(\omega)) \in A\}) = \int_A \pi_{\Xi_1, \ldots, \Xi_n}(x_1, \ldots, x_n) dx_1 \ldots dx_n.$$

In particular,

$$\int_{\mathbf{R}^n} \pi_{\Xi_1,\dots,\Xi_n}(x_1,\dots,x_n)dx_1\dots dx_n = 1.$$

A collection of random variables $\underline{Y}$ of length $m+n$, containing m discrete variables and n continuous variables, is a random vector if it satisfies the following: there is an ordering σ of $1,\dots,m+n$ such that $(Y_{\sigma(1)},\dots,Y_{\sigma(m)})$ is a discrete random vector and $(Y_{\sigma(m+1)},\dots,Y_{\sigma(m+n)})$ is a continuous random vector. Furthermore, let $\tilde{C}$ denote the state space for $(Y_{\sigma(1)},\dots,Y_{\sigma(m)})$, then for each $(y_1,\dots,y_m)\in\tilde{C}$, there is a Riemann integrable function $\pi_{Y_{\sigma(m+1)},\dots,Y_{\sigma(m+n)}|y_1,\dots,y_m}^{(\sigma)}: \mathbf{R}^n\to\mathbf{R}_+$ such that for any set $A\in\mathbf{R}^n$ such that $\mathbf{1}_{A\cap[N,N]^n}$ is Riemann integrable for each $N\in\mathbf{Z}_+$,

$$p(\{\omega|(Y_{\sigma(1)},\dots,Y_{\sigma(m)}) = (y_1,\dots,y_m), (Y_{\sigma(m+1)},\dots,Y_{\sigma(m+n)})\in A\})$$

$$= p_{Y_{\sigma(1)},\dots,Y_{\sigma(m)}}(y_1,\dots,y_m)\int_A \pi_{Y_{\sigma(m+1)},\dots,Y_{\sigma(m+n)}|y_1,\dots,y_m}^{(\sigma)}(x_1,\dots,x_n)dx_1\dots dx_n.$$

and such that

$$\int_{\mathbf{R}^n} \pi_{Y_{\sigma(m+1)},\dots,Y_{\sigma(m+n)}|y_1,\dots,y_m}^{(\sigma)}(x_1,\dots,x_n)dx_1\dots dx_n = 1.$$

Definition 1.5 (Marginal Distribution) Let $X = (X_1,\dots,X_n)$ be a discrete random vector, with joint probability function $p_{X_1,\dots,X_n}$. The probability distribution for $(X_{j_1},\dots,X_{j_m})$, where $m\le n$ and $1\le j_1 < \dots < j_m \le n$ is known as the *marginal distribution*, and the marginal probability function is defined as

$$p_{X_{j_1},\dots,X_{j_m}}(x_1,\dots,x_m) = \sum_{(y_1,\dots,y_n)|(y_{j_1},\dots,y_{j_m})=(x_{j_1},\dots,x_{j_m})} p_{X_1,\dots,X_n}(y_1,\dots,y_n).$$

In particular, for two discrete variables X and Y taking values in the spaces $\mathcal{C}_X$ and $\mathcal{C}_Y$ respectively, with joint probability function $p_{X,Y}$, the marginal probability function for the random variable X is defined as

$$p_X(x) = \sum_{y\in\mathcal{C}_Y} p_{X,Y}(x,y)$$

and the marginal probability function for the random variable Y is defined as

$$p_Y(y) = \sum_{x\in\mathcal{C}_X} p_{X,Y}(x,y).$$

If $\Xi = (\Xi_1,\dots,\Xi_n)$ is a continuous random vector, with joint probability density function $\pi_{\Xi_1,\dots,\Xi_n}$, then the *marginal density function* for $\Xi_{j_1},\dots,\Xi_{j_m}$, where $\{j_1,\dots,j_m\}\subset\{1,\dots,n\}$ is defined as

$$\pi_{\Xi_{j_1},\dots,\Xi_{j_m}}(x_1,\dots,x_m) = \int_{\mathbf{R}^{n-m}} \pi_{\Xi_1,\dots,\Xi_n}(y_1,\dots,y_n|(y_{j_1},\dots,y_{j_m})$$

$$= (x_1,\dots,x_m)) \prod_{k\notin(j_1,\dots j_m)} dy_k. \qquad \square$$

Categorical random variables   In this text, the sample space $\mathcal{X}$ will contain a finite, or countably infinite, number of outcomes, while usually the parameter space $\tilde{\Theta} \subseteq \mathbf{R}^n$, where n is the number of parameters. Most of the discrete variables arising will be *categorical*, in the sense that the outcomes are classified in several categories. For example, suppose an urn contains 16 balls, of which four are white, six are green, five are blue and one is red. Pick a ball at random and let C denote the colour of the ball, then C is an example of a discrete random variable. The probability distribution of a categorical variable C is denoted by p_C. Here, for example, $p(\{C = \text{green}\}) = p_C(\text{green}) = \frac{6}{16} = \frac{3}{8}$;

$$p_C = \frac{\begin{array}{cccc} \text{white} & \text{green} & \text{blue} & \text{red} \end{array}}{\begin{array}{cccc} \frac{1}{4} & \frac{3}{8} & \frac{5}{16} & \frac{1}{16} \end{array}}$$

The set-up described above is adequate for this text, where only two types of random variables are considered; continuous (which, by definition, have a Riemann integrable density function) and discrete variables. Statements of uncertainty made within this framework are consistent and coherent. Any probability function p over $\mathcal{F}$ that is to provide a quantitative assessment of the probability of an event, which is also to be mathematically coherent over a constructive algebra of events, must satisfy the axioms listed above. Any set function over an algebra of sets that satisfies these axioms will provide a mathematically coherent measure of uncertainty.

1.3 The Bayes update of probability

The *prior* probability is the probability distribution p over $\mathcal{F}$ before any relevant data is obtained. The prior probability is related to background information, modelling assumptions, or simply a function introduced for mathematical convenience, which is intended to express a degree of vagueness. It could be written $p^{(K)}(.)$, where K designates what could be called the *context* or a *frame of knowledge* [9]. The notation $p(.|K)$ is often employed, although this is a little misleading since, formally, conditional probability is only defined in terms of an initial distribution. Here, the *initial* distribution is based on K. Since the notation is now standard, it will be employed in this text, although caution should be observed.

Consider a prior distribution, denoted by $p(.)$. Suppose the experimenter, or agent, has the information that $B \in \mathcal{F}$ holds and desires to update the probabilities based on this piece of information. This update involves introducing a new probability distribution p^* on $\mathcal{F}$. Suppose also that $p(B) > 0$. Since it is now known that B is a certainty, the update requires

$$p^*(B) = 1$$

so that $p^*(B^c) = 0$, where $B^c = \Omega \backslash B$ is the complement of B in Ω. The updated probability is constructed so that the *ratio* of probabilities for any $B_1 \subset B$ and $B_2 \subset B$ does not change. That is, for $B_i \subset B$, $i = 1, 2$,

$$\frac{p^*(B_1)}{p^*(B_2)} = \frac{p(B_1)}{p(B_2)}.$$

For arbitrary $A \in \mathcal{F}$, the axioms yield

$$p^*(A) = p^*(A \cap B) + p^*(A \cap B^c).$$

But since $p^*(B^c) = 0$, it follows that $p^*(A \cap B^c) \leq p^*(B^c) = 0$. In other words, it follows that for arbitrary $A \in \mathcal{F}$, since $p^*(B) = 1$,

$$p^*(A) = \frac{p^*(A \cap B) + p^*(A \cap B^c)}{p^*(B)} = \frac{p^*(A \cap B)}{p^*(B)} = \frac{p(A \cap B)}{p(B)}.$$

The transformation $p \rightarrow p^*$ is known as the *Bayes update of probability*. When the evidence is obtained is precisely that an event $B \in \mathcal{F}$ *within the algebra* has happened, Bayes' rule may be used to update the probability distribution. It is customary to use the notation $p(A|B)$ to denote $p^*(A)$. Hence

$$p(A|B) \overset{def}{=} p^*(A) = \frac{p(A \cap B)}{p(B)}. \tag{1.1}$$

This is called the conditional probability of A given B. This characterization of the conditional probability $p(A|B)$ follows [10]. Further discussion may be found in [11]. This is the update used when the evidence is precisely that an event $B \in \mathcal{F}$ has occurred. Different updates are required to incorporate knowledge that cannot be expressed in this way. This is discussed in [12].

From the definition of $p(A|B)$, the trivial but important identity

$$p(A|B)p(B) = p(A \cap B) \tag{1.2}$$

follows for any $A, B \in \mathcal{F}$.

The Bayes factor   Bayes' rule simply states that for any two events A and C,

$$p(A|C) = \frac{p(C|A)p(A)}{p(C)}.$$

If event C represents new evidence, and $p^*(.) = p(.|C)$ represents the updated probability distribution, then for any two events A and B, Bayes' rule yields:

$$\frac{p^*(A)}{p^*(B)} = \frac{p(C|A)}{p(C|B)} \frac{p(A)}{p(B)}.$$

The factor $\frac{p(C|A)}{p(C|B)}$ therefore updates the ratio $\frac{p(A)}{p(B)}$ to $\frac{p^*(A)}{p^*(B)}$. Note that

$$\frac{p(C|A)}{p(C|B)} = \frac{p^*(A)/p^*(B)}{p(A)/p(B)}.$$

This leads to the following definition, which will be used later.

Definition 1.6 *Let p and q denote two probability distributions over an algebra $\mathcal{A}$. For any two events $A, B \in \mathcal{A}$, the Bayes factor $F_{q,p}(A; B)$ is defined as*

$$F_{q,p}(A, B) := \frac{q(A)/q(B)}{p(A)/p(B)}. \tag{1.3}$$

Here p plays the role of a probability before updating and q plays the role of an updated probability. The Bayes factor indicates whether or not the new information has increased the *odds* of an event A relative to B.

Bayes' rule applied to random variables   Let X and Y be two discrete random variables. Let p_X, p_Y, $p_{X|Y}$ and $p_{Y|X}$ denote the probability mass functions of X, Y, X given Y and Y given X respectively. It follows directly from Bayes' rule that for all x, y

$$p_{X|Y}(x|y) = \frac{p_{Y|X}(y|x)p_X(x)}{p_Y(y)}. \tag{1.4}$$

If X and Y are continuous random variables with density functions π_X and π_Y, then the conditional probability density function of X, given $Y = y$ is

$$\pi_{X|Y}(x|y) = \frac{\pi_{Y|X}(y|x)\pi_X(x)}{\pi_Y(y)}.$$

If X is a discrete random variable and Θ is a continuous random variable with state space $\tilde{\Theta}$, where $p_{X|\Theta}(.|\theta)$ is the conditional probability function for X given $\Theta = \theta$ and Θ has density function π_Θ, then Bayes' rule in this context gives

$$\pi_{\Theta|X}(\theta|x) = \frac{p_{X|\Theta}(x|\theta)\pi_\Theta(\theta)}{\int_{\tilde{\Theta}} p_{X|\Theta}(x|\theta)\pi_\Theta(\theta)d\theta} = \frac{p_{X|\Theta}(x|\theta)\pi_\Theta(\theta)}{p_X(x)}. \tag{1.5}$$

1.4   Inductive learning

The task of *inductive learning* is, roughly stated, to find a *general* law, based of a *finite* number of particular examples. Without further information, a law established in this way cannot be a certainty, but necessarily has a level of uncertainty attached to it. The assessment uses the idea that future events similar to past events will cause similar outcomes, so that outcomes from past events may be used to predict outcomes from future events. There is a *subjective* component in the assessment of the uncertainty, which enters through the *prior* distribution. This is the assessment made before any particular examples are taken into consideration.

In any real situation within the engineering sciences, a mathematical model is only ever at best a *model* and can never present a *full* description of the situation that is being modelled. In many situations, the information is not presented in terms of absolute certainties to which deductive logic may be applied and 'inductive learning', as described above, is the only way to proceed.

For machine learning of situations arising in the engineering sciences, the machine learns inductively from past examples, from which it makes predictions of future behaviour and takes appropriate decisions; for example, deciding whether a paper mill is running abnormally and, if it is, shutting it down and locating the error.

In the following discussion, the second person singular pronoun, 'You', will be used to denote the person, or machine,[2] with well defined instructions, analysing some uncertain statements and making predictions based on this analysis.

---

[2] A more precise description of the machine is the hypothetical artificial intelligence robot 'Robbie' in [13].

The Bayes update rule is now applied to learning from experience. The update is based on the definition of the conditional probability $p(E|A)$ of an event E given an event A. The rule for calculating the conditional probability of A given E (or, given that E 'occurs') is known as *Bayes' rule* and is given by

$$p(A|E) = \frac{p(E|A)p(A)}{p(E)}. \tag{1.6}$$

Equation (1.6) follows immediately from the definitions introduced in Equations (1.1) and (1.2). Here E is a mnemonic notation for *evidence*.

The formula (1.6) was for a long time known more widely as 'inverse probability' rather than as 'Bayes' rule'. The quantity $p(A|E)$ does not always need to be computed using this formula; sometimes it is arrived at by other means. For example, the Jeffrey's update rule, as will be seen later, uses a particular set of conditional probabilities as *fundamental* and derives the other probabilities from it.

1.4.1   Bayes' rule

A more instructive form of Bayes' rule is obtained by considering a finite exhaustive set of mutually exclusive hypotheses $\{H_i\}_{i=1}^m$. The law of total probability gives for any E

$$p(E) = \sum_{i=1}^m p(H_i)p(E|H_i) \tag{1.7}$$

and Equation (1.6) yields for any H_i

$$p(H_i|E) = \frac{p(E|H_i)p(H_i)}{\sum_{i=1}^m p(H_i)p(E|H_i)}. \tag{1.8}$$

Here $p(H_i)$ is the initial or *prior* probability for a hypothesis, before any evidence E is obtained. The prior probability will have a subjective element, depending on background information and how You interpret this information. As discussed earlier, this background information is often denoted by the letter K. In computer science, this is often referred to as *domain knowledge* [14]. The prior is therefore often denoted $p(H_i|K)$. As discussed earlier, this is misleading, because no application of the formula given in Equation (1.1) has been made. The notation is simplified by dropping K.

The quantity in $p(H_i|E)$ in Equation (1.8) denotes how $p(H_i)$ is *updated* to a new probability, called the *posterior probability*, based on this new evidence. This update is the key concept of Bayesian learning from experience.

There is a basic question: why should probability calculus be used when performing the inductive logic in the presence of uncertainty described above? This is connected with the notion of *coherence*. The probability calculus outlined above ensures that the uncertainty statements 'fit together' or 'cohere' and it is the only way to ensure that mathematical statements of uncertainty are consistent. To ensure coherence, inductive logic should therefore be expressed through probability. Inference about uncertain events H_i from an observed event E will be mathematically coherent if and only if they are made by computing $p(H_i|E)$ in the way described above. All calculations, in order to achieve coherence, must be made within this probability calculus. Hence, to make coherent statements about different events, Your calculations for learning by experience have to satisfy Bayes' rule.

The introduction of an arbitrary prior distribution may appear, at first sight, to be ad hoc. The important point here is that all modelling contains some ad hoc element in the choice of the model. The strength of the Bayesian approach is that once the prior distribution is declared, this contains the modelling assumptions and the ad hoc element becomes transparent. The ad hoc element is always present and is stated clearly in the Bayesian approach.

When an event E occurs that belongs to the well defined algebra of events, over which there are well defined probabilities, the Bayes rule for updating the probability of an event A from $p(A)$ to $p(A|E)$ is applied. There are situations where the evidence E may be given in terms of observations that are less precise (that is, E is not an event that clearly belongs to the algebra), but nevertheless an update of the the probability function $p(\cdot)$ is required. Jeffrey's rule and Pearl's method of virtual evidence can be useful in these situations.

1.4.2   Jeffrey's rule

Suppose that the new evidence implies that You form an exhaustive set of r mutually exclusive hypotheses $(G_i)_{i=1}^r$ which, following the soft evidence have probabilities $p^*(G_i)$. The question is how to update the probability of any event $A \in \mathcal{F}$. Note that You cannot use Bayes' rule (Equation (1.6)), since the evidence has not been expressed in terms of a well defined event E for which the prior probability value is known. Jeffrey's rule may be applied to the situation where it may be assumed that for all events $A \in \mathcal{F}$, the probabilities $p(A|G_i)$ remain unchanged. It is only the assessment of the the mutually exclusive hypotheses $(G_i)_{i=1}^r$ that changes; no new information is given about the the relevance of $(G_i)_{i=1}^r$ to *other* events.

Definition 1.7 (Jeffrey's Update) *The Jeffrey's rule for computing the update of the probability for any event A, is given by*

$$p^*(A) = \sum_{i=1}^r p^*(G_i)p(A|G_i). \tag{1.9}$$

This is discussed in [15]. Jeffrey's rule provides a consistent probability function, such that $p^*(A|G_i) = p(A|G_i)$ for all i. Equation (1.9) is therefore an expression of the rule of total probability (Definition 1.7).   □

1.4.3   Pearl's method of virtual evidence

In the situation considered by Pearl, new evidence gives information on a set of mutually exclusive and exhaustive events $G_1, \ldots, G_n$, but is not specified as a set of new probabilities for these events. Instead, for each of the events $G_1, \ldots, G_n$, the ratio $\lambda_j = \frac{p(A|G_j)}{p(A|G_1)}$, for $j = 1, \ldots, n$ is given for an event A. That is, λ_j represents the likelihood ratio that the event A occurs given that G_j occurs, compared with G_1. Note that $\lambda_1 = 1$.

Definition 1.8 *Let p denote a probability distribution over a countable space $\mathcal{X}$ (Definition 1.2) and let $G_1, \ldots, G_n \in \mathcal{F}$ be a mutually exclusive (that is $G_i \cap G_j = \phi$ for all* *i ≠ j) and exhaustive (that is $\cup_{j=1}^{n} G_j = \Omega$) events, where $p(G_j) = p_j$. Set $\lambda_j = \frac{p(A|G_j)}{p(A|G_1)}$ for j = 1, . . . , n. Then, for each $x \in \mathcal{X}$, the* Pearl update $\tilde{p}$ *is defined as*

$$\tilde{p}(\{x\}) = p(\{x\}) \frac{\lambda_j}{\sum_{j=1}^{n} \lambda_j p_j} \qquad x \in G_j, \qquad j = 1, \ldots, n. \tag{1.10}$$

It is clear that this provides a well defined probability distribution.

Jeffrey's rule and Pearl's method for virtual evidence will be discussed further in Section 3.2. They are methods that, under some circumstances, enable evidence to be incorporated that does not fit directly into the framework of the chosen statistical model.

1.5 Interpretations of probability and Bayesian networks

Loosely speaking, 'classical statistics' proposes a probability distribution over the event space, where the probability distribution is a member of a parametric family, where the value of the parameters are unknown. The parameters are *estimated*, and the estimates are used to obtain an *approximation* to the 'true' probability, which is unknown. In *Bayesian* probability, the lack of knowledge of the parameter is expressed though a probability distribution over the parameter space, to feature Your personal assessment of the probability of where the parameter may lie. For this reason, Bayesian probability is also referred to as *personal* probability, or *epistemological* probability. Built into a Bayesian probability is Your a priori state of knowledge, understanding and assessment concerning the model and the source of data. So, loosely speaking, classic statistics yields an approximation to an objectively 'true' probability. The 'true' probability is fixed, but unknown, and the *estimate* of this probability may differ between researchers. Bayesian statistics yields an *exact* computation of a *subjective* probability. It is the *probability itself*, rather than the estimate of the probability, that may differ between researchers.

Bayesian networks are frequently implemented as information processing components of Expert Systems (in artificial intelligence) [14], where personal and epistemological probability provides a natural framework for the machine to learn from its experience. P. Cheeseman in [13] and [16] argues that personal probability as the calculus of plausible reasoning is the natural paradigm for artificial intelligence.

J. Williamson [4] and other authors distinguish between *subjective* and *objective* Bayesian probability. Consider two different learning agents, called (for convenience) Robbie$_\alpha$ and Robbie$_\beta$. These are two hardware copies of Robbie in [13], with some different internal representations (i.e. different ways of assessing prior information). A Bayesian probability is said to be *objective* if, with the same background information, the two agents will agree on the probabilities. Bayesian probabilities are *subjective* if the two different learning agents, Robbie$_\alpha$ and Robbie$_\beta$ may disagree about the probabilities, even though they share the same background knowledge, but without either of then being provably wrong.

It seems rational to require that *subjective* probabilities should be calibrated with the external world. This does not follow from requirements of coherence; it is rather a principle of inference that is imposed in any serious statistical study. For this, one often cites *Lewis's principal principle*, stated in [17]. This is the assertion that Your subjective probability for an event, conditional upon the knowledge of a *physical* probability of that event, should equal the physical probability. In terms of the formula (where ch denotes the physical chance),

$$p(A \mid ch\,(A) = x) = x.$$

There are several theoretical foundations for reconciling subjective opinion to objective probability. One of the more prominent was given by R.J. Aumann.[3] He proved the following fact. If two agents have same priors, and if their posteriors for an event A are *common knowledge* (i.e. Robbie$_\alpha$ and Robbie$_\beta$ know the posterior, and Robbie$_\alpha$ knows that Robbie$_\beta$ knows and that Robbie$_\beta$ knows that Robbie$_\alpha$ knows that Robbie$_\beta$ knows and so on, ad infinitum), then the two posterior probability masses will be identical [18].

This relies on the rational idea that if someone presents an opinion different from Yours, then this is an important piece of information which should induce You to revise Your opinion. The result by Aumann quoted above implies that there will be a process of revision, which will continue until objective probability (an equilibrium of consensus) is reached. It can be shown that this will happen in a finite number of steps.

In probabilistic theories of causation there is a set of variables V, over which there is a probability distribution p. The causal relationships between the variables in V are the object of study. In [19], the variables are indexed by a time parameter and causality is reduced to probability. The Bayesian network approach is different; in addition to the probability distribution p, the variables are nodes of a directed acyclic graph $\mathcal{G}$, where the edges represent direct causal relationships between the variables in V. This requires, as pointed out by D. Freedman and P. Humphreys [20], that we already know the causal structure obtained, for example, by exercise of common sense and knowledge of the variables V. Both p and $\mathcal{G}$ are required [21]. Additional assumptions are therefore required to infer the direction of cause to effect relationships between variables; there is no form of Bayesian coherence from which they may be inferred. The role of a directed graph is to represent information about dependence between variables V. In particular, the graph may be used to indicate what would happen to values of some variables under changes to other variables that are called interventions; namely, the variable is forced to take a particular value irrespective of the state of the rest of the system. The graph will also indicate how the various different distributions for subsets of V are consistently connected to each other to yield p.

1.6   Learning as inference about parameters

Consider a random row vector $\underline{X} = (X_1, \ldots, X_d)$, denoting d attributes. Suppose that n independent observations are made on $\underline{X}$ and the values obtained are recorded in a matrix $\mathbf{x}$, where x_{jk} denotes the jth recorded value of attribute k. The $n \times d$ matrix $\mathbf{x}$ may be regarded as an instantiation of the $n \times d$ random matrix $\mathbf{X}$, where each row of $\mathbf{X}$ is an independent copy of $\underline{X}$. Suppose that the *evidence* $\mathbf{x}$ is to be used to update the probability function for another collection of m variables, $(Y_1, \ldots, Y_m)$ from $p_{Y_1, \ldots, Y_m}(.)$ to $p_{Y_1, \ldots, Y_m | \mathbf{x}}(.|\mathbf{x})$.

A fundamental special case, discussed in [22], is that of computing the predictive probability for the next observation in the univariate setting. That is, $d = 1$. Here, the

---

[3] 2005 Laureate of the Sveriges Riksbank Prize in Economic Sciences in Memory of Alfred Nobel.

matrix $\mathbf{X}$ is an $n \times 1$ matrix and may therefore be considered as a column vector $\underline{X}_{(n)} = (X_1, \ldots, X_n)^t$, where X_j, $j = 1, \ldots, n$ are independent identically distributed. Here, $\mathbf{x} = \underline{x}_{(n)} = (x_1, \ldots, x_n)^t$, a vector of n observed values. The random vector $\underline{Y}$ is simply X_{n+1}. The problem is to compute the conditional probability distribution of Y given $\underline{X}_{(n)} = \underline{x}_{(n)}$. The connection between $\underline{x}_{(n)}$ and y is described by a probability $p_{Y|\underline{X}_{(n)}}(y|\underline{x}_{(n)})$ of Y given $\underline{X}_{(n)} = \underline{x}_{(n)}$. Let $\underline{X}_{(n+1)} = (X_1, \ldots, X_{n+1})^t$ and $\underline{x}_{(n+1)} = (x_1, \ldots x_{n+1})^t$. You compute

$$p_{X_{n+1}|\underline{X}_{(n)}}(x_{n+1}|\underline{x}_{(n)}) = \frac{p_{\underline{X}_{(n+1)}}(\underline{x}_{(n+1)})}{p_{\underline{X}_{(n)}}(\underline{x}_{(n)})}. \tag{1.11}$$

The progression from here requires a mathematical model containing *parameters* denoted θ such that, given θ, the random variables $X_1, X_2, \ldots$ are independent and identically distributed (i.i.d.). That is, with notation $(.|\theta)$ to denote the parameter fixed as θ, there is a decomposition

$$p_{\underline{X}_{(n)}}(\underline{x}_{(n)}|\theta) = \prod_1^n q_{X_i}(x_i|\theta).$$

The family of probability functions $q_{X_i}(.|\theta)$, $\theta \in \tilde{\Theta}$ (where $\tilde{\Theta}$ is the space of all permissible values of the unknown parameter θ) and the parameter θ need to be specified. Since the value of θ is unknown, the Bayesian approach is to consider a *probability distribution* over $\tilde{\Theta}$. Thus, θ may be regarded as a realisation of a random variable Θ. You need to specify a prior distribution over $\tilde{\Theta}$. This discussion confines itself to the case where Θ is considered to be a *continuous* random variable and hence the prior distribution is described by a probability density function $\pi_\Theta : \tilde{\Theta} \to \mathbf{R}_+$. The prior predictive distribution may then be written as

$$p_{\underline{X}_{(n)}}(\underline{x}_{(n)}) = \int_{\tilde{\Theta}} \prod_1^n q_{X_i}(x_i|\theta)\pi(\theta)d\theta. \tag{1.12}$$

Definition 1.9 (Prior Predictive Probability Distribution) *The prior distribution $p_{\underline{X}_{(n)}}$ for the collection of random variables $\underline{X}_{(n)}$ (for which $\underline{x}_{(n)}$ is an observation) is known as the* Prior Predictive Probability Distribution.

B. De Finetti showed in [23] that if the x_js are infinitely exchangeable (and lie in a reasonable space), then the structure for $p_{\underline{X}_{(n)}}(.)$ given by Equation (1.12) is the only possible one.

Inserting (1.12) in the right hand side of (1.11) yields

$$p_{X_{n+1}}(x_{n+1}|\underline{x}_{(n)}) = \frac{\int_{\tilde{\Theta}} \prod_1^{n+1} q_{X_i}(x_i|\theta)\pi_\Theta(\theta)d\theta}{\int_{\tilde{\Theta}} \prod_1^n q_{X_i}(x_i|\theta)\pi_\Theta(\theta)d\theta}$$

$$= \int_{\tilde{\Theta}} q_{X_{n+1}}(x_{n+1}|\theta) \frac{\prod_1^n q_{X_i}(x_i|\theta)\pi_\Theta(\theta)}{\int_{\tilde{\Theta}} \prod_1^n q_{X_i}(x_i|\theta)\pi_\Theta(\theta)d\theta} d\theta.$$

The conditional probability density of Θ given $\underline{X}_{(n)} = \underline{x}_{(n)}$ may be obtained by Bayes' rule:

$$\pi_{\Theta|\underline{X}_{(n)}}\left(\theta \mid \underline{x}_{(n)}\right) = \frac{\prod_1^n q_{X_i}(x_i|\theta)\pi_{\Theta}(\theta)}{\int_{\tilde{\Theta}} \prod_1^n q_{X_i}(x_i|\theta)\pi_{\Theta}(\theta)d\theta}. \tag{1.13}$$

It follows directly that

$$p_{X_{n+1}}(x_{n+1}|\underline{x}_{(n)}) = \int_{\tilde{\Theta}} q_{X_{n+1}}(x_{n+1}|\theta)\pi_{\Theta|\underline{X}_{(n)}}\left(\theta \mid \underline{x}_{(n)}\right)d\theta. \tag{1.14}$$

In Section 1.9, explicit examples of evaluations of Equation (1.14) are given.

The probability density function π_{Θ}, placed over $\tilde{\Theta}$ before any data is observed is known as the prior density; the probability density function $\pi_{\Theta|\underline{X}_{(n)}}$ defined by Equation (1.13) is known as the *posterior* probability density. Equation (1.14) shows how Bayesian learning about X_{n+1} is based on learning about θ from $\underline{x}_{(n)}$. *Bayesian statistical inference* is the term used to denote Bayesian learning of the posterior distribution of a set of parameters.

The meaning of causality for K. Pearson, see Chapter 4 in [2], seems to be expressible by Equation (1.14), as he writes 'that a certain sequence has occurred and recurred in the past is a matter of experience to which we give expression in the concept of causation, that it will recur in the future is a matter of belief to which we give expression in the concept of probability.'

1.7 Bayesian statistical inference

The aim of learning is to predict the nature of future data based on past experience [22]. One constructs a probabilistic model for a situation where the model contains unknown parameters. The parameters are only a *mechanism* to help estimate *future* behaviour; they are not an end in themselves.

As stated, the 'classical' approach regards a parameter as fixed. It is unknown and has to be estimated, but it is not considered to be a random variable. One therefore computes approximations to the unknown parameters, and uses these to compute an approximation to the probability density. The parameter is considered to be fixed and unknown, because there is usually a basic assumption that in ideal circumstances, the experiment could be repeated infinitely often and the estimating procedure would return a precise value for the parameter. That is, if one increases the number of replications indefinitely, the *estimate* of the unknown parameter converges, with probability one, to the true value. This is known as the 'frequentist' interpretation of probability.

Basic to the 'frequentist' interpretation is the assumption that an experiment may be repeated, in an identical manner, an indefinite number of times. With the classical approach, the sample space Ω and the event space $\mathcal{A}$ of subsets of Ω have to be defined in advance. This makes incorporation of 'soft evidence' or 'virtual evidence,' that will be considered later in the text, harder; these are situations where the information obtained cannot be expressed as one of the well defined events of the event space. Then, the probability distribution is interpreted as follows: for each $A \in \mathcal{A}$, $p(A)$ is interpreted as the limit, *from observed data* that would be obtained if the experiment could be repeated independently, under identical circumstance, arbitrarily often. That is,

$$p(A) = \lim_{n \to +\infty} \frac{N(n, A)}{n},$$

where $N(n, A)$ denotes the number of times event A has been observed from n replications of the experiment.

This interpretation is intuitively appealing, but there is room for caution, since the infinite independent replications are *imagined*, and therefore the convergence of relative frequencies to a limit is hypothetical; the imagined infinite sequence of replications under *exactly* the same conditions is a textbook construction and abstraction. In concrete terms, it is supposed that there are many sources, each with large numbers of data so that the 'empirical' distribution can approximate the limit with arbitrary precision. Despite the hypothetical element in the formulation, the 'frequentist' interpretation of probability follows basic human common sense; the probability distribution is interpreted as the long run average. The 'long run average' interpretation assumes prior knowledge; when an agent like Robbie is to compute probability for its actions, it cannot be instructed to wait for the result in an infinite outcome of experiments and, indeed, it cannot run in 'real time' if it is expected to wait for a large number of outcomes.

Once the existence of a probability measure p over $(\Omega, \mathcal{A})$ has been established, which may be interpreted in the sense of 'long run averages', it is then a matter of computation to prove that the parameter estimates $\hat{\theta}_n$ based on n observations converge with probability 1, provided a sensible estimating procedure is used, to a parameter value θ.

As discussed earlier, the *Bayesian* approach takes the view that since the parameter is unknown, it is a random variable as far as You are concerned. A probability distribution, known as the *prior distribution*, is put over the *parameter space*, based on a prior assessment of where the parameter may lie. One then carries out the experiment and, using the data available, which is necessarily a *finite* number of data, one uses the Bayes rule to compute the *posterior distribution* in Equation (1.13), which is the updated probability distribution over the parameter space.

The posterior distribution over the parameter space is then used to compute the probability distribution for future events, based on past experience. Unlike the classical approach, this is an exact distribution, but it contains a subjective element. The subjective element is described by the prior distribution.

In Bayesian statistics, the computation of the posterior distribution usually requires numerical methods, and Markov chain Monte Carlo methods seem to be the most efficient. This technique is 'frequentist', in the sense that it relies upon an arbitrarily large supply of independent random numbers to obtain the desired precision. From an engineering point of view, there are efficient pseudo-random number generators that supply arbitrarily large sequences of 'random' numbers of very good quality. That is, there are tests available to show whether a sequence 'behaves' like an observation of a sequence of suitable independent random numbers.

Both approaches to statistical inference have an arbitrary element. For the classical approach, one sees this in the choice of sample space. The sample space is, to use H. Jeffreys' [24] vivid description, 'the class of observations that might have been obtained, but weren't'. For some experiments, the sample space is a clear and well defined object, but for others, there is an arbitrary element in the choice of the sample space. For example, an experiment may be set up with n plants, but some of the plants may die before the results are established.

Alternative hypotheses   There is *no distinction* within the Bayesian approach between the various values of the parameter except in the prior $\pi(\theta)$. The view is one of contrast between various values of θ. Consider the case where the parameter space consists of just two values, (θ_0, θ_1). Dropping subscripts where they are clearly implied, Bayes' rule for data x gives

$$\pi(\theta_0|x) = \frac{p(x|\theta_0)\pi(\theta_0)}{p(x)}$$

and

$$\pi(\theta_1|x) = \frac{p(x|\theta_1)\pi(\theta_1)}{p(x)}.$$

It follows that

$$\frac{\pi(\theta_0|x)}{\pi(\theta_1|x)} = \frac{p(x|\theta_0)\pi(\theta_0)}{p(x|\theta_1)\pi(\theta_1)}. \tag{1.15}$$

The *likelihood ratio* for two different parameter values is the ratio of the likelihood functions for these parameter values; denoting the likelihood ratio by LR,

$$LR(\theta_0, \theta_1; x) = \frac{p(x|\theta_0)}{p(x|\theta_1)}.$$

The *prior odds ratio* is simply the ratio $\pi(\theta_0)/\pi(\theta_1)$ and the *posterior odds ratio* is simply the ratio $\pi(\theta_0|x)/\pi(\theta_1|x)$. An odds ratio of greater than 1 indicates support for the parameter value in the numerator.

Equation (1.15) may be rewritten as

$$\text{posterior odds} \quad = \quad LR \times \text{prior odds}.$$

The data affect the change of assessment of probabilities through the likelihood ratio, comparing the probabilities of data on θ_0 and θ_1. This is in contrast with a sampling theory, or tail area significance test, where only the null hypothesis (say θ_0) is considered by the user of the test.

In 'classical' statistics, statements about parameters may be made through *confidence intervals*. It is important to note that a confidence interval for θ is *not* a probability statement about θ, because in classical statistics θ is *not* a random variable. It is a fixed, though unknown, value. The confidence interval is derived from probability statements about x the observation, namely from $p(x|\theta)$.

There is no axiomatic system that leads to confidence *measures*, while the axioms of probability are well defined. Operations strictly in accord with the calculus of probability give coherent conclusions. Ideas outside the probability calculus may give anomalies.

The next two sections give a detailed examination of two probability distributions that are often central to the analysis of Bayesian networks. Section 1.8 discusses binary variables, while Section 1.9 discusses multinomial variables. The distributions discussed in Section 1.8 are a useful special case of those discussed in Section 1.9.

1.8 Tossing a thumb-tack

The discussion of the thumb-tack is taken from D. Heckerman [25].

If a thumb-tack is thrown in the air, it will come to rest either on its point (0) or on its head (1). Suppose the thumb-tack is flipped n times, making sure that the physical properties of the thumb-tack and the conditions under which it is flipped remain stable over time. Let $\underline{x}_{(n)}$ denote the sequence of outcomes

$$\underline{x}_{(n)} = (x_1, \ldots, x_n)^t.$$

Each trial is a *Bernoulli trial* with probability θ of success (obtaining a 1). This is denoted by

$$X_i \sim Be(\theta), \qquad i = 1, \ldots, n.$$

Using the Bayesian approach, the parameter θ is be regarded as the outcome of a random variable, which is denoted by Θ. The outcomes are *conditionally independent, given θ*. This is denoted by

$$X_i \perp X_j | \Theta, \qquad i \neq j.$$

When $\Theta = \theta$ is given, the random variables $X_1, \ldots, X_n$ are independent, so that

$$p_{\underline{X}_{(n)}}(\underline{x}_{(n)} | \theta) = \prod_{l=1}^{n} \theta^{x_l}(1 - \theta)^{1-x_l} = \theta^k (1 - \theta)^{n-k}$$

where $k = \sum_{l=1}^{n} x_l$.

The problem is to estimate θ, finding the value that is best for $\underline{x}_{(n)}$. The Bayesian approach is, starting with a prior density $\pi_\Theta(.)$ over the parameter space $\tilde{\Theta} = [0, 1]$, to find the posterior density $\pi_{\Theta | \underline{X}_{(n)}}(.| \underline{x}_{(n)})$.

$$\pi_{\Theta | \underline{X}_{(n)}}(\theta | \underline{x}_{(n)}) = \frac{p_{\underline{X}_{(n)} | \Theta}(\underline{x}_{(n)} | \theta) \pi_\Theta(\theta)}{p_{\underline{X}_{(n)}}(\underline{x}_{(n)})} = \frac{p_{\underline{X}_{(n)} | \Theta}(\underline{x}_{(n)} | \theta) \pi_\Theta(\theta)}{\int p_{\underline{X}_{(n)} | \Theta}(\underline{x}_{(n)} | \phi) \pi_\Theta(\phi) d\phi}.$$

Let π_Θ be the uniform density on $[0, 1]$. This represents that initially You have no preference concerning θ; all values are equally plausible.[4] The choice of prior may seem arbitrary, but following the computations below, it should be clear that, from a large class of priors, the final answer does not depend much on the choice of prior if the thumb-tack is thrown a large number of times.

---

[4] As previously stated, the prior distribution contains the 'ad hoc' element. The results obtained from any statistical analysis are only reliable if there is sufficient data so that any inference will be robust under a rather general choice of prior.

There are well known difficulties with the statement that a uniform prior represents no preference concerning the value of θ. If the prior density for Θ is uniform, then the prior density of Θ^2 will *not* be uniform, so 'no preference' for values of Θ indicates that there is a *distinct* preference among possible initial values of Θ^2. If $\pi_1(x) = 1$ for $0 < x < 1$ is the density function for Θ and π_2 is the density function for Θ^2, then $\pi_2(x) = \frac{1}{2x^{1/2}}$ for $0 < x < 1$.

With the uniform prior,

$$\int_0^1 p_{\underline{X}_{(n)}|\Theta}(\underline{x}_{(n)}|\theta)\pi_\Theta(\theta)d\theta = \int_0^1 \theta^k(1-\theta)^{n-k}d\theta = \frac{k!(n-k)!}{(n+1)!}. \tag{1.16}$$

This may be computed using integration by parts, as follows. Set

$$I_{n,k} = \int_0^1 \theta^k(1-\theta)^{n-k}d\theta,$$

then

$$I_{n,0} = \int_0^1 (1-\theta)^n d\theta = \frac{1}{n+1}.$$

Using integration by parts,

$$I_{n,k} = \left[-\frac{\theta^k(1-\theta)^{n-k+1}}{n-k+1}\right]_{\theta=0}^1 + \frac{k}{n-k+1}I_{n,k-1} = \frac{k}{n-k+1}I_{n,k-1}.$$

From this,

$$I_{n,k} = \frac{k!}{n(n-1)\dots(n-k+1)}\frac{1}{(n+1)} = \frac{k!(n-k)!}{(n+1)!}.$$

This is an example of the *Beta integral*. The *posterior distribution* is therefore a *Beta density*

$$\pi_{\Theta|\underline{X}_{(n)}}(\theta|\underline{x}^{(n)}) = \begin{cases} \frac{(n+1)!}{k!(n-k)!}\theta^k(1-\theta)^{n-k} & 0 \le \theta \le 1 \\ 0 & \text{otherwise.} \end{cases} \tag{1.17}$$

It should be apparent that, in this case, there would have been tremendous difficulties carrying out the integral if the prior had been anything other than the uniform, or a member of the Beta family. The computational aspects are, or were, prior to the development of Markov chain Monte Carlo (McMC) methods [26], the main drawback to the Bayesian approach.

The Beta distribution is not restricted to integer values; the *Euler gamma function* is necessary to extend the definition to positive real numbers.

Definition 1.10 (Euler Gamma function) *The* Euler Gamma function $\Gamma(\alpha) : (0, +\infty) \to (0, +\infty)$ *is defined as*

$$\Gamma(\alpha) = \int_0^\infty x^{\alpha-1}e^{-x}dx. \tag{1.18}$$

The Euler Gamma function satisfies the following properties.

Lemma 1.1 *For all* $\alpha > 0$, $\Gamma(\alpha+1) = \alpha\Gamma(\alpha)$. *If* n *is an integer satisfying* $n \ge 1$, *then*

$$\Gamma(n) = (n-1)!$$

Proof of Lemma 1.1 Note that $\Gamma(1) = \int_0^\infty e^{-x} dx = 1$. For all $\alpha > 0$, integration by parts gives

$$\Gamma(\alpha + 1) = \int_0^\infty x^\alpha e^{-x} dx = \alpha \Gamma(\alpha). \tag{1.19}$$

The result follows directly. ☐

Definition 1.11 (Beta Density) *The Beta density Beta(α, β) with parameters $\alpha > 0$ and $\beta > 0$ is defined as the function*

$$\psi(t) = \begin{cases} \frac{\Gamma(\alpha+\beta)}{\Gamma(\alpha)\Gamma(\beta)} t^{\alpha-1}(1-t)^{\beta-1} & t \in [0, 1] \\ 0 & t \notin [0, 1] \end{cases} \tag{1.20}$$

The following results show that the Beta density is a probability density function for all real $\alpha > 0$ and $\beta > 0$.

Lemma 1.2 *Set*

$$B(\alpha, \beta) = \int_0^1 t^{\alpha-1}(1-t)^{\beta-1} dt.$$

Then

$$B(\alpha, \beta) = \frac{\Gamma(\alpha)\Gamma(\beta)}{\Gamma(\alpha+\beta)}.$$

Proof of Lemma 1.2 Directly from the definition of the Gamma function, using the substitutions $u = a^2$ and $v = b^2$ and, at the end of the argument $\cos^2 \theta = t$ so that $\frac{dt}{d\theta} = -2\cos\theta\sin\theta$,

$$\Gamma(\alpha)\Gamma(\beta) = \int_0^\infty \int_0^\infty e^{-u} u^{\alpha-1} e^{-v} v^{\beta-1} du dv$$

$$= 4 \int_0^\infty \int_0^\infty e^{-(a^2+b^2)} a^{2(\alpha-1)} b^{2(\beta-1)} ab\, da\, db$$

$$= \int_{-\infty}^\infty \int_{-\infty}^\infty e^{-(a^2+b^2)} |a|^{2\alpha-1} |b|^{2\beta-1} da\, db$$

$$= \int_0^{2\pi} \int_0^\infty e^{-r^2} r^{2(\alpha+\beta)-2} |\cos\theta|^{2\alpha-1} |\sin\theta|^{2\beta-1} r\, dr\, d\theta$$

$$= \frac{1}{2} \left(\int_0^{2\pi} |\cos\theta|^{2\alpha-1} |\sin\theta|^{2\beta-1} d\theta \right) \int_0^\infty e^{-u} u^{(\alpha+\beta)-1} du$$

$$= \left(2 \int_0^{\pi/2} (\cos\theta)^{2(\alpha-1)} (\sin\theta)^{2(\beta-1)} \cos\theta \sin\theta\, d\theta \right) \Gamma(\alpha+\beta)$$

$$= \left(\int_0^1 t^{\alpha-1} (1-t)^{\beta-1} dt \right) \Gamma(\alpha+\beta)$$

$$= B(\alpha, \beta)\Gamma(\alpha+\beta).$$

The result follows directly. ☐

Corollary 1.1 *Let ψ denote the Beta density, defined in Equation (1.20), then $\int_0^1 \psi(\theta)d\theta = 1$.*

Proof of Corollary 1.1 This is a direct consequence of Lemma 1.2. □

It follows that, for binomial sampling, updating may be carried out very easily for any prior distribution within the Beta family. Suppose the prior distribution π_0 is the $B(\alpha, \beta)$ density function, n trials are observed, with k taking the value 1 and $n - k$ taking the value 0. Then

$$\pi_{\Theta|X_{(n)}}(\theta|x_{(n)}) = \frac{p_{X_{(n)}|\Theta}(x_{(n)}|\theta)\pi_\Theta(\theta)}{p_{X_{(n)}}(x_{(n)})}$$

$$= \frac{\Gamma(\alpha + \beta)}{\Gamma(\alpha)\Gamma(\beta)p_{X_{(n)}}(x_{(n)})}\theta^{\alpha+k-1}(1-\theta)^{\beta+n-k-1} = c\theta^{\alpha+k-1}(1-\theta)^{\beta+n-k-1}.$$

Since $\int_0^1 \pi_{\Theta|X_{(n)}}(\theta|x_{(n)})d\theta = 1$, it follows from Lemma 1.2, that

$$\pi_{\Theta|X_{(n)}}(\theta|x_{(n)}) = \begin{cases} \frac{\Gamma(\alpha+\beta+n)}{\Gamma(\alpha+k)\Gamma(\beta+n-k)}\theta^{\alpha+k}(1-\theta)^{\beta+n-k} & \theta \in (0,1) \\ 0 & \theta \notin (0,1). \end{cases}$$

so that $\pi_{\Theta|X_{(n)}}(\theta|x_{(n)})$ is a $B(\alpha + k, \beta + n - k)$ density. □

Recall the definition of the maximum likelihood estimate: it is the value of θ that maximizes $p(x_{(n)}|\theta) = \theta^k(1-\theta)^{n-k}$. It is well known that

$$\hat{\theta}_{MLE}\left(x_{(n)}\right) = \frac{k}{n}.$$

The same pattern of thought can be applied to maximize the posterior density.

Definition 1.12 (Maximum Posterior Estimate) *The maximum posterior estimate, $\hat{\theta}_{MAP}$, is the value of θ which maximizes the posterior density $\pi_{\Theta|x_{(n)}}(\theta|x_{(n)})$.*

When the posterior density is $B(k + \alpha, n - k + \beta)$, an easy computation gives

$$\hat{\theta}_{MAP} = \frac{\alpha + k}{\alpha + \beta + n}.$$

Note that when the prior density is uniform, as in the case above, the MAP and MLE are exactly the same. The parameter, of course, is not an end in itself. The parameter ought to be regarded as a means to computing the predictive probability. The posterior is used to compute this; c.f. (1.14) above.

The predictive probability for the next toss   Recall that the 'parameter' is, in general, an artificial introduction, to help compute $p_{X_{n+1}|X_{(n)}}(x_{n+1}|x_{(n)})$. Suppose that $\pi(\theta|x_{(n)})$ has a $B(\alpha + k, \beta + n - k)$ distribution. The *predictive probability for the next toss*, for $a = 0$ or 1, is given by

$$p_{X_{n+1}|X_{(n)}}(a|x_{(n)}) = \int_0^1 p_{X_{n+1}}(a|\theta)\pi_{\Theta|x_{(n)}}(\theta|x_{(n)})d\theta.$$

Since $p_{X_{n+1}}(1|\theta) = \theta$, it follows (using equation (1.19)) that

$$p_{X_{n+1}|\underline{X}_{(n)}}(1|\underline{x}_{(n)}) = \frac{\Gamma(\alpha + \beta + n)}{\Gamma(\alpha + k)\Gamma(\beta + n - k)} \int_0^1 \theta^{(\alpha+k+1)}(1 - \theta)^{\beta+n-k} d\theta$$

$$= \frac{\Gamma(\alpha + \beta + n)}{\Gamma(\alpha + k)\Gamma(\beta + n - k)} \frac{\Gamma(\alpha + k + 1)\Gamma(\beta + n - k)}{\Gamma(\alpha + \beta + n + 1)}$$

$$= \frac{\alpha + k}{\alpha + \beta + n}.$$

In particular, note that the *uniform* prior, $\pi_0(\theta) = 1$ for $\theta \in (0, 1)$, is the $B(1, 1)$ density function, so that for binomial sampling with a uniform prior, the predictive probability is

$$p_{X_{n+1}|\underline{X}_{(n)}}(1|\underline{x}_{(n)}) = \frac{k + 1}{n + 2};$$

$$p_{X_{n+1}|\underline{X}_{(n)}}(0|\underline{x}_{(n)}) = \frac{n + 1 - k}{n + 2}. \tag{1.21}$$

This distribution, or more precisely $\frac{k+1}{n+2}$, is known as the Laplace rule of succession. A combinatorial derivation for it is given in [27].

Reconciling subjective predictive probabilities   The example of *agreeing to disagree*, referred to in the preceding, is due to R.J. Aumann, in [18]. Suppose two agents, Robbie_α and Robbie_β, both toss a thumb-tack once without communicating the outcome to each other. Both Robbie_α and Robbie_β have the same uniform prior on θ. Suppose Robbie_α and Robbie_β communicate the value of their respective predictive (posterior) probabilities as

$$p(\{X_{n+1} = 1\}|\text{Robbie}_\alpha) = \frac{2}{3}; \qquad p(\{X_{n+1} = 1\}|\text{Robbie}_\beta) = \frac{1}{3}.$$

Note that in the conditional probabilities above Robbie_α and Robbie_β actually refer to respective states of knowledge. Now, since both the number of tosses by each agent and the predictive probabilities held by the two agents is their common knowledge, they can revise their opinions by (1.21) to

$$p(\{X_{n+1} = 1\}|\text{Robbie}_\alpha, \text{Robbie}_\beta) = \frac{1 + 1}{2 + 2} = \frac{1}{2}.$$

This holds as Robbie_α and Robbie_β deduce by (1.21) that exactly one outcome of the two tosses was 1 (and the other was 0). The revision would not hold if the number of tosses was not common knowledge.

1.9   Multinomial sampling and the Dirichlet integral

Consider the case of multinomial sampling, where an experiment can take one of k outcomes, labelled $C_1, \ldots, C_k$. Suppose that $p(X = C_j) = \theta_j$, so that $\theta_1 + \ldots + \theta_k = 1$.

Consider n independent trials, $X_1, \ldots, X_n$. The notation $\mathbf{1}_A$, to denote the indicator function of a set A, will be used; that is

$$\mathbf{1}_A(x) = \begin{cases} 1 & x \in A \\ 0 & x \notin A. \end{cases}$$

Let $\mathbf{1}_{C_i}(x) = 1$ if $x = C_i$ and 0 otherwise. Set

$$Y_i = \sum_{j=1}^{n} \mathbf{1}_{C_i}(X_j).$$

Then Y_i denotes the number of trials that result in outcome C_i. Note that

$$Y_1 + \ldots + Y_k = n.$$

Then $(Y_1, \ldots, Y_k)$ is said to have a *multinomial* distribution and

$$p_{Y_1, \ldots, Y_k}(x_1, \ldots, x_k) = \frac{n!}{x_1! x_2! \ldots x_{k-1}! x_k!} \theta_1^{x_1} \ldots \theta_k^{x_k},$$

where the expression in front of the $\theta_1^{x_1} \ldots \theta_k^{x_k}$ is the multinomial coefficient.

In the *Bayesian* approach, a *prior distribution* is put over $\theta_1, \ldots, \theta_k$. Then, using the observations, this is updated using Bayes' rule to a posterior probability distribution over $\theta_1, \ldots, \theta_k$.

A particularly convenient family of distributions to use is the *Dirichlet* family, defined as follows.

Definition 1.13 (Dirichlet Density) *The Dirichlet density* $Dir(a_1, \ldots, a_k)$ *is the function*

$$\pi(\theta_1, \ldots, \theta_k) = \begin{cases} \frac{\Gamma(a_1 + \ldots + a_k)}{\prod_{j=1}^{k} \Gamma(a_k)} (\prod_{j=1}^{k} \theta_j^{a_j - 1}) & \theta_j \geq 0, \sum_{j=1}^{k} \theta_j = 1, \\ 0 & otherwise, \end{cases} \tag{1.22}$$

where Γ denotes the Euler Gamma function, given in Definition 1.10. The parameters $(a_1, \ldots, a_k)$ are all strictly positive and are known as *hyper parameters*.

This density, and integration with respect to this density function, are to be understood in the following sense. Since $\theta_k = 1 - \sum_{j=1}^{k-1} \theta_j$, it follows that π may be written as $\pi(\theta_1, \ldots, \theta_k) = \tilde{\pi}(\theta_1, \ldots, \theta_{k-1})$, where

$$\tilde{\pi}(\theta_1, \ldots, \theta_{k-1})$$

$$= \begin{cases} \frac{\Gamma(a_1 + \ldots + a_k)}{\prod_{j=1}^{k} \Gamma(a_k)} \left(\prod_{j=1}^{k-1} \theta_j^{a_j - 1}\right) \left(1 - \sum_{j=1}^{k-1} \theta_j\right)^{a_k - 1} & \theta_j \geq 0, \sum_{j=1}^{k-1} \theta_j \leq 1, \\ 0 & otherwise. \end{cases} \tag{1.23}$$

Clearly, when $k = 2$, this reduces to the *Beta density*. The following results show that the Dirichlet density is a probability density function.

Lemma 1.3 *Set*

$$D(a_1, \ldots, a_k) = \int_0^1 \int_0^{1-x_1} \int_0^{1-(x_1+x_2)} \cdots \int_0^{1-\sum_{j=1}^{k-2} x_j}$$

$$\left(\prod_{j=1}^{k-1} x_j^{a_j-1} \right) \left(1 - \sum_{j=1}^{k-1} x_j \right)^{a_k-1} dx_{k-1} \ldots dx_1.$$

Then

$$D(a_1, \ldots, a_k) = \frac{\prod_{j=1}^n \Gamma(a_j)}{\Gamma\left(\sum_{j=1}^k a_j \right)}.$$

Proof of Lemma 1.3 Straight from the definition of the Euler Gamma function, using the substitutions $x_j^2 = u_j$,

$$\prod_{j=1}^n \Gamma(a_j) = \int_0^\infty \cdots \int_0^\infty e^{-\sum_{j=1}^k u_j} \prod_{j=1}^k u_j^{a_j-1} du_1 \ldots du_k$$

$$= 2^k \int_0^\infty \cdots \int_0^\infty e^{-\sum_{j=1}^k x_j^2} \prod_{j=1}^k x_j^{2a_j-1} dx_1 \ldots dx_k$$

$$= \int_{-\infty}^\infty \cdots \int_{-\infty}^\infty e^{-\sum_{j=1}^k x_j^2} \prod_{j=1}^k |x_j|^{2a_j-1} dx_1 \ldots dx_k.$$

Now let $r = \sqrt{\sum_{j=1}^k x_j^2}$ and $z_j = \frac{x_j}{r}$ for $1 \le j \le k-1$. Using $x_j = rz_j$ for $j = 1, \ldots, k-1$ and $x_k = r\sqrt{1 - \sum_{j=1}^{k-1} z_j^2}$, the computation of the Jacobian easy and is left as an exercise:

$$J((x_1, \ldots, x_k) \to (r, z_1, \ldots, z_{k-1})) = \frac{r^{k-1}}{\sqrt{1 - \sum_{j=1}^{k-1} z_j^2}}.$$

Then

$$\prod_{j=1}^n \Gamma(a_j) = \int_0^\infty e^{-r^2} r^{2(\sum_{j=1}^k a_j)-k} r^{k-1} dr$$

$$\times \int_{-1}^1 \int_{-(1-z_1^2)}^{1-z_1^2} \cdots \int_{-(1-\sum_{j=1}^{k-2} z_j^2)}^{1-\sum_{j=1}^{k-2} z_j^2} \left(\prod_{j=1}^{k-1} z_j^{2a_j-1} \right) \left(1 - \sum_{j=1}^{k-1} z_j^2 \right)^{a_k-1/2}$$

$$\times \frac{1}{\sqrt{1 - \sum_{j=1}^{k-1} z_j^2}} \prod_{j=1}^{k-1} dz_j.$$

$$= \Gamma \left(\sum_{j=1}^{k} a_j \right)$$

$$\times \int_0^1 \int_0^{1-z_1^2} \cdots \int_0^{1-\sum_{j=1}^{k-2} z_j^2} \left(\prod_{j=1}^{k-1} z_j^{2(a_j-1)} \right) \left(1 - \sum_{j=1}^{k-1} z_j^2 \right)^{a_k-1} \prod_{j=1}^{k-1} 2z_j \, dz_j$$

$$= \Gamma \left(\sum_{j=1}^{k} a_j \right) D(a_1, \ldots, a_k)$$

and the result follows. □

Theorem 1.1 *The function $\tilde{\pi}(\theta_1, \ldots, \theta_{k-1})$ defined by Equation (1.23) satisfies*

$$\int_0^1 \int_0^{1-\theta_1} \cdots \int_0^{1-\sum_{j=1}^{k-2} \theta_j} \tilde{\pi}(\theta_1, \ldots, \theta_{k-1}) d\theta_{k-1} \ldots d\theta_1 = 1,$$

hence the Dirichlet density (Definition 1.13) is a well defined probability density function.

Proof This follows directly from the lemma. □

Properties of the Dirichlet density Theorem 1.1 shows that the Dirichlet density is a probability density function.

Another very important property is that the Dirichlet densities $\text{Dir}(a_1, \ldots, a_k)$: $a_1 > 0, \ldots, a_k > 0$ form a family of distributions that is *closed under sampling*. Consider a prior distribution $\pi_\Theta \sim \text{Dir}(a_1, \ldots, a_k)$ and suppose that an observation $\underline{x} := (x_1, \ldots, x_k)$ is made on $\underline{Y} := (Y_1, \ldots, Y_k)$ based on n independent trials (i.e. $x_1 + \ldots + x_k = n$). Let $\pi_{\Theta|\underline{Y}}$ denote the posterior distribution. Then, using Bayes' rule,

$$\pi_{\Theta|\underline{Y}}(\theta_1, \ldots, \theta_k) = \frac{\pi_\Theta(\theta_1, \ldots, \theta_{k-1}) p_{\underline{Y}}(\underline{x}|\theta_1, \ldots, \theta_k)}{p_{\underline{Y}}(\underline{x})}.$$

It follows that

$$\pi_{\Theta|\underline{Y}}(\theta_1, \ldots, \theta_k) = \frac{1}{p_{\underline{Y}}(\underline{x})} \frac{n!}{x_1! x_2! \ldots x_{k-1}! x_k!} \theta_1^{a_1+x_1-1} \ldots \theta_k^{a_k+x_k-1},$$

where $\theta_k = 1 - \sum_{j=1}^{k-1} \theta_j$.

Since the posterior density is a *probability* density, belonging to the Dirichlet family, it follows that the constant

$$\frac{1}{p_{\underline{Y}}(\underline{x})} \frac{n!}{x_1! x_2! \ldots x_{k-1}! x_k!} = \frac{\Gamma(a_1 + \cdots + a_k + x_1 + \cdots + x_k)}{\prod_{j=1}^{k} \Gamma(a_j + x_j)}$$

and hence that

$$\pi_{\Theta|\underline{Y}}(\theta_1, \ldots, \theta_k) \sim \text{Dir}(a_1 + x_1, \ldots, a_k + x_k).$$

The results in this section were perhaps first found by G. Lidstone [28].

Later in the text, the Dirichlet density will be written exclusively as a function of k variables, $\pi_\Theta(\theta_1, \ldots, \theta_k)$, where there are $k - 1$ independent variables and $\theta_k = 1 - \sum_{j=1}^{k-1} \theta_j$.

A question is how to select the hyper parameters $a_1, \ldots, a_k$ for the prior distribution. The choice of $a_1 = \ldots = a_k = \frac{1}{k}$ was suggested by W. Perks in [29].

Definition 1.14 (Conjugate Prior) *A prior distribution from a family that is closed under sampling is known as a* conjugate prior.

In [30], I.J. Good proved that exchangeability and *sufficientness* of samples implied that the prior is necessarily Dirichlet, if $k > 2$. The notion of sufficientness was originally defined by W.E. Johnson and I.J. Good. Loosely speaking, it means that the conditional probability of seeing case i of k possible in the next sample given n past samples, depends only on n, the number of times you have seen i in the past *and NOT on the other cases*.

Notes   Full accounts of the coherence argument may be found, for example, in [4], [31] and [32]. An introduction to inductive logic is given in [33].

The monograph [34] includes a thorough presentation of the topics of statistical inference and Bayesian machine learning. The papers [13] and [16] argue for subjective probability as the appropriate inference and language procedures for artificial intelligence agents, see also [14]. The book [35] provides a clear introduction to the application of Bayesian methods in artificial intelligence.

The work [36] by Thomas Bayes (1702–1761) and Richard Price was published posthumously in 1763. This paper makes difficult reading for a modern mathematician. Consequently, there is a considerable literature investigating the question of what Bayes actually proved, see, e.g. [22, 37–39] and the references therein. There is, however, a wide consensus that [36] does contain Equation (1.17).

For this, Bayes deals with billiard balls. Suppose You throw one billiard ball o (orange) on a square table (e.g. a billiard table without pockets) and measure the shortest distance from the side of the table, when the side of the table is scaled in size to 1. Let this value be denoted by p. Then You throw n balls W (white) on the table and note the number of white balls, say k, to the left of the orange ball. Then it is understood that Bayes computed the distribution of p given k given by Equation (1.17).

In this setting the uniform prior distribution on p is based on a physical understanding that is verifiable by repeated experimentation.

There is even the question of whether Bayes was the *first* to discover the results attributed to him. This is discussed in [40]. Another up-to-date report on the life and thinking of the Reverend Thomas Bayes, by D.R. Bellhouse [41], also discusses the question of whether he was the first to prove these results. The author has discovered some previously unknown documents. The paper points out that the canonical picture of Bayes is not proved to be an image of him.[5]

An alternative procedure on the billiard table is that $n + 1$ balls W are thrown on the table. One of them is then selected at random to play the role of the orange ball, and k,

---

[5] see http://www-history.mcs.st-andrews.ac.uk/PictDisplay/Bayes.html.

the number of balls to the left of the orange ball, is counted. Then You have a uniform distribution $\frac{1}{n+1}$ on the values of k.

It has been argued that Bayes demonstrated that a prior density $\pi(\theta)$ satisfying the equality

$$\frac{n!}{k!(n-k)!} \int_0^1 \theta^k (1-\theta)^{n-k} \pi(\theta) d\theta = \frac{1}{n+1} \qquad (1.24)$$

for all $0 \le k \le n$ and all n must be the uniform density. It may be checked rather easily using Equation (1.16) that the uniform density indeed satisfies this equality. F.H. Murray in [42] observed that Equation (1.24) implies for $k = n$ that

$$\int_0^1 \theta^n \pi(\theta) d\theta = \frac{1}{n+1}, \qquad (1.25)$$

which means that all the moments of $\pi(\theta)$ are given. Murray then went on to show that these moments determine a *unique* distribution, which is in fact the uniform distribution.[6]

The probability in Equation (1.24) is a uniform distribution on the number of successes in n Bernoulli trials with an unknown parameter. Hence Bayes (or Murray) has shown that *the uniform distribution on the number of successes is equivalent to the uniform density on the probability of success*. But this probability on the number of successes is a predictive probability on observables. This understanding of the Bayesian inference due to Thomas Bayes is different from many standard recapitulations of it, as pointed out in [39].

The ultimate question raised by reading of [36] is, 'what is it all about?'. In other words, what was the problem that Bayes was actually dealing with?

It is hardly credible that Bayes, a clergyman, should have studied this as a mere curious speculation, and even less that scoring at a billiard room should have been at the forefront of his mind. Richard Price writes in [36],

> ... the problem ... mentioned [is] necessary to be solved in order to provide a sure foundation for all our reasoning concerning past facts, and what is likely to be hereafter ...

For a layman in the history of philosophy the argument in [37] and [43] may carry a convincing power: Bayes and Price developed an inductive logic as a response to the critical and, in particular, anti-clerical objections to induction, causation and miracles advanced by David Hume [3] in his book of 1748; the famous philosopher and scholar was a contemporary of Bayes and Price.

Further evidence that this consideration may have prompted Bayes to develop a mathematical framework for inductive logic is seen from his theological interests. In 1731, he published the following paper: 'Divine Benevolence, or an Attempt to Prove That the Principal End of the Divine Providence and Government is the Happiness of His Creatures'.

---

[6] The *moment problem* is a classic problem; whether or not the moments of a distribution uniquely characterize the distribution. The technique usually employed is to check whether the *Carlemann conditions* are satisfied. For this problem, Murray showed *directly* that the moments uniquely determined the distribution.

The models that were later to be called Bayesian networks were introduced into artificial intelligence by J. Pearl, in the article [44]. Within the artificial intelligence literature, this is a seminal article, which concludes with the following statement: *The paper demonstrates that the centuries-old Bayes formula still retains its potency for serving as the basic belief revising rule in large, multi hypotheses, inference systems.*

1.10 Exercises: Probabilistic theories of causality, Bayes' rule, multinomial sampling and the Dirichlet density

1. This exercise considers the statistical notion of association due to G.U. Yule, who used it in a classical statistical study to demonstrate the positive effect of innoculation against cholera.

   Here the *association* between two events A and B, denoted by $\alpha(A, B)$, is defined as

   $$\alpha(A, B) = p(A \cap B) - p(A) \cdot p(B).$$

   (a) Show that

   $$\alpha(A, B) = -\alpha\left(A, B^c\right),$$

   where B^c is the complement of B.

   (b) Show that

   $$\alpha(A, B) = \alpha\left(A^c, B^c\right).$$

   Comment: Association is clearly symmetric. That is, for any two events A and B, $\alpha(A, B) = \alpha(B, A)$. It does not seem reasonable to claim that a decrease in cholera causes an increase in the number of innoculations. In this case it is common sense to conclude that there is an underlying causal relation, where innoculation (say B) causes a decreased prevalence of cholera (A), although without a controlled experiment, it is not possible to conclude that there is not a hidden factor C that both causes cholera and makes innoculation less likely.

2. **On a probabilistic theory of causality** Following the theory of causality due to P. Suppes [19] an event B_s is *defined* as a *prima facie cause*[7] of the event A_t if and only if the following three statements hold:

   - $s < t$,

   - $p(B_s) > 0$

   - $p(A_t \mid B_s) > p(A_t)$.

   Here the parameter is considered as a time parameter, and $s < t$ means that B_s occurs prior to A_t; a cause occurs before an effect.

   An event B_s is defined as a *prima facie negative cause* of an event A_t [19] if and only if the following three statements hold:

   - $s < t$,

   - $p(B_s) > 0$

---

[7] **Prima facie** is a Latin expression meaning 'on its first appearance', or 'by first instance'. Literally the phrase translates as first face, 'prima' first, 'facie' face. It is used in modern legal English to signify that on first examination, a matter appears to be self-evident from the facts. In common law jurisdictions, 'prima facie' denotes evidence that (unless rebutted) would be sufficient to prove a particular proposition or fact.

- $p(A_t \mid B_s) < p(A_t)$.

Intuitively, a negative cause is an event that prevents another event from happening. For example, the theory and practice of preventive medicine focuses on certain types of negative causation. In the problems the indices s, t are dropped for ease of writing.

(a) Show that if B^c is a prima facie negative cause of A, then B is a prima facie cause of A.

(b) Show that if B is a prima facie cause of A, then B^c is a prima facie cause of A^c. Also, show that if B is a prima facie *negative* cause of A, then B^c is a prima facie negative cause of A^c.

(c) Recall the definition of association from Exercise 1. Show that if B is a prima facie cause of A, then $\alpha(A, B) > 0$ and that if B is a prima facie negative cause of A, then $\alpha(A, B) < 0$.

3. **On odds and the weight of evidence** Let p be a probability distribution over a space $\mathcal{X}$. The *odds* of an event $A \subseteq \mathcal{X}$ given $B \subseteq \mathcal{X}$ under p, denoted by $O_p(A \mid B)$, is defined as

$$O_p(A \mid B) = \frac{p(A \mid B)}{p(A^c \mid B)}. \tag{1.26}$$

The odds ration will play an important role in Chapter 7, which considers sensitivity analysis. Next, the *weight of evidence E* in favour of an event A given B, denoted by $W(A : E \mid B)$, is defined as

$$W(A : E \mid B) = \log \frac{O_p(A \mid B \cap E)}{O_p(A \mid B)}. \tag{1.27}$$

Show that if $p(E \cap A^c \cap B) > 0$, then

$$W(A : E \mid B) = \log \frac{p(E \mid A \cap B)}{p(E \mid A^c \cap B)}. \tag{1.28}$$

4. **On a generalized odds and the weight of evidence** Let p denote a probability distribution over a space $\mathcal{X}$ and let $H_1 \subseteq \mathcal{X}$, $H_2 \subseteq \mathcal{X}$, $G \subseteq \mathcal{X}$ and $E \subseteq \mathcal{X}$. The *odds of H_1 compared to H_2 given G*, denoted by $O_p(H_1/H_2 \mid G)$, is defined as

$$O_p(H_1/H_2 \mid G) = \frac{p(H_1 \mid G)}{p(H_2 \mid G)}. \tag{1.29}$$

The *generalized weight of evidence* is defined by

$$W(H_1/H_2 : E \mid G) = \log \frac{O_p(H_1/H_2 \mid G \cap E)}{O_p(H_1/H_2 \mid G)}. \tag{1.30}$$

Show that if $p(H_1 \cap G \cap E) > 0$ and $p(H_2 \cap G \cap E) > 0$ then

$$W(H_1/H_2 : E \mid B) = \log \frac{p(E \mid H_1 \cap G)}{p(E \mid H_2 \cap G)}. \tag{1.31}$$

Clearly this is just a log likelihood ratio and these notions are another expression for posterior odds = likelihood ratio × prior odds.

5. In [45], I.J. Good discusses the causes of an event that are necessary and sufficient from probabilistic view point. For example, let E is the event of being hit by a car and F the event of going for a walk. Then F tends to be a necessary cause of E. The quantitites $Q_{suf}(E : F \mid U)$ and $Q_{nec}(E : F \mid U)$ are defined to measure the probabilistic tendency of an event F to be a sufficient and/or necessary cause, respectively, for an event E with background information U, by the weights of evidence discussed in the preceding exercise. They are defined respectively by

$$Q_{suf}(E : F \mid U) = W\left(F^c : E^c \mid U\right) \tag{1.32}$$

and

$$Q_{nec}(E : F \mid U) = W(F : E \mid U). \tag{1.33}$$

In view of the preceding definitions, Q_{suf} may be read as the weight of evidence against F provided by non-occurrence of E. Similarly, Q_{nec} is the the weight of evidence in favour of F given by occurrence of E. Both quantities are computed, to borrow a philosophical phrase, 'given the state of universe U just before F occurred'.

(a) If $p(E^c \mid F \cap U) > 0$, show that

$$Q_{suf}(E : F \mid U) = \log \frac{p(E^c \mid F^c \cap U)}{p(E^c \mid F \cap U)}. \tag{1.34}$$

(b) If $p(E \mid F^c \cap U) > 0$, show that

$$Q_{nec}(E : F \mid U) = \log \frac{p(E \mid F \cap U)}{p(E \mid F^c \cap U)}. \tag{1.35}$$

6. This exercise considers a few more properties of Q_{suf} and Q_{nec}. Following Exercise 4 above, set

$$Q_{nec}(E : F_1/F_2 \mid U) = W(F_1/F_2 : E \mid U)$$

which is the *necessitivity* of E of F_1 against F_2 and

$$Q_{suf}(E : F_2/F_1 \mid U) = W\left(F_1/F_2 : E^c \mid U\right),$$

which is the *sufficientivity* of E of F_1 against F_2.

(a) Show that $Q_{suf}(E : F \mid U) < 0$, if and only if $Q_{nec}(E : F \mid U) < 0$. Compare with *prima facie negative cause* in Exercise 2.

(b) Show that

$$Q_{nec}(E : F_1/F_2 \mid U) = Q_{suf}\left(E^c : F_2/F_1 \mid U\right).$$

This is called a *probabilistic contraposition*.

(c) Show that

$$Q_{nec} (E : F \mid U) = Q_{suf} \left(E^c : F^c \mid U \right).$$

This may interpreted along the following lines. Going for a walk F tends to be a necessary cause for being hit by a vehicle E, whereas staying home tends to be a sufficient cause for not being hit by a vehicle. (Note that cars and aircraft are known to have crashed into houses.) Both Q_{nec} and Q_{suf} should have high values in this case.

7. Let $\underline{X} = (X_1, \ldots, X_n)^t$ be an exchangeable sample of Bernoulli trials and let $T = \sum_{j=1}^{n} X_j$. Show that there is a probability density function π such that

(a)

$$p_T(t) = \int_0^1 \binom{n}{t} \theta^t (1 - \theta)^{n-t} \pi(\theta) d\theta, \quad t = 0, 1, \ldots, n$$

(b)

$$E[T] = n \int_0^1 \theta \pi(\theta) d\theta.$$

You may use the result of DeFinetti.

8. Consider a sequence of n independent, identically distributed Bernoulli trials, with unknown parameter θ, the 'success' probability. For a *uniform prior* over θ, show that the posterior density for θ, if the sequence has k successes, is

$$\pi_{\Theta|\underline{x}} (\theta \mid \underline{x}) = \begin{cases} \frac{(n+1)!}{k!(n-k)!} \cdot \theta^k (1 - \theta)^{n-k} & 0 \leq \theta \leq 1 \\ 0 & \text{elsewhere.} \end{cases} \quad (1.36)$$

9. Consider the thumb-tack experiment and the conditional independence model for the problem and the uniform prior density for θ. What is $P_{X_{n+1}|\underline{X}_{(n)}} (\text{head}|\underline{x}_{(n)})$, where $\underline{x}_{(n)}$ denotes the outcome of the first n throws?

10. Consider multinomial sampling, where θ_j is the probability that category j is obtained, with prior density $\pi_\Theta(\theta_1, \ldots, \theta_L)$ is the Dirichlet prior $Dir(\alpha q_1, \ldots, \alpha q_L)$ with $\sum_{j=1}^{L} q_j = 1$, defined by

$$\pi_\Theta(\theta) = \begin{cases} \frac{\Gamma(\alpha)}{\prod_{j=1}^{L} \Gamma(\alpha q_j)} \prod_{j=1}^{L} \theta_j^{\alpha q_j - 1} & \theta_1 + \ldots + \theta_L = 1, 0 \leq \theta_i \leq 1 \\ 0 & \text{elsewhere.} \end{cases}$$

Show that for multinomial sampling, with the Dirichlet prior, the posterior density $p_{\Theta|\underline{x}} (\theta|\underline{x}; \alpha)$ is the Dirichlet density

$$Dir (n_1 + \alpha q_1, \ldots, n_L + \alpha q_L),$$

which is shorthand for

$$\pi_{\Theta|\underline{X}_{(n)}} \left(\theta|\underline{x}_{(n)}; \alpha\underline{q} \right) = \frac{\Gamma (n + \alpha)}{\prod_{i=1}^{L} \Gamma (\alpha q_i + n_i)} \prod_{i=1}^{L} \theta_i^{n_i + \alpha q_i - 1}, \quad (1.37)$$

where $\underline{q} = (q_1, \ldots, q_L)$.

11. A useful property of the Dirichlet density is that the predictive distribution of X_{n+1} may be computed explicitly by integrating $p_{X_{n+1}|\Theta}(.\mid\theta)$ with respect to the posterior distribution containing the stored experience $\underline{x}_{(n)}$. Using the previous exercise, show that

$$p_{X_{n+1}|\underline{X}_{(n)}}\left(x_i\mid\underline{x}_{(n)}\right) = \int_{S_L}\theta_i\pi\left(\theta_1,\ldots,\theta_L|\underline{x};\alpha\underline{q}\right)d\theta_1\ldots d\theta_L = \frac{n_i+\alpha q_i}{n+\alpha}. \quad (1.38)$$

12. Let $\Theta=(\Theta_1,\ldots,\Theta_L)$ be a continuous random vector with $Dir(\alpha_1,\ldots,\alpha_L)$ distribution. Compute $Var(\Theta_i)$.

13. Prove the *Laplace rule of succession*. Namely, let $\{X_1,\ldots,X_{n+1}\}$ be independent, identically distributed Bernoulli random variables, where $p_{X_i}(1)=1-p_{X_i}(0)=\theta$ and $\theta\sim U(0,1)$. Then the Laplace rule of succession states that

$$p(\{X_{n+1}=1\}|\{X_1+\ldots+X_n=s\}) = \frac{s+1}{n+2}.$$

14. Let $\underline{V}=(V_1,\ldots,V_K)$ be a continuous random vector, with

$$\underline{V}\sim Dir(a_1,\ldots,a_K),$$

and set

$$U_i = \frac{V_ix_i^{-1}}{\sum_{i=1}^K V_ix_i^{-1}}, \quad ,i=1,\ldots,K,$$

where $\underline{x}=(x_1,\ldots,x_K)$ is a vector of positive real numbers; that is, $x_i>0$ for each $i=1,\ldots,K$. Show that $\underline{U}=(U_1,\ldots,U_K)$ has density function

$$\frac{\Gamma\left(\sum_{i=1}^k a_i\right)}{\prod_{i=1}^K\Gamma(a_i)}\prod_{i=1}^K u_i^{a_i-1}\left(\frac{1}{\sum_{i=1}^K u_ix_i}\right)^{\sum_{i=1}^K a_i}\prod_{i=1}^K x_i^{a_i}.$$

This density is denoted

$$\underline{U}\sim S\left(\underline{a},\underline{x}\right).$$

This is due to J.L. Savage [46]. Note that the Dirichlet density is obtained as a special case when $x_i=c$ for $i=1,\ldots,K$.

15. The next two examples illustrate how the Savage distribution of the previous exercise can arise in Bayesian analysis, for updating an objective distribution over the subjective assessments of a probability distribution by several different researchers, faced with a common set of data. Consider several researchers studying an unknown quantity X, where X can take values in $\{1,2,\ldots,K\}$. Each researcher has his own initial assessment of the probability distribution $\underline{V}=(V_1,\ldots,V_K)$ for the value that X takes. That is, for a particular researcher,

$$V_i=p_X(i), \quad i=1,\ldots,K.$$

It is assumed that

$$\underline{V} \sim \text{Dir}\,(a_1, \ldots, a_K)\,.$$

Each researcher observes the same set of data with the common likelihood function

$$l_i = p\,(\text{data}|\{X = i\})\,, \quad i = 1, \ldots, K.$$

The coherent posterior probability of a researcher is

$$U_i = p\,(\{X = i\} \mid \text{data})\,, \quad i = 1, 2, \ldots, K.$$

Let $\underline{U} = (U_1, \ldots, U_K)$. Prove that

$$\underline{U} \sim S\,(\underline{a}, \underline{l}^{-1})\,,$$

where $\underline{a} = (a_1, \ldots, a_K)$ and $\underline{l}^{-1} = \left(l_1^{-1}, \ldots, l_K^{-1}\right)$. This is due to J.M. Dickey [47].

16. Show that the family of distributions $S\,(\underline{a}, \underline{l}^{-1})$ is closed under updating of the opinion populations. In other words, if

$$\underline{V} \sim S\,(\underline{a}, \underline{z})\,,$$

before the data is considered, then

$$\underline{U} \sim S\,(\underline{a}, \underline{z} \times \underline{l}^{-1})\,,$$

after the data update, where

$$\underline{z} \times \underline{l}^{-1} = \left(z_1 l_1^{-1}, \ldots, z_K l_K^{-1}\right).$$

2

Conditional independence, graphs and d-separation

2.1 Joint probabilities

Consider a random vector $\underline{X} = (X_1, \ldots, X_d)$, defined on a state space $\mathcal{X} = \mathcal{X}_1 \times \ldots \times \mathcal{X}_d$, where $\mathcal{X}_j$ is the state space for X_j, where $\mathcal{X}_j = \{x_j^{(1)}, \ldots, x_j^{(j_k)}\}$ for $j = 1, \ldots, d$. In principle, the *joint probability function* $p_{X_1, \ldots, X_d}$ contains full information about the d random variables $X_1, \ldots, X_d$. But, although mathematically complete, this is not usually such a useful description in practice; the important features of the distribution may not be immediately clear in a table that has $\prod_{j=1}^{d} k_j$ elements, if this number is large. Furthermore, in many situations, the elementary building blocks will be low order conditional probabilities, each defined over small groups of variables.

Definition 2.1 (Independence) *Two discrete random variables X and Y are* independent *if and only if*

$$p_{X,Y} = p_X p_Y.$$

A collection of d random variables $\{X_1, \ldots, X_d\}$ *is said to be* jointly independent *if for any random vector* $(X_{i_1}, \ldots, X_{i_m})$ *where* $i_j \neq i_k$ *for all* $j \neq k$,

$$p_{X_{i_1}, \ldots, X_{i_m}} = \prod_{j=1}^{m} p_{X_{i_j}}.$$

Bayesian Networks: An Introduction   T. Koski, J. Noble
© 2009 John Wiley & Sons, Ltd

In practice, dependence between variables, or independence, can often be detected or understood even if the precise numerical values of the joint probability distribution are unavailable. Likewise, the relationships of conditional dependence (for example, the conditional distribution of X and Y given Z and the distribution of Z) often provide more convenient basic building blocks than the joint probability function $p_{X,Y,Z}$, since the conditional probability distribution is often easier to assess.

In many cases, qualitative dependencies among variables may be asserted relatively easily, before making numerical assignments for the relevant probabilities. This is simply an assertion as to whether or not two sets of variables are conditionally independent given another set of variables. Such a dependence structure may be modelled by a *directed acyclic graph*, where the nodes of the graph represent random variables.

In many problems, the directed acyclic graph may be interpreted in the following way: a directed edge between two variables may be used to indicate the modelling assumption that there is a *direct causal connection* between the two variables, the cause to effect relationship indicated by the direction of the arrow. Lack of any arrow indicates that there is no *direct* causal relation between the variables. The dependence structure between different variables in the network is described by the structure of the directed acyclic graph which, under certain circumstances, may have a causal interpretation. This leads to the notion of d-*separation* of variables (or *directed* separation), which will be introduced later.

It will be shown that d-separation characterises the conditional independence statements that can be inferred from a given DAG. A DAG where all conditional independence statements may be inferred from d-separation 'faithfully' represents the probability distribution. That is, when the representation is 'faithful', there are no artificial dependencies that have to be considered simply through an unfortunate choice of parametrization. In situations where there is a causal structure between the variables, it can, in many situations, be modelled by a faithful DAG. This idea is expanded in [48].

2.2   Conditional independence

Conditional independence (CI) is the key probabilistic notion in Bayesian networks. The following gives a quick summary of some basic properties of CI.

Characterizations of CI   Let $(\underline{X}, \underline{Y}, \underline{Z})$ be three discrete random vectors, with joint probability function $p_{\underline{X},\underline{Y},\underline{Z}}$. Let $\mathcal{X}_X$, $\mathcal{X}_Y$ and $\mathcal{X}_Z$ denote the state spaces for $\underline{X}$, $\underline{Y}$ and $\underline{Z}$ respectively. The vectors $\underline{X}$ and $\underline{Y}$ are said to be conditionally independent given $\underline{Z}$ if for all $(\underline{x}, \underline{y}, \underline{z}) \in \mathcal{X}_X \times \mathcal{X}_Y \times \mathcal{X}_Z$,

$$p_{\underline{X},\underline{Y},\underline{Z}}(\underline{x}, \underline{y}, \underline{z}) = p_{\underline{X}|\underline{Z}}(\underline{x}|\underline{z}) p_{\underline{Y}|\underline{Z}}(\underline{y}|\underline{z}) p_{\underline{Z}}(\underline{z}).$$

This will be indicated by the notation

$$\underline{X} \perp \underline{Y} | \underline{Z}.$$

The notation $\underline{X} \perp \underline{Y}$ denotes that $\underline{X}$ and $\underline{Y}$ are independent; that is, $p_{\underline{X},\underline{Y}}(\underline{x}, \underline{y}) = p_{\underline{X}}(\underline{x}) p_{\underline{Y}}(\underline{y})$ for all $(\underline{x}, \underline{y}) \in \mathcal{X}_X \times \mathcal{X}_Y$. This may be considered as $\underline{X} \perp \underline{Y} | \phi$, where ϕ denotes the empty vector. Similarly, for a set $V = \{X_1, \ldots, X_d\}$ of random variables, and three subsets $A \subset V$, $B \subset V$, $C \subset V$, the notation $A \perp B|C$ denotes that the variables in A are independent of the variables in B once the variables in set C are instantiated. $A \perp B$ means that the variables in A are independent of those in B and could be written $A \perp B|\phi$, where ϕ denotes the empty set.

Theorem 2.1 *The following are all equivalent to* $\underline{X} \perp \underline{Y}|\underline{Z}$:

1) *For all* $(\underline{x}, \underline{y}, \underline{z}) \in \mathcal{X}_Z \times \mathcal{X}_Y \times \mathcal{X}_Z$ *such that* $p_{\underline{Y}|\underline{Z}}(\underline{y}|\underline{z}) > 0$ *and* $p_{\underline{Z}}(\underline{z}) > 0$,

$$p_{\underline{X}|\underline{Y},\underline{Z}}(\underline{x}|\underline{y}, \underline{z}) = p_{\underline{X}|\underline{Z}}(\underline{x}|\underline{z}).$$

2) *Then there exists a function* $a : \mathcal{X}_X \times \mathcal{X}_Z \to [0, 1]$ *such that for all* $(\underline{x}, \underline{y}, \underline{z}) \in \mathcal{X}_X \times \mathcal{X}_Y \times \mathcal{X}_Z$ *satisfying* $p_{\underline{Y}|\underline{Z}}(\underline{y}|\underline{z}) > 0$ *and* $p_{\underline{Z}}(\underline{z}) > 0$,

$$p_{\underline{X}|\underline{Y},\underline{Z}}(\underline{x}|\underline{y}, \underline{z}) = a(\underline{x}, \underline{z})$$

3) *There exist functions* $a : \mathcal{X}_X \times \mathcal{X}_Z \to \mathbf{R}$ *and* $b : \mathcal{X}_Y \times \mathcal{X}_Z \to \mathbf{R}$ *such that for all* $(\underline{x}, \underline{y}, \underline{z}) \in \mathcal{X}_X \times \mathcal{X}_Y \times \mathcal{X}_Z$ *satisfying* $p_{\underline{Z}}(\underline{z}) > 0$,

$$p_{\underline{X},\underline{Y}|\underline{Z}}(\underline{x}, \underline{y}|\underline{z}) = a(\underline{x}, \underline{z})b(\underline{y}, \underline{z})$$

4) *For all* $(\underline{x}, \underline{y}, \underline{z}) \in \mathcal{X}_X \times \mathcal{X}_Y \times \mathcal{X}_Z$ *such that* $p_{\underline{Z}}(\underline{z}) > 0$,

$$p_{\underline{X},\underline{Y},\underline{Z}}(\underline{x}, \underline{y}, \underline{z}) = \frac{p_{\underline{X},\underline{Z}}(\underline{x}, \underline{z})p_{\underline{Y},\underline{Z}}(\underline{y}, \underline{z})}{p_{\underline{Z}}(\underline{z})}.$$

5) *There exist functions* $a : \mathcal{X}_X \times \mathcal{X}_Z \to \mathbf{R}$ *and* $b : \mathcal{X}_Y \times \mathcal{X}_Z \to \mathbf{R}$ *such that*

$$p_{\underline{X},\underline{Y},\underline{Z}}(\underline{x}, \underline{y}, \underline{z}) = a(\underline{x}, \underline{z})b(\underline{y}, \underline{z}).$$

Proof of Theorem 2.1

CI $\Longrightarrow$ 1) This is proved as follows: firstly,

$$p_{\underline{X},\underline{Y},\underline{Z}}(\underline{x}, \underline{y}, \underline{z}) = p_{\underline{X}|\underline{Y},\underline{Z}}(\underline{x}|\underline{y}, \underline{z})p_{\underline{Y}|\underline{Z}}(\underline{y}|\underline{z})p_{\underline{Z}}(\underline{z}).$$

Recall that CI is defined as

$$p_{\underline{X},\underline{Y},\underline{Z}}(\underline{x}, \underline{y}, \underline{z}) = p_{\underline{X}|\underline{Z}}(\underline{x}|\underline{z})p_{\underline{Y}|\underline{Z}}(\underline{y}|\underline{z})p_{\underline{Z}}(\underline{z}) \qquad \forall (\underline{x}, \underline{y}, \underline{z}) \in \mathcal{X}_X \times \mathcal{X}_Y \times \mathcal{X}_Z.$$

By equating these two expressions, it follows that if CI holds, then for all $(\underline{x}, \underline{y}, \underline{z})$ such that $p_{\underline{Z}}(\underline{z}) > 0$, either $p_{\underline{Y}|\underline{Z}}(\underline{y}|\underline{z}) = 0$ or $p_{\underline{X}|\underline{Z}}(\underline{x}|\underline{z}) = p_{\underline{X}|\underline{Y},\underline{Z}}(\underline{x}|\underline{y}, \underline{z})$ so CI $\Longrightarrow$ 1).

1) $\Longrightarrow$ 2) The first characterization of CI implies the second by taking $a(\underline{x}, \underline{z}) = p_{\underline{X}|\underline{Z}}(\underline{x}|\underline{z})$.

2) $\Longrightarrow$ 3) The second implies the third by taking

$$p_{\underline{X},\underline{Y}|\underline{Z}}(\underline{x}, \underline{y}|\underline{z}) = p_{\underline{X}|\underline{Y},\underline{Z}}(\underline{x}|\underline{y}, \underline{z})p_{\underline{Y}|\underline{Z}}(\underline{y}|\underline{z}),$$

where, from 2), $a(\underline{x}, \underline{z}) = p_{\underline{X}|\underline{Y},\underline{Z}}(\underline{x}|\underline{y}, \underline{z})$. The result follows by taking $b(\underline{y}, \underline{z}) = p_{\underline{Y}|\underline{Z}}(\underline{y}|\underline{z})$.

3) $\Longrightarrow$ **4)** The third implies the fourth as follows: assume there are two functions a and b such that

$$p_{\underline{X},\underline{Y}|\underline{Z}}(\underline{x},\underline{y}|\underline{z}) = a(\underline{x},\underline{z})b(\underline{y},\underline{z}). \qquad (2.1)$$

Set $A(\underline{z}) = \sum_{\underline{x} \in \mathcal{X}_X} a(\underline{x},\underline{z})$ and $B(\underline{z}) = \sum_{\underline{y} \in \mathcal{X}_Y} b(\underline{y},\underline{z})$. By summing over $\mathcal{X}_Y$ on both sides of Equation (2.1), it follows that

$$p_{\underline{X}|\underline{Z}}(\underline{x}|\underline{z}) = B(\underline{z})a(\underline{x},\underline{z}) \qquad (2.2)$$

and by summing over $\mathcal{X}_X$ on both sides of Equation (2.1), it follows that

$$p_{\underline{Y}|\underline{Z}}(\underline{y}|\underline{z}) = A(\underline{z})b(\underline{y},\underline{z}).$$

It follows, from summing over $\mathcal{X}_X$ on both sides of Equation (2.2), that $B(\underline{z})A(\underline{z}) = 1$. From this, it follows directly that

$$p_{\underline{X},\underline{Y}|\underline{Z}}(\underline{x},\underline{y}|\underline{z}) = a(\underline{x},\underline{z})b(\underline{y},\underline{z}) = B(\underline{z})a(\underline{x},\underline{z})A(\underline{z})b(\underline{y},\underline{z}) = p_{\underline{X}|\underline{Z}}(\underline{x}|\underline{z})p_{\underline{Y}|\underline{Z}}(\underline{y}|\underline{z}).$$

This, incidentally, shows that 3) implies CI. If 3) holds, then since $a(\underline{x},\underline{z})b(\underline{y},\underline{z}) = p_{\underline{X}|\underline{Z}}(\underline{x}|\underline{z})p_{\underline{Y}|\underline{Z}}(\underline{y}|\underline{z})$, it follows that

$$p_{\underline{X},\underline{Y},\underline{Z}}(\underline{x},\underline{y},\underline{z}) = p_{\underline{X},\underline{Y}|\underline{Z}}(\underline{x},\underline{y}|\underline{z})p_{\underline{Z}}(\underline{z})$$

$$= a(\underline{x},\underline{z})b(\underline{y},\underline{z})p_{\underline{Z}}(\underline{z}) = p_{\underline{X}|\underline{Z}}(\underline{x}|\underline{z})p_{\underline{Y}|\underline{Z}}(\underline{y}|\underline{z})p_{\underline{Z}}(\underline{z}) = \frac{p_{\underline{X},\underline{Z}}(\underline{x},\underline{z})p_{\underline{Y},\underline{Z}}(\underline{y},\underline{z})}{p_{\underline{Z}}(\underline{z})},$$

and therefore 3) $\Longrightarrow$ 4) is proved.

4) $\Longrightarrow$ **5)** This is proved by taking (for example) $a(\underline{x},\underline{z}) = p_{\underline{X}|\underline{Z}}(\underline{x}|\underline{z})$ and $b(\underline{y},\underline{z}) = p_{\underline{Y}|\underline{Z}}(\underline{y}|\underline{z})p_{\underline{Z}}(\underline{z})$.

5) $\Longrightarrow$ **CI** This is proved as follows: 5) gives

$$p_{\underline{X},\underline{Y},\underline{Z}}(\underline{x},\underline{y},\underline{z}) = p_{\underline{X}|\underline{Y},\underline{Z}}(\underline{x}|\underline{y},\underline{z})p_{\underline{Y}|\underline{Z}}(\underline{y}|\underline{z})p_{\underline{Z}}(\underline{z}) = a(\underline{x},\underline{z})b(\underline{y},\underline{z}). \qquad (2.3)$$

Set $C(\underline{z}) = \sum_{\underline{x} \in \mathcal{X}_X} a(\underline{x},\underline{z})$ and $D(\underline{z}) = \sum_{\underline{y} \in \mathcal{X}_Y} b(\underline{y},\underline{z})$. It follows, by summing over $\mathcal{X}_X$ first and then summing over $\mathcal{X}_Y$ in Equation (2.3), that $C(\underline{z})D(\underline{z}) = p_{\underline{Z}}(\underline{z})$. Set

$$\tilde{a}(\underline{x},\underline{z}) = \frac{a(\underline{x},\underline{z})}{A(\underline{z})}p_{\underline{Z}}(\underline{z})$$

and

$$\tilde{b}(\underline{y},\underline{z}) = \frac{b(\underline{y},\underline{z})}{B(\underline{z})}.$$

It follows that

$$p_{\underline{X},\underline{Y},\underline{Z}}(\underline{x},\underline{y},\underline{z}) = p_{\underline{X}|\underline{Y},\underline{Z}}(\underline{x}|\underline{y},\underline{z})p_{\underline{Y}|\underline{Z}}(\underline{y}|\underline{z})p_{\underline{Z}}(\underline{z}) = \tilde{a}(\underline{x},\underline{z})\tilde{b}(\underline{y},\underline{z})$$

and summing over $\mathcal{X}_X$ gives

$$p_{\underline{Y}|\underline{Z}}(\underline{y}|\underline{z})p_{\underline{Z}}(\underline{z}) = \tilde{b}(\underline{y},\underline{z})p_{\underline{Z}}(\underline{z}).$$

Therefore, for $p_{\underline{Z}}(\underline{z}) > 0$, $\tilde{b}(\underline{y}, \underline{z}) = p_{\underline{Y}|\underline{Z}}(\underline{y}|\underline{z})$. Similarly, it follows that $\tilde{a}(\underline{x}, \underline{z}) = p_{\underline{X}|\underline{Y},\underline{Z}}(\underline{x}|\underline{y}, \underline{z})p_{\underline{Z}}(\underline{z})$. For $p_{\underline{Z}}(\underline{z}) > 0$, it follows that $p_{\underline{X}|\underline{Y},\underline{Z}}(\underline{x}|\underline{y}, \underline{z}) = p_{\underline{X}|\underline{Z}}(\underline{x}|\underline{z})$, so that $\tilde{a}(\underline{x}, \underline{z}) = p_{\underline{X}|\underline{Z}}(\underline{x}|\underline{z})p_{\underline{Z}}(\underline{z})$ and

$$p_{\underline{X},\underline{Y},\underline{Z}}(\underline{x}, \underline{y}, \underline{z}) = p_{\underline{X}|\underline{Z}}(\underline{x}|\underline{z})p_{\underline{Y}|\underline{Z}}(\underline{y}|\underline{z})p_{\underline{Z}}(\underline{z})$$

thus proving CI. The proof of Theorem 2.1 is complete.   □

2.3 Directed acyclic graphs and d-separation

A graphical model is a representation of a collection of the components of a random vector $\underline{X} = (X_1, \ldots, X_d)$ as *nodes* of a *graph* $\mathcal{G} = (V, E)$, where important aspects of the conditional independence structure between the variables may be inferred from the structure of the graph, and in some cases the whole conditional independence structure is described by the graph.

2.3.1 Graphs

This section introduces some of the necessary graph theory. The remainder is presented in Chapter 4.

Definition 2.2 (Graph, Simple Graph) *A graph $\mathcal{G} = (V, E)$ consists of a finite set of nodes V and an edge set E, where each edge is contained in $V \times V$. The edge set therefore consists of ordered pairs of nodes.*

Let $V = \{\alpha_1, \ldots, \alpha_d\}$. A graph is said to be simple *if E does not contain any edges of the form (α_j, α_j) (that is a loop from the node to itself) and any edge $(\alpha_j, \alpha_k) \in E$ appears exactly once. That is, multiple edges are not permitted.*

For any two distinct nodes α and $\beta \in V$, the ordered pair $(\alpha, \beta) \in E$ if and only if there is a directed edge from α to β. An undirected *edge will be denoted $\langle \alpha, \beta \rangle$. In terms of directed edges,*

$$\langle \alpha, \beta \rangle \in E \Leftrightarrow (\alpha, \beta) \in E \quad and \quad (\alpha, \beta) \in E.$$

For a simple graph that may contain both directed and undirected edges, the edge set E may be decomposed as $E = D \cup U$, where $D \cap U = \phi$, the empty set. The sets U and D are defined by

$$\langle \alpha, \beta \rangle \in U \Leftrightarrow (\alpha, \beta) \in E \quad and \quad (\beta, \alpha) \in E.$$
$$(\alpha, \beta) \in D \Leftrightarrow (\alpha, \beta) \in E \quad and \quad (\beta, \alpha) \notin E.$$

For the definitions of 'path', 'trail' and 'cycle', an undirected edge will be considered as a single edge.

All the graphs considered in this text will be simple graphs and the term 'graph' will be used to mean 'simple graph'. If $(\alpha_i, \alpha_j) \in D$, this is denoted by an arrow going from α_i to α_j. If $\langle \alpha_i, \alpha_j \rangle \in U$, this is denoted by an edge between the two variables α_i and α_j.

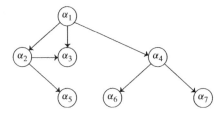

Figure 2.1 Example of a graph illustrating Definition 2.3.

Figure 2.1 gives an example of a graph that will be used to illustrate the definitions that follow. The node set is

$$V = \{\alpha_1, \alpha_2, \alpha_3, \alpha_4, \alpha_5, \alpha_6, \alpha_7\}$$

and the edge set is

$$E = \{(\alpha_1, \alpha_2), (\alpha_1, \alpha_3), (\alpha_1, \alpha_4), (\alpha_2, \alpha_3), (\alpha_2, \alpha_5), (\alpha_3, \alpha_5), (\alpha_4, \alpha_6), (\alpha_4, \alpha_7)\}.$$

The node α_4, for example, has neighbours $\alpha_1, \alpha_6, \alpha_7$, since the edge set contains (α_1, α_4), (α_4, α_6) and (α_4, α_7).

Definition 2.3 (Parent, Child, Directed and Undirected Neighbour, Family) *Consider a graph $\mathcal{G} = (V, E)$, where $V = \{\alpha_1, \ldots, \alpha_d\}$ and let $E = D \cup U$, where D is the set of directed edges and U the set of undirected edges. These are defined by*

$$(\alpha, \beta) \in D \Leftrightarrow (\alpha, \beta) \in E, \quad (\beta, \alpha) \notin E$$

$$\langle \alpha, \beta \rangle \in U \Leftrightarrow (\alpha, \beta) \in E, \quad (\beta, \alpha) \in E.$$

Let $\alpha_j, \alpha_k \in \mathcal{G}$. If $(\alpha_j, \alpha_k) \in D$, then α_k is referred to as a child *of α_j and α_j as a* parent *of α_k.*

For any node $\alpha \subseteq V$, the set of parents *is defined as*

$$\Pi(\alpha) = \{\beta \in V \mid (\beta, \alpha) \in D\} \tag{2.4}$$

and the set of children *is defined as*

$$Ch(\alpha) = \{\beta \in V \mid (\alpha, \beta) \in D\}. \tag{2.5}$$

For any subset $A \subseteq V$, the set of parents of A is defined as

$$\Pi(A) = \cup_{\alpha \in A}\{\beta \in V \backslash A \mid (\beta, \alpha) \in D\}. \tag{2.6}$$

The set of directed neighbours *of a node α is defined as*

$$N_{(d)}(\alpha) = \Pi(\alpha) \cup Ch(\alpha)$$

and the set of undirected neighbours *of α as*

$$N_{(u)}(\alpha) = \{\beta \in V \mid \langle \alpha, \beta \rangle \in U\}. \tag{2.7}$$

For any subset $A \subseteq V$, the set of undirected neighbours *of A is defined as*

$$N_{(u)}(A) = \cup_{\alpha \in A}\{\beta \in V \backslash A \mid \langle \alpha, \beta \rangle \in U\}. \tag{2.8}$$

For a node α, the set of neighbours $N(\alpha)$ is defined as

$$N(\alpha) = N_{(u)}(\alpha) \cup N_{(d)}(\alpha).$$

The family *of a node β is the set containing the node β together with its parents and undirected neighbours. It is denoted:*

$$F(\beta) = \{\beta\} \cup \Pi(\beta) \cup N_{(u)}(\beta) = \{\text{family of } \beta\}.$$

When $\mathcal{G}$ is undirected, this reduces to $F(\beta) = \{\beta\} \cup N(\beta)$.

When the variables have a clear indexing set, for example, the variables of the set $V = \{\alpha_1, \ldots, \alpha_d\}$ are clearly indexed by the set $\tilde{V} = \{1, \ldots, d\}$, the notation Π_j will also be used to denote the parent set $\Pi(\alpha_j)$ of variable α_j. Similarly with children, family and neighbour.

The notation $\alpha \sim \beta$ will be used to denote that $\alpha \in N(\beta)$; namely, that α and β are neighbours. Note that $\alpha \in N(\beta) \implies \beta \in N(\alpha)$.

For example, in Figure 2.1, $\Pi(\alpha_1) = \Pi_1 = \phi$ where ϕ denotes the empty set, $\Pi(\alpha_2) = \Pi_2 = \{\alpha_1\}$, $\Pi(\alpha_3) = \Pi_3 = \{\alpha_2, \alpha_1\}$, $\Pi(\alpha_4) = \Pi_4 = \{\alpha_1\}$, $\Pi(\alpha_5) = \Pi_5 = \{\alpha_2, \alpha_3\}$, $\Pi(\alpha_6) = \Pi_6 = \{\alpha_4\}$ and $\Pi(\alpha_7) = \Pi_7 = \{\alpha_4\}$.

In this text, a directed edge (α_j, α_k) is indicated by a pointed arrow from α_j to α_k; that is, from the parent to the child. In the graph in Figure 2.1, α_4 has a single parent α_1 and two children α_6 and α_7.

Definition 2.4 (Directed, Undirected Graph) *If all edges of a graph are undirected, then the graph $\mathcal{G}$ is said to be* undirected. *If all edges are directed, then the graph is said to be* directed. *The* undirected version of a graph $\mathcal{G}$, denoted by $\tilde{\mathcal{G}}$, is obtained by replacing the directed edges of $\mathcal{G}$ by undirected edges.

The graph in Figure 2.1 is a *directed* graph.

Definition 2.5 (Trail) *Let $\mathcal{G} = (V, E)$ be a graph, where $E = D \cup U$; $D \cap U = \phi$, D denotes the directed edges and U the undirected edges. A* trail τ *between two nodes $\alpha \in V$ and $\beta \in V$ is a collection of nodes $\tau = (\tau_1, \ldots, \tau_m)$, where $\tau_i \in V$ for each $i = 1, \ldots, m$, $\tau_1 = \alpha$ and $\tau_m = \alpha$ and such that for each $i = 1, \ldots, m - 1$, $\tau_i \sim \tau_{i+1}$. That is, for each $i = 1, \ldots, m - 1$, either $(\tau_i, \tau_{i+1}) \in D$ or $(\tau_{i+1}, \tau_i) \in D$ or $\langle \tau_i, \tau_{i+1} \rangle \in U$.*

For example, in the graph in Figure 2.1, there is a trail $\tau = (\alpha_3, \alpha_1, \alpha_4, \alpha_7)$ between α_3 and α_7, since the edges (α_1, α_3), (α_1, α_4), (α_4, α_7) are contained in the edge set.

Definition 2.6 (Sub-graph, Induced Sub-graph) *Let $A \subseteq V$ and $E_A \subseteq E \cap A \times A$. Then $\mathcal{F} = (A, E_A)$ is a sub graph of $\mathcal{G}$.*

If $A \subset V$ and $E_A = E \cap A \times A$, then $\mathcal{G}_A = (A, E_A)$ is the sub-graph induced by A.

Note that in general it is possible for a sub-graph to contain the same nodes, but fewer edges, but the sub-graph *induced* by the same node set will have the same edges.

Definition 2.7 (Connected Graph, Connected Component) *A graph is said to be con-nected if between any two nodes $\alpha_j \in V$ and $\alpha_k \in V$ there is a trail. A connected com-ponent of a graph $\mathcal{G} = (V, E)$ is an induced sub-graph $\mathcal{G}_A$ such that $\mathcal{G}_A$ is connected and such that if $A \neq V$, then for any two nodes $(\alpha, \beta) \in V \times V$ such that $\alpha \in A$ and $\beta \in V \backslash A$, there is no trail between α and β.*

It is clear that the graph in Figure 2.1 is connected.

Definition 2.8 (Path, Directed Path) *Let $\mathcal{G} = (V, E)$ denote a simple graph, where $E = D \cup U$. That is, $D \cap U = \phi$, D denotes the directed edges and U denotes the undirected edges. A path of length m from a node α to a node β is a sequence of distinct nodes $(\tau_0, \ldots, \tau_m)$ such that $\tau_0 = \alpha$ and $\tau_m = \beta$ such that $(\tau_{i-1}, \tau_i) \in E$ for each $i = 1, \ldots, m$. That is, for each $i = 1, \ldots, m$, either $(\tau_{i-1}, \tau_i) \in D$, or $\langle \tau_{i-1}, \tau_i \rangle \in U$.*

The path is a directed path if $(\tau_{i-1}, \tau_i) \in D$ for each $i = 1, \ldots, m$. That is, there are no undirected edges along the directed path.

It follows that a *trail* in $\mathcal{G}$ is a sequence of nodes that form a path in the undirected version $\tilde{\mathcal{G}}$.

Unlike a trail, a directed path $(\tau_0, \ldots, \tau_m)$ requires that the directed edge $(\tau_i, \tau_{i+1}) \in D$ for all $i = 0, \ldots, m - 1$. Therefore, in Figure 2.1, there is no *path* between α_3 and α_7, although there is a trail between these two nodes.

Definition 2.9 (Descendant, Ancestor) *Let $\mathcal{G} = (V, E)$ be a graph. A node α is a descen-dant of a node β if and only if there is a directed path from α to β. A node γ is an ancestor of a node α if and only if there is a directed path from γ to α.*

Let $E = U \cup D$, where U denotes the undirected edges and D denotes the directed edges. The set of descendants $D(\alpha)$ of a node α is defined as

$$D(\alpha) = \{\beta \in V \mid \exists \tau = (\tau_0, \ldots, \tau_k) : \tau_0 = \alpha, \ \tau_k = \beta, \ (\tau_j, \tau_{j+1}) \in D, \ j = 0, 1, \ldots, k\}. \tag{2.9}$$

The set of ancestors $A(\alpha)$ of a node α is defined as

$$A(\alpha) = \{\beta \in V \mid \exists \tau = (\tau_0, \ldots, \tau_k) : \tau_0 = \beta, \ \tau_k = \alpha, \ (\tau_j, \tau_{j+1}) \in D, \ j = 0, 1, \ldots, k\}. \tag{2.10}$$

In both cases, the paths are directed; they consist of directed edges only; they do not contain undirected edges.

In Figure 2.1, *all* the nodes $\alpha_2, \alpha_3, \alpha_4, \alpha_5, \alpha_6, \alpha_7$ are descendants of α_1, while α_3 and α_5 are the descendants of α_2.

Definition 2.10 (Cycle) *Let $\mathcal{G} = (V, E)$ be a graph. An m-cycle in $\mathcal{G}$ is a sequence of distinct nodes*

$$\tau_0, \ldots, \tau_{m-1}$$

such that $\tau_0, \ldots, \tau_{m-1}, \tau_0$ is a path (Definition 2.9).

Definition 2.11 (Directed Acyclic Graph (DAG)) *A graph $\mathcal{G} = (V, E)$ is said to be a directed acyclic graph if each edge is directed (that is, $\mathcal{G}$ is a simple graph such that for each pair $(\alpha, \beta) \in V \times V$, $(\alpha, \beta) \in E \Longrightarrow (\beta, \alpha) \notin E$) and for any node $\alpha \in V$ there does not exist any set of distinct nodes $\tau_1, \ldots, \tau_m$ such that $\alpha \neq \tau_i$ for all $i = 1, \ldots, m$ and $(\alpha, \tau_1, \ldots, \tau_m, \alpha)$ forms a directed path. That is, there are no m-cycles in $\mathcal{G}$ for any $m \geq 1$.*

The graph in Figure 2.1 is a directed acyclic graph.

Definition 2.12 (Tree) *A tree is a graph $\mathcal{G} = (V, E)$ that is* connected *and such that for any node $\alpha \in V$, there is no trail between α and α and for any two nodes α and β in V with $\alpha \neq \beta$, there is a unique trail. A* leaf *of a tree is a node that is connected to exactly one other node.*

The graph in Figure 2.1 is *not* a tree; $\tau = (\alpha_1, \alpha_2, \alpha_3, \alpha_1)$ is a trail from α_1 to α_1. Figure 2.2 gives an example of a tree.

Definition 2.13 (Forest) *A forest is a graph where all its connected components (Definition 2.7) are trees.*

This is illustrated in Figure 2.3. Each connected component in Figure 2.3 is a tree.

2.3.2 Directed acyclic graphs and probability distributions

Now consider a random vector $\underline{X} = (X_1, \ldots, X_d)$. Throughout the text, when the nodes of the graph represent random variables, they will be labelled by random variables that

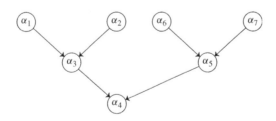

Figure 2.2 Example of a tree.

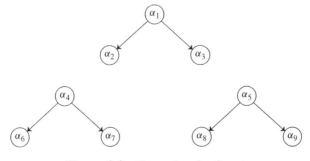

Figure 2.3 Example of a forest.

they represent. That is, $V = \{X_1, \ldots, X_d\}$ will denote the node set for the graph, which is a set of random variables. A directed acyclic graph $\mathcal{G} = (V, E)$ may be used to model assumptions that certain variables have direct causal relations on others. If a variable X_i is considered to have a direct causal effect on variable X_j, then $(X_i, X_j) \in E$. For a variable X_i, $\Pi_i = \{X_j | (X_j, X_i) \in E\}$. In situations where immediate associations between variables are considered to be causal, Π_i, the parent set for variable X_i, is the complete set of variables in the model whose values are considered to have a direct cause on the value taken by X_i.

A variable can have any number of *states*. For example, (red, green, blue, brown) (four states), or number of children in a family $(0, 1, 2, 3, 4, 5, 6, > 6)$ (eight states). In this text, attention is restricted to variables with a finite number of possible states. A variable is in exactly one of its states, which may or may not be known.

Factorization of a probability distribution   For *any* collection of random variables $(X_1, \ldots, X_d)$, it *always* holds, as a straightforward consequence of the definition of conditional probability, that the probability function $p_{X_1,\ldots,X_d}$ may be written as

$$p_{X_1,\ldots,X_d} = p_{X_1} p_{X_2|X_1} p_{X_3|X_1,X_2} \cdots p_{X_d|X_1,\ldots X_{d-1}}.$$

By reordering the variables, it therefore holds that for any ordering σ of $(1, \ldots, d)$,

$$p_{X_1,\ldots,X_d} = p_{X_{\sigma(1)}} p_{X_{\sigma(2)}|X_{\sigma(1)}} p_{X_{\sigma(3)}|X_{\sigma(1)},X_{\sigma(2)}} \cdots p_{X_{\sigma(d)}|X_{\sigma(1)},\ldots X_{\sigma(d-1)}}.$$

This way of writing a probability distribution is referred to as a *factorization*. A *directed acyclic graph* may be used to indicate that certain variables are conditionally independent of other variables, thus indicating how a factorization may be simplified.

Definition 2.14 (Factorization Along a Directed Acyclic Graph) *A probability function $p_{X_1,\ldots,X_d}$ over the variables $X_1, \ldots, X_d$ is said to* factorize along a directed acyclic graph $\mathcal{G}$ if the following holds: there is an ordering $X_{\sigma(1)}, \ldots, X_{\sigma(d)}$ of the variables such that

- $\Pi(X_{\sigma(1)}) = \Pi_{\sigma(1)} = \phi$; that is, $X_{\sigma(1)}$ has no parents.
- For each j, $\Pi(X_{\sigma(j)}) = \Pi_{\sigma(j)} \subset \{X_{\sigma(1)}, \ldots, X_{\sigma(j-1)}\}$.
- $p_{X_{\sigma(j)}|X_{\sigma(1)},\ldots,X_{\sigma(j-1)}} = p_{X_{\sigma(j)}|\Pi_{\sigma(j)}}$.

For each ordering of the variables, there is a directed acyclic graph that indicates how to factorize the probability distribution and those variables that may be excluded in the conditioning.

In some of the applications, the DAGs of interest will be *trees* (Definition 2.12) and *forests* (Definition 2.13).

Definition 2.15 (Instantiated) *When the state of variable is known, the variable is said to be* instantiated.

Within a directed acyclic graph, there are three basic ways in which two variables can be connected via a third variable and the whole graph is built up from these connections. They are the *chain*, *fork* and *collider* connections respectively.

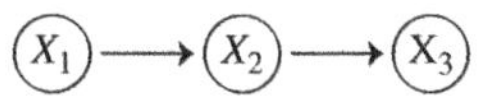

$$(X_1) \longrightarrow (X_2) \longrightarrow (X_3)$$

Figure 2.4   Chain connection: X_2 is a chain node.

Chain connections   Consider a situation with three random variables (X_1, X_2, X_3), where X_1 influences X_2, which in turn influences X_3, as in Figure 2.4, but there is no *direct* influence from X_1 to X_3. If the state of X_2 is *unknown*, then information about X_1 will influence the probability distribution of X_2, which then influences the probability distribution of X_3. Similarly, information about X_3 will influence the probability distribution of X_1 through X_2.

If state X_2 is *known*, then the channel is *blocked* and X_1 and X_3 become *independent* given X_2. The DAG indicates that the probability distribution of (X_1, X_2, X_3) may be factorized as:

$$p_{X_1,X_2,X_3} = p_{X_1} p_{X_2|X_1} p_{X_3|X_2}.$$

If $X_2 = x_2$ is known, then

$$p_{X_1,X_2,X_3}(., x_2, .) = (p_{X_1}(.) p_{X_2|X_1}(x_2|.))(p_{X_3|X_2}(.|x_2))$$

and so, following characterization 5) of conditional independence from Theorem 2.1, the variables X_1 and X_3 are conditionally independent, given X_2.

From the DAG, the variables X_1 and X_3 are said to be d-*separated given* X_2. The full definition of *d*-separation is given in Definition 2.18, found later.

Fork connections   A *fork* is illustrated in Figure 2.5. Influence can pass between all the children of X_1 unless the state of X_1 is known. If X_1 is known, then the variables X_2 and X_3 are said to be d-*separated given* X_1. Evidence may be transmitted through a fork node unless it is instantiated.

The DAG for the fork indicates that the probability distribution may be factorized as

$$p_{X_1,X_2,X_3} = p_{X_1} p_{X_2|X_1} p_{X_3|X_1}.$$

This implies that if the state of X_1 is known, then

$$p_{X_2,X_3|X_1}(., .|x_1) = p_{X_2|X_1}(.|x_1) p_{X_3|X_1}(.|x_1),$$

so that X_2 and X_3 are conditionally independent, following characterization 3) from the characterizations of conditional independence listed in the statement of Theorem 2.1.

Example 2.1   The directed acyclic graph (Figure 2.6) illustrates the following situation, described by Albert Engström (1869–1940), a Swedish cartoonist. During a convivial gathering there is talk of the unhygienic aspect of using galoshes. One of those present chimes in: 'Yes, I've also noticed this. Every time I've woken up with my galoshes on, I've had a headache.'

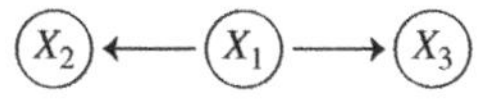

$$(X_2) \longleftarrow (X_1) \longrightarrow (X_3)$$

Figure 2.5   Fork connection: X_1 is a fork node.

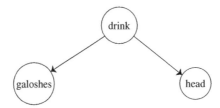

Figure 2.6   Causal relationship between galoshes, head and drink.

But does the footwear that he adopts while sleeping really influence the state of his head the next morning? The causal model represented by the graph indicates that if there is no information about his activities of the previous evening, then the presence of galoshes on his feet when he awakes indicates that his head will not be in such good shape. On the other hand, if there is full information concerning the activities of the previous evening, then the state of the feet gives no further information about the state of the head.   □

Collider connections   Consider the graph in Figure 2.7. If nothing is known about X_1 except that which may be inferred from knowledge about its parents, then the parents are independent; information about one of them will not affect the assessment of the probability values of the others. Information about one possible *cause* of an event does not give any further information about other possible causes, if there is no information as to whether or not the event actually happened. But if there is any information concerning the event, then information about one possible *cause* may give information about the other causes. In Figure 2.7, if it is known that the event $\{X_1 = x_1\}$ has occurred, and it is considered a priori that both $\{X_2 = x_2\}$ and $\{X_3 = x_3\}$ make the event $\{X_1 = x_1\}$ more likely, then the information that the event $\{X_2 = x_2\}$ has occurred will, in general, decrease the probability that the event $\{X_3 = x_3\}$ has occurred.

Information may only be transmitted through a collider if information has been received either about the variable in the connection or about one of its descendants.

The factorization of the distribution p_{X_1,X_2,X_3} corresponding to the DAG for the collider is

$$p_{X_1,X_2,X_3} = p_{X_2} p_{X_3} p_{X_1|X_2,X_3}.$$

Clearly, from the characterizations of conditional independence, X_2 and X_3 are *not* conditionally independent given X_1, but if X_1 is unknown, then X_2 and X_3 are independent; for any $(x_2^{(i)}, x_3^{(j)}) \in \mathcal{X}_2 \times \mathcal{X}_3$, where $\mathcal{X}_1$, $\mathcal{X}_2$, $\mathcal{X}_3$ are the state spaces for the random

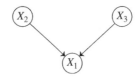

Figure 2.7   Collider connection: X_1 is a collider node.

variables X_1, X_2 and X_3 respectively.

$$p_{X_2,X_3}(x_2^{(i)}, x_3^{(j)}) = \sum_{y \in \mathcal{X}_1} p_{X_2}(x_2^{(i)}) p_{X_3}(x_3^{(j)}) p_{X_1|X_2,X_3}(y|x_2^{(i)}, x_3^{(j)})$$

$$= p_{X_2}(x_2^{(i)}) p_{X_3}(x_3^{(j)}) \sum_{y \in \mathcal{X}_1} p_{X_1|X_2,X_3}(y|x_2^{(i)}, x_3^{(j)})$$

$$= p_{X_2}(x_2^{(i)}) p_{X_3}(x_3^{(j)}).$$

For a chain or a fork, blocking requires the chain or fork variable respectively to be *instantiated*. Opening in the case of a collider holds for any information at all on any of the descendant variables. Information may be transmitted between nodes of a graph along an *active trail*, defined below.

Definition 2.16 (*S*-Active Trail) *Let* $\mathcal{G} = (V, E)$ *be a directed acyclic graph. Let* $S \subset V$ *and let* $X, Y \in V \backslash S$. *A trail* τ *between the two variables* X *and* Y *is said to be* S-*active if*

1. *Every collider node in* τ *is in* S, *or has a descendant (Definition 2.9) in* S.

2. *Every other node is outside* S.

Definition 2.17 (Blocked Trail) *A trail between* X *and* Y *that is not* S-*active is said to be blocked by* S.

The following definition is basic for all that follows in this chapter.

Definition 2.18 (*D*-separation) *Let* $\mathcal{G} = (V, E)$ *be a directed acyclic graph, where* $V = \{X_1, \ldots, X_d\}$ *is a collection of random variables. Let* $S \subset V$ *such that all the variables in* S *are instantiated and all the variables in* $V \backslash S$ *are not instantiated. Two distinct variables* X_i *and* X_j *not in* S *are* d-*separated by* S *if all trails between* X_i *and* X_j *are blocked by* S.

Let C *and* D *denote two* sets *of variables. If every trail from any variable in* C *to any variable in* D *is blocked by* S, *then the sets* C *and* D *are said to be* d-*separated by* S. *This is written*

$$C \perp D \,\|_{\mathcal{G}}\, S. \tag{2.11}$$

The set S *blocks every path between* C *and* D.

The terminology d-separation is short for *directed* separation. The insertion of the letter '*d*' points out that this is not the standard use of the term 'separation' found in graph theory. The term 'separation' will be introduced and used in the standard way in the later discussion about decomposable graphs (Chapter 4).

In common language, d-separation may be regarded in terms of *irrelevance*. In other words, if A and B are d-separated by a set S then, once the variables in the set S are known, new information on one of the sets A or B does not change the probability distribution of the other. This has been seen in some basic cases above; a general proof will be given later.

Definition 2.19 (*d*-connected) *If two variables X and Y are not* d-*separated, they are said to be* d-*connected.*

A procedure for determining *d*-separation   The following procedure may be used to check whether a set of variables S *d*-separates a set C from a set D.

1. Find all trails connecting the variables in C to the variables in D.

2. Check for each trail, until an active trail is found:

    (a) If there a chain or fork node in S on the trail, then the trail is *not* active.

    (b) If there is a *collider* node on the trail, then check whether any of its descendants are in S. If not, then the trail is not active.

    (c) Otherwise, the trail is active.

3. If an active trail was found, then C and D are not *d*-separated by S. If *none* of the trails are active, they are *d*-separated by S.

To declare that sets are *not* *d*-separated, it is only necessary to find a single active trail. To declare that sets *are* *d*-separated, it is necessary to show that *all* trails are not active.

It is clear that if all the parents of a variable X and all the children of X and all the variables sharing a child with X are instantiated, then X is *d*-separated from the rest of the network. This set of variables is known as the *Markov blanket* of the variable X:

Definition 2.20 (Markov Blanket) *The* Markov blanket *of a variable X is the set consisting of the parents of X, the children of X and the variables sharing a child with X.*

Example 2.2   Consider the DAG given in Figure 2.8. The Markov blanket of the variable X_1 is the set of nodes $\{X_2, X_3, X_4, X_6\}$. Let $S = \{X_2, X_3, X_4, X_6\}$, then $X_1 \perp X_5 \|_{\mathcal{G}} S$.

There are several interesting algorithmic applications of Markov blankets, e.g. to selection of variables in pattern recognition; see [49].

2.4   The Bayes ball

The *Bayes ball* provides a convenient method for deciding whether or not two nodes are *d*-separated. The idea was introduced by R. Schachter [50] and is illustrated in Figure 2.9. Variables are *d*-connected if the Bayes ball can be passed between them employing the following rule. The uninstantiated (hidden) nodes are represented by circles, the instantiated nodes as squares. This notation will be used throughout the book.

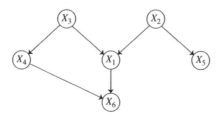

Figure 2.8   DAG for example of a Markov blanket.

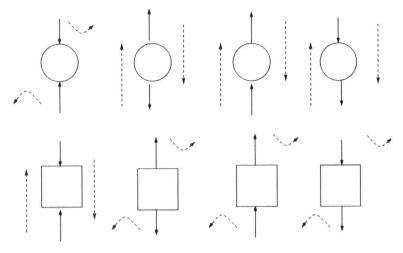

Figure 2.9 Bayes ball.

Consider the three types of connection in a DAG: chain, collider and fork.

- For the *chain* connection illustrated in Figure 2.4, the Bayes ball algorithm indicates that *if* the node is instantiated, *then* the ball does not move from X_1 to X_3 through X_2. The communication in the trail is *blocked*. If the node is *not* instantiated, then communication is possible.

- For the *fork* connection illustrated in Figure 2.5, the algorithm states that *if* node X_1 is instantiated, then again communication between X_2 and X_3 is *blocked*. If the node is *not* instantiated, then communication is possible.

- For the *collider* connection illustrated in Figure 2.7, the Bayes ball algorithm states that the ball *does* move from X_2 to X_3 if node X_1 is instantiated. *Instantiation of X_1 opens communication between the parents.*

For a collider node X_1, instantiation of any of the descendants of X_1 *also* opens communication. If node X_1 is uninstantiated, *and* none of its descendants is instantiated, then there is no communication.

Explaining away   The ways that the Bayes ball may move in a collider are the opposite of those for a chain or fork. In Figure 2.7, the nodes X_2 and X_3 are independent when X_1 is not instantiated, but a link emerges after instantiation. This pattern of reasoning is known as *explaining away*. That is, if there are two possible causes for an event and if one possible cause is known to have happened, then the other possible cause is less likely to have happened.

2.4.1 Illustrations

The following examples illustrate situations that may be modelled using random variables with causal relations between them, how these may be expressed as DAGs and inferences that may be drawn.

Example 2.3   Consider a situation where a desktop computer is powered by an electricity supply. The light source operates from the same electricity supply. If the computer is turned on and nothing happens, then the problem may be the result either of a hardware malfunction in the computer, or of a fault in the electricity supply.

If the lights are not working, this gives information about electricity failure, since it is a likely cause of light failure. If the computer does not respond when the electricity is switched on, it could be a result of a failure in the electricity supply, or a problem with the computer hardware. A computer malfunction as a cause for lack of response is less likely if it is known that there is an electricity failure. A network describing the situation is shown in Figure 2.10.

This is an example of *explaining away*, as described above. If the 'effect' variable is instantiated, then any particular cause is less likely if another possible cause is known to have taken place.

Example 2.4   The DAG in Figure 2.11 has been instantiated in X_3, X_4, X_7 and X_8, the set of neighbours of X_5. In this graph, the variable X_5 is *d*-connected to X_6, through X_8; the connection is a *collider* and X_8 is instantiated.

Example 2.5   The DAG in Figure 2.12 has been instantiated in X_2, X_3, X_4. The node X_6 *is d*-separated from *all* the uninstantiated nodes.

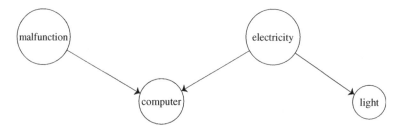

Figure 2.10   Light, Computer, Electricity.

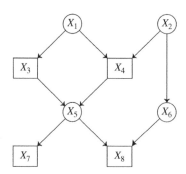

Figure 2.11   Figure for Example 2.

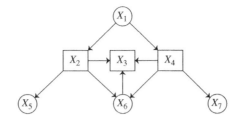

Figure 2.12 Figure for Example 2.5.

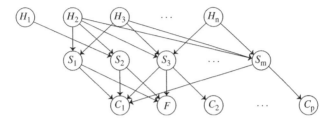

Figure 2.13 Bayesian network for Root Cause Analysis.

Example 2.6 (Root Cause Analysis) The following example is taken from [51], p. 1999. In large scale and complex industrial processes, the process operator has to isolate the cause of a failure by analyzing the signals from many sensors.

The *root causes* in the hardware, $(H_i)_{i=1}^n$ are the parent variables, which are the causes of various *symptoms*, $(S_j)_{j=1}^m$. The word 'symptoms' refers to changes in the process operation conditions, which affect the equipment performance or the final output. These in turn cause either *failure F* or *confirming events* $(C_k)_{k=1}^p$ that confirm that the process is abnormal. From the *confirming events* that occur, the objective is to deduce which hardware problem is the root cause. A directed acyclic graph for Root Cause Analysis is given in Figure 2.13.

Root Cause Analysis is a situation where the concept of *explaining away* appears; symptoms connected with a root cause may make it less likely that other root causes (which are not supported by confirming events) are present.

2.5 Potentials

Let $V = \{X_1, \ldots, X_d\}$ denote a collection of random variables, where variable X_j has state space $\mathcal{X}_j = (x_j^{(1)}, \ldots, x_j^{(k_j)})$ for $j = 1, \ldots, d$. Let $\mathcal{X} = \times_{j=1}^d \mathcal{X}_j$ denote the state space for $\underline{X}$. Let $\tilde{V} = \{1, \ldots, d\}$ denote the indexing set for the variables. For $D \subset \tilde{V}$, where $D = \{j_1, \ldots, j_m\}$, let $\mathcal{X}_D = \times_{j \in D} \mathcal{X}$ and let $\underline{X}_D = (X_{j_1}, \ldots, X_{j_m})$. Let $\underline{x} \in \mathcal{X}$ denote a generic element of $\mathcal{X}$ and let $\underline{x}_D = (x_{j_1}, \ldots, x_{j_m}) \in \mathcal{X}_D$, when $\underline{x} = (x_1, \ldots, x_d) \in \mathcal{X}$.

Furthermore, for $W \subset V$, let $\tilde{W}$ denote the indexing set for W. The notation $\mathcal{X}_W$ will also be used to denote $\mathcal{X}_{\tilde{W}}$, $\underline{X}_W$ to denote $\underline{X}_{\tilde{W}}$ and $\underline{x}_W$ to denote $\underline{x}_{\tilde{W}}$.

Definition 2.21 (Potential) *A potential ϕ over a domain $\mathcal{X}_D$ is defined as a non negative function $\phi : \mathcal{X}_D \to \mathbf{R}_+$. The space $\mathcal{X}_D$ is known as the* domain *of the potential. If the domain is the state space of a random vector $\underline{X}_D$, then the random vector $\underline{X}_D$ may also be referred to as the domain of the potential.*

In this setting, a potential over a domain $\mathcal{X}_D$ has $\prod_{j \in D} k_j$ entries. For $W \subset V$, the domain of a potential $\mathcal{X}_W$ may also be denoted by the collection of random variables W.

For example, let $\{X_1, X_2\} \subset V$ and let ϕ denote the joint probability distribution of (X_1, X_2), defined by

	$x_2^{(1)}$	$x_2^{(2)}$	$x_2^{(3)}$
$x_1^{(1)}$	0.05	0.10	0.05
$x_1^{(2)}$	0.15	0.00	0.25
$x_1^{(3)}$	0.10	0.20	0.10

Here $\mathcal{X}_1 = (x_1^{(1)}, x_1^{(2)}, x_1^{(3)})$ and $\mathcal{X}_2 = (x_2^{(1)}, x_2^{(2)}, x_2^{(3)})$. The potential ϕ is a function of two variables. The *domain* of the potential ϕ is $\mathcal{X}_1 \times \mathcal{X}_2$, which may also be denoted by (X_1, X_2). A pair $c = (x_1^{(i)}, x_2^{(j)})$ is called a *configuration* of (X_1, X_2). If, for example $c = (x_1^{(1)}, x_2^{(3)})$, then

$$\phi(c) = \phi((x_1^{(1)}, x_2^{(3)})) = 0.05.$$

Consider two potentials over (X_1, X_2); ϕ and ϕ', given by

$$\phi = \begin{array}{c|ccc} & x_2^{(1)} & x_2^{(2)} & x_2^{(3)} \\ \hline x_1^{(1)} & a_1 & a_2 & a_3 \\ x_1^{(2)} & b_1 & b_2 & b_3 \\ x_1^{(3)} & c_1 & c_2 & c_3 \end{array} \quad \text{and} \quad \phi' = \begin{array}{c|ccc} & x_2^{(1)} & x_2^{(2)} & x_2^{(3)} \\ \hline x_1^{(1)} & a_1' & a_2' & a_3' \\ x_1^{(2)} & b_1' & b_2' & b_3' \\ x_1^{(3)} & c_1' & c_2' & c_3' \end{array}$$

These will be used to illustrate the definitions of multiplication and division of potentials.

Definition 2.22 (Addition of Potentials) *Two potentials ϕ and ϕ' defined over the same domain $\mathcal{X}_D$ may be added together. Their sum is defined as the coordinate-wize sum; for each $\underline{x}_D \in \mathcal{X}_D$,*

$$(\phi + \phi')(\underline{x}_D) = \phi(\underline{x}_D) + \phi'(\underline{x}_D).$$

Definition 2.23 (Multiplication of Potentials) *Two potentials ϕ and ϕ' may be multiplied together to yield the potential $\phi.\phi'$ if they are both defined over the same domain. Multiplication of potentials is defined by multiplying each entry in the configuration.*

Hence, in the example above,

$X \backslash Y$	y_1	y_2	y_3
x_1	$a_1 a_1'$	$a_2 a_2'$	$a_3 a_3'$
x_2	$b_1 b_1'$	$b_2 b_2'$	$b_3 b_3'$
x_3	$c_1 c_1'$	$c_2 c_2'$	$c_3 c_3'$

$\phi.\phi' \leftrightarrow$

$\square$

Definition 2.24 (Division of Potentials) *Two potentials ϕ and ϕ' may be divided if they are defined over the same domain. The division of a potential ϕ by ϕ' to give the potential ϕ/ϕ' is defined by coordinate-wize division where the definition is that $b = 0 \Longrightarrow \frac{a}{b} = 0$.*

In the example above, provided that none of the entries of potential ϕ' are zero, the potential $\frac{\phi}{\phi'}$ is given by

$$\phi/\phi' \leftrightarrow$$

X\Y	y_1	y_2	y_3
x_1	a_1/a_1'	a_2/a_2'	a_3/a_3'
x_2	b_1/b_1'	b_2/b_2'	b_3/b_3'
x_3	c_1/c_1'	c_2/c_2'	c_3/c_3'

□

Potentials over different domains If potential ϕ_1 is defined over domain $\mathcal{X}_{D_1}$ and potential ϕ_2 is defined over domain $\mathcal{X}_{D_2}$, then multiplication and division of potentials may be defined by first extending both potentials to the domain $\mathcal{X}_{D_1 \cup D_2}$.

Definition 2.25 (Extending the Domain) *Let the potential ϕ be defined on a domain $\mathcal{X}_D$, where $D \subset \tilde{W} \subseteq \tilde{V}$. Then ϕ, defined over a domain $\mathcal{X}_D$, is extended to the domain $\mathcal{X}_{\tilde{W}}$ in the following way. For each $\underline{x}_{\tilde{W}} \in \mathcal{X}_{\tilde{W}}$,*

$$\phi(\underline{x}_{\tilde{W}}) = \phi(\underline{x}_D),$$

where $\underline{x}_D$ is the projection of $\underline{x}_{\tilde{W}}$ onto $\mathcal{X}_D$, using the definition of $\underline{x}_D$ (and hence $\underline{x}_{\tilde{W}}$) from the beginning of Section 2.5. In other words, the extended potential depends on $\underline{x}_{\tilde{W}}$ only through $\underline{x}_D$.

Definition 2.26 (Addition, Multiplication and Division of Potentials over Different Domains) *Addition, multiplication and division of potentials over different domains is defined as first, extending the domains of definition, using Definition 2.25, so that they are defined over the same domain, and then using Definition 2.22 for adding, Definition 2.23 for multiplication and Definition 2.24 for division.*

Multiplication of potentials may be expressed in the following terms: the product $\phi_1.\phi_2$ of potentials ϕ_1 and ϕ_2, defined over domains $\mathcal{X}_{D_1}$ and $\mathcal{X}_{D_2}$ is defined as

$$(\phi_1.\phi_2)(\underline{x}_{D_1 \cup D_2}) = \phi_1(\underline{x}_{D_1 \cup D_2})\phi_2(\underline{x}_{D_1 \cup D_2}),$$

where ϕ_1 and ϕ_2 have *first* been extended to $\mathcal{X}_{D_1 \cup D_2}$.

Let D_ϕ denote the index set for the domain variables of a potential ϕ. Multiplication has the following properties:

1. $D_{\phi_1.\phi_2} = D_{\phi_1} \cup D_{\phi_2}$,

2. (Commutative Law): $\phi_1.\phi_2 = \phi_2.\phi_1$

3. (Associative law): $(\phi_1.\phi_2).\phi_3 = \phi_1.(\phi_2.\phi_3)$.

4. (Existence of unit) The number 1 is a potential over the empty domain and $1\phi = \phi$ for all potentials ϕ. The unit potential is denoted by 1.

Example 2.7   Let

$X\backslash Y$	y_1	y_2
$\phi_{XY} \leftrightarrow \quad x_1$	a_1	a_2
x_2	a_3	a_4

and

$X\backslash Z$	z_1	z_2
$\phi'_{XZ} \leftrightarrow \quad x_1$	b_1	b_2
x_2	b_3	b_4

Following Definition 2.26, the product $\phi_{XY}.\phi_{XZ}$ is

$$[\phi_{XY}.\phi_{XZ}](x, y, z) \overset{def}{=} \phi_{XY}(x, y)\phi_{XZ}(x, z).$$

Thus

$X\backslash Y\backslash Z$	y_1	y_1
$\phi_{XY}\phi_{XZ} \leftrightarrow \quad x_1$	(a_1b_1, a_1b_2)	(a_2b_1, a_2b_2)
x_2	(a_3b_1, a_3b_4)	(a_4b_3, a_4b_4)

Now consider *marginalizing* the potential $\phi_{XY}\phi_{XZ}$ over the variable Z. The term *marginalizing* simply means summing over the state space $\mathcal{X}_Z$. This produces a potential with domain $\mathcal{X}_X \times \mathcal{X}_Y$;

$X\backslash Y$	y_1	y_2
$\sum_{\mathcal{X}_Z}\phi_{XY}\phi_{XZ} \leftrightarrow \quad x_1$	$a_1b_1 + a_1b_2$	$a_2b_1 + a_2b_2$
x_2	$a_3b_1 + a_3b_4$	$a_4b_3 + a_4b_4$

The entry for configuration (x_1, y_2) is

$$[\sum_Z \phi_{XY}.\phi_{XZ}](x_1, y_2) = \phi_{XY}.\phi_{XZ}(x_1, y_2, z_1) + \phi_{XY}\phi_{XZ}(x_1, y_2, z_2) = a_2b_1 + a_2b_2.$$

Marginalization   The operation of marginalization is now considered more generally. Let $V = \{X_1, \ldots, X_d\}$ denote a set of d random variables, with indexing set $\tilde{V} = \{1, \ldots, d\}$. Let $U \subseteq W \subseteq V$ and let ϕ be a potential defined over $\mathcal{X}_W$. The expression $\sum_{\mathcal{X}_{W\backslash U}} \phi$ denotes the *margin* (or the *sum margin*) of ϕ over $\mathcal{X}_U$ and is defined for $\underline{x}_U \in \mathcal{X}_U$ by

$$\left(\sum_{W\backslash U} \phi\right)(\underline{x}_U) = \left(\sum_{\underline{z} \in \mathcal{X}_{W\backslash U}} \phi\right)(\underline{z}, \underline{x}_U),$$

where the arguments have been rearranged so that those corresponding to W appear first, $\underline{z} \in \mathcal{X}_W$ is the projection of $(\underline{z}, \underline{x}_U) \in \mathcal{X}$ onto $\mathcal{X}_W$ and $\underline{x}_U \in \mathcal{X}_U$ the projection of $(\underline{z}, \underline{x}_U) \in \mathcal{X}$ onto $\mathcal{X}_U$. The following notation is also used:

$$\phi^{\downarrow U} = \left(\sum_{W\backslash U} \phi\right).$$

The marginalization operation obeys the following rules:

1. The Commutative Law: for any two sets of variables $U \subset V$ and $W \subset V$,

$$(\phi^{\downarrow U})^{\downarrow W} = (\phi^{\downarrow W})^{\downarrow U}.$$

2. The Distributive Law:
   If $\mathcal{X}_{D_1}$ is the domain of ϕ_1 and $D_1 \subseteq \tilde{V}$, then $(\phi_1 \phi_2)^{\downarrow D_1} = \phi_1 (\phi_2)^{\downarrow D_1}$.

Joint probability distributions   Consider three variables, X_1, X_2, X_3, with state spaces $\mathcal{X}_1$, $\mathcal{X}_2$ and $\mathcal{X}_3$ respectively, where $\mathcal{X}_1 = (x_1^{(1)}, x_1^{(2)})$, $\mathcal{X}_2 = (x_2^{(1)}, x_2^{(2)}, x_2^{(3)})$ and $\mathcal{X}_3 = (x_3^{(1)}, x_3^{(2)}, x_3^{(3)})$, with joint probability function p_{X_1, X_2, X_3}. The joint probability function is a three-way potential. In the potential below, the entry at position (i, j) is the triple

$$(p_{X_1, X_2, X_3}(x_1^{(i)}, x_2^{(j)}, x_3^{(1)}), p_{X_1, X_2, X_3}(x_1^{(i)}, x_2^{(j)}, x_3^{(2)}), p_{X_1, X_2, X_3}(x_1^{(i)}, x_2^{(j)}, x_3^{(3)})).$$

	$x_2^{(1)}$	$x_2^{(2)}$	$x_2^{(3)}$
$x_1^{(1)}$	$(0, 0.05, 0.05)$	$(0.05, 0.05, 0)$	$(0.05, 0.05, 0.05)$
$x_1^{(2)}$	$(0.1, 0.1, 0)$	$(0.1, 0, 0.1)$	$(0.2, 0, 0.05)$

In this example, $p_{X_1, X_2, X_3}(x_1^{(2)}, x_2^{(3)}, x_3^{(1)}) = 0.2$. The distribution of (for example) X_1 and X_3, is found by *marginalizing* over the unwanted variables; here, by summing over X_2. This gives

	$x_3^{(1)}$	$x_3^{(2)}$	$x_3^{(3)}$
$x_1^{(1)}$	0.10	0.15	0.10
$x_1^{(2)}$	0.40	0.10	0.15

The conditional probability distribution of X_2 given $X_1 = x_1^{(2)}$ and $X_3 = x_3^{(1)}$, for example, is computed using

$$p_{X_2 | X_1, X_3}(. | x_1^{(2)}, x_3^{(1)}) = \frac{p_{X_1, X_2, X_3}(x_1^{(2)}, .., x_3^{(1)})}{p_{X_1, X_3}(x_1^{(2)}, x_3^{(1)})} = (0.25, 0.25, 0.5).$$

The entire potential giving the conditional probability distribution of X_2 given X_1 and X_3 may be computed by marginalizing over X_2 to give the potential p_{X_1, X_3} and then by division of potentials to give the potential

$$p_{X_2 | X_1, X_3} = \frac{p_{X_1, X_2, X_3}}{p_{X_1, X_3}}.$$

2.6 Bayesian networks

The formal definition of a *Bayesian network* is given in Definition 2.27. It consists of a *graphical model* (namely, the directed acyclic graph) together with the corresponding *probability potentials*.

The result of the following lemma is clear and is inserted for completeness. It is necessary for the formal definition and construction of a Bayesian network. The statement of the lemma follows p.22 in [52].

Lemma 2.1 (Shafer) *For any DAG with a finite number of nodes $\alpha_1, \ldots, \alpha_d$ there is an ordering $(\alpha_{\sigma(1)}, \ldots, \alpha_{\sigma(d)})$ of the nodes (not necessarily unique) such that the parents of $\alpha_{\sigma(i)}$ are a subset of $\{\alpha_{\sigma(1)} \ldots, \alpha_{\sigma(i-1)}\}$. That is, by renaming the nodes as $\beta_j = \alpha_{\sigma(j)}$, $j = 1, \ldots, d$, the parents of β_j are a subset of $\{\beta_1, \ldots, \beta_{j-1}\}$ for each $j = 1, \ldots, d$.*

The statement of the lemma does not require the DAG to be connected.

Proof In any DAG with $d > 1$ nodes, it is possible to find at least one node that has no children. This is easily proved by contradiction. Assume there are no such nodes. Then, since there are only a finite number of nodes, it possible to move from any given node to one of its children. After repeating this $d + 1$ times, a node already visited is encountered again. This implies that there is a cycle, and hence a contradiction.

For $k = 1, \ldots, d$ do the following: In the DAG with $d - k + 1$ nodes, choose a node without a child, call it j_k. Let $\sigma(d - k + 1) = j_k$. Remove the node α_{j_k} from the node set and all edges (α_i, α_{j_k}) (i.e. those with a directed arrow pointing towards α_{j_k} from the edge set. This leaves a DAG with $d - k$ nodes.

This produces an ordering of the nodes $(\alpha_{\sigma(1)}, \ldots, \alpha_{\sigma(d)})$ that satisfies the given condition. □

For any joint probability distribution over a set of variables, with a given ordering for the variables, there is a directed acyclic graph over which the probability distribution may be factorized, where for each node X_j, $\Pi_j \subseteq \{X_1, \ldots, X_{j-1}\}$. The directed acyclic graph is induced by the ordering of the nodes $X_1, \ldots, X_d$ and any other ordering of the nodes $X_{\sigma(1)}, \ldots, X_{\sigma(d)}$ will induce a different directed acyclic graph along which the probability distribution may be satisfied such that $\Pi_{\sigma(j)} \subseteq \{X_{\sigma(1)}, \ldots, X_{\sigma(j-1)}\}$. In many situations of interest, additional causal modelling assumptions are to be incorporated, which may be modelled into a Bayesian network rather easily, by choosing an ordering of the variables to reflect them; for each $i = 1, \ldots, d$, X_i is chosen in such a way that all the variables that have a causal effect on X_i are contained in the set of variables $X_1, \ldots, X_{i-1}$. For a variable X_i, the *parent* set Π_i will be the subset of $\{X_1, \ldots, X_{i-1}\}$ containing those variables that, according to the model, may have a *direct* causal effect on X_i. The model needs an a priori assessment of the conditional probability potentials $p_{X_i|\Pi_i}$ for $i = 1, \ldots, d$. The prior probability distribution over the whole set of variables is obtained by multiplying these together.

The factorization $p_X = \prod_{j=1}^{d} p_{X_j|\Pi_j}$ expresses certain explicit modelling assumptions of conditional independence between variables and, if there are assumptions of direct causal relations between variables, these are also expressed by the factorization. Further conditional independence relations implied by these modelling assumptions may be inferred by checking whether nodes of the graph are d-separated. A graph for which the *only* conditional independence relations are those that are given by *d*-separation within the graph is said to be *faithful* to the probability distribution. When a graph is faithful to a probability distribution, the variables that are *d*-connected within the graph are associated. That is, when 'faithfulness' holds, associations suggested by the graph are real associations and not an accident of the parametrization.

Definition 2.27 (Bayesian Network) *A Bayesian network is a pair* $(\mathcal{G}, p)$, *where* $\mathcal{G} = (V, E)$ *is a directed acyclic graph with node set* $V = \{1, \ldots, d\}$ *for some* $d \in \mathbb{N}$, E *is the edge set, and* p *is either a probability distribution or a family of probability distributions, indexed by a parameter set* Θ, *over* d *discrete random variables,* $\{X_1, \ldots, X_d\}$. *The pair* $(\mathcal{G}, p)$ *satisfies the following criteria:*

- *For each* $\theta \in \Theta$, $p(.|\theta)$ *is a probability function with the same state space* $\mathcal{X}$, *where* $\mathcal{X}$ *has a finite number of elements. That is, for each* $\theta \in \Theta$, $p(.|\theta) : \mathcal{X} \to [0, 1]$ *and* $\sum_{\underline{x} \in \mathcal{X}} p(\underline{x}|\theta) = 1$.

- *For each node* $X_v \in V$ *with no parent variables, there is assigned a potential denoted by* p_{X_v}, *giving the probability distribution of the random variable* X_v. *To each variable* $X_v \in V$ *with a non empty parent set* $\Pi_v = (X_{b_1^{(v)}}, \ldots, X_{b_m^{(v)}})$, *there is assigned a potential*

$$p_{X_v|\Pi_v}$$

containing the conditional probability function of X_v *given the variables* $\{X_{b_1^{(v)}}, \ldots, X_{b_m^{(v)}}\}$. *If* X_v *has no parents, set* $\Pi_v = \phi$, *the empty set, so that* $p_{X_v} = p_{X_v|\Pi_v}$. *The joint probability function* p *may be factorized using the potentials* $p_{X_v|\Pi_v}$ *thus defined:*

$$p_{X_1,\ldots,X_d} = \prod_{v=1}^{d} p_{X_v|\Pi_v}.$$

- *The factorization is* minimal *in the sense that for an ordering of the variables such that* $\Pi_j \subseteq \{X_1, \ldots, X_{j-1}\}$, Π_j *is the smallest set of variables such that* $X_j \perp \Pi_j^c | \Pi_j$. *That is,*

$$\Pi_j = \cap\{A \subseteq \{X_1, \ldots, X_{j-1}\} \quad \text{such that} \quad X_j \perp A^c | A\}.$$

Example 2.8 Consider the Bayesian network with four variables X, Y, Z, W given in Figure 2.14.

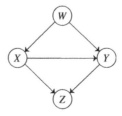

Figure 2.14 A graph on four variables.

The graph in Figure 2.14 represents a probability distribution which may be factorized according to

$$p_{X,Y,Z,W} = p_{Z|X,Y} \, p_{Y|X,W} \, p_{X|W} \, p_W .$$

The joint distribution has been expressed in terms of conditional probabilities which have been given. Furthermore, using the notation of Lemma 2.1, the variables may be renamed as $W = \beta_1$, $X = \beta_2$, $Y = \beta_3$ and $Z = \beta_4$, giving the variables an ordering that has the property described in Lemma 2.1. The parent set of $W = \beta_1$ is empty.

Vorobev's example   It is important for the calculus developed in this text that there are no *feedback* cycles and that the probability distributions factorizes along a directed *acyclic* graph. In the context of causal probability calculus, this is interpreted as a requirement that circular reasoning is not permitted, where an event A has a causal influence on an event B, while at the same time B has a causal influence on an event A. The following example, due to N. Vorobev [53], illustrates how things can go wrong if cycles are permitted. This is a fundamental paper in the field, but rarely cited.

Consider the *cyclic* graph in Figure 2.15. Three two-dimensional *marginal* distributions are specified, for a joint probability of three variables, which is to be factorized according to the directed graph given by Figure 2.15.

The potentials for the *joint* distributions are:

		x_2	
		0	1
x_1	0	1/2	0
	1	0	1/2

		x_3	
		0	1
x_2	0	0	1/2
	1	1/2	0

		x_1	
		0	1
x_3	0	1/2	0
	1	0	1/2

Here the potentials yield the marginals $p_{X_1}(0, 1) = (\frac{1}{2}, \frac{1}{2})$, $p_{X_2}(0, 1) = (\frac{1}{2}, \frac{1}{2})$ and $p_{X_3}(0, 1) = (\frac{1}{2}, \frac{1}{2})$. If these potentials are used to compute the joint distribution p_{X_1,X_2,X_3}, then, factorizing along the graph,

$$p_{X_1,X_2,X_3}(0, 0, 0) = p_{X_1|X_3}(0|0) p_{X_3|X_2}(0|0) p_{X_2}(0)$$

$$= p_{X_1|X_3}(0|0) p_{X_2,X_3}(0, 0)$$

$$= \frac{p_{X_1,X_3}(0, 0)}{p_{X_3}(0)} \times 0$$

$$= 1 \times 0 = 0,$$

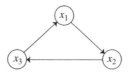

Figure 2.15   A cyclic graph.

but

$$p_{X_1,X_2,X_3}(0,0,0) = p_{X_2|X_1}(0|0)p_{X_1|X_3}(0|0)p_{X_3}(0)$$
$$= p_{X_2|X_1}(0|0)p_{X_1,X_3}(0,0)$$
$$= \frac{p_{X_1,X_2}(0,0)}{p_{X_1}(0)} \times \frac{1}{2}$$
$$= 1 \times \frac{1}{2} = \frac{1}{2}$$

and hence a contradiction is reached if cyclic graphs are permitted.   □

Example 2.9   Consider the following three variables: Electricity Failure (E) (1/0), Malfunction (M) (1/0), Computer Breakdown (C) (1/0) where 1 denotes 'yes' and 0 denotes 'no'. Suppose the causal relation is given by the DAG in Figure 2.16. Then the specified probabilities for the DAG are $p_{C|E,M}$, p_E and p_M. The factorization of the joint probability along the DAG is

$$p_{C,E,M} = p_{C|E,M}\, p_E\, p_M.$$

For this DAG, E and M are independent. Suppose that

$$p_E(1) = 0.1, \qquad p_M(1) = 0.2$$
$$p_{C|M,E}(1|1,1) = 1, \qquad p_{C|M,E}(1|0,1) = 1,$$
$$p_{C|M,E}(1|1,0) = 0.5, \qquad p_{C|M,E}(1|0,0) = 0.$$

This gives all the information necessary for a Bayesian network.

Suppose the computer is turned on and nothing happens. Then $C = 1$ has been *instantiated*. $p_C(1)$ is computed by marginalizing over E and M;

$$p_C(1) = \sum_{e,m} p_{C|E,M}(1|e,m)p_E(e)p_M(m) = 0.19.$$

The joint distribution of E and M given $C = 1$ may be computed via Bayes' theorem,

$$p_{M,E|C}(m,e|1) = \frac{p_{C|M,E}(1|m,e)p_M(m)p_E(e)}{p_C(1)}$$

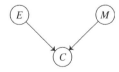

Figure 2.16   Electricity, Malfunction, Breakdown.

which gives

		e	
		1	0
m	1	$\dfrac{0.02}{0.19}$	$\dfrac{0.09}{0.19}$
	0	$\dfrac{0.08}{0.19}$	0

The marginal probabilities may be computed. For example:

$$p_{M|C}(1|1) = \sum_e p_{M,E|C}(1, e|1) = 0.58$$

$$p_{E|C}(1|1) = \sum_m p_{E,M|C}(1, m|1) = 0.53.$$

From this, it is clear that

$$p_{M|C}(1|1)p_{E|C}(1|1) \neq p_{M,E|C}(1, 1|1).$$

That is, M and E are not conditionally independent given C.   □

Example 2.10   Continuing the previous example, suppose that, having observed a computer failure, $C = 1$, it is also observed that the lights in the room have gone off, $L = 1$. The situation is described by the DAG in Figure 2.17, where square nodes represent the instantiated variables.

Only one more conditional probability, $p_{L|E}$ is required, in addition to the ones given above, to create a Bayesian network. Suppose that

$$p_{L|E}(1|1) = 1, \quad p_{L|E}(1|0) = 0.2.$$

Then
$$p_L(1) = p_{L|E}(1|1)p_E(1) + p_{L|E}(1|0)p_E(0) = 0.28.$$

Having first observed $C = 1$ and then $L = 1$, the previous posterior distribution $p_{M,E|C}(m, e|1)$ becomes the new prior and Bayes' rule gives

$$p_{M,E|C,L}(m, e|1, 1) = \frac{p_{C,L|M,E}(1, 1|m, e)p_{M,E}(m, e)}{p_{C,L}(1, 1)}$$

$$= \frac{p_{C|M,E}(1|m, e)p_{L|E}(1|e)p_M(m)p_E(e)}{\sum_{e,m} p_{C|E,M}(1|e, m)p_{L|E}(1|e)p_E(e)p_M(m)}$$

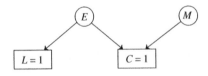

Figure 2.17   Electricity, Malfunction, Breakdown, Lights.

The computation is as follows:

$$p_M(m) = \frac{\begin{array}{cc} 1 & 0 \end{array}}{\begin{array}{cc} 0.2 & 0.8 \end{array}}$$

$$p_E(e) = \frac{\begin{array}{cc} 1 & 0 \end{array}}{\begin{array}{cc} 0.1 & 0.9 \end{array}}$$

so that

$$p_M(m)p_E(e) = \begin{array}{c|cc} M\backslash E & 1 & 0 \\ \hline 1 & 0.02 & 0.18 \\ 0 & 0.08 & 0.72 \end{array}$$

$$p_{C|M,E}(1|m,e) = \begin{array}{c|cc} M\backslash E & 1 & 0 \\ \hline 1 & 1 & 0.5 \\ 0 & 1 & 0 \end{array}$$

so that

$$p_M(m)p_E(e)p_{C|M,E}(1|m,e) = \begin{array}{c|cc} M\backslash E & 1 & 0 \\ \hline 1 & 0.02 & 0.09 \\ 0 & 0.08 & 0 \end{array}$$

The potential $p_{L|E}$ is

$$p_{L|E}(1|e) = \frac{\begin{array}{cc} 1 & 0 \end{array}}{\begin{array}{cc} 1 & 0.2 \end{array}}$$

and multiplication of potentials gives

$$p_M(m)p_E(e)p_{C|M,E}(1|m,e)p_{L|E}(1|e) = \begin{array}{c|cc} M\backslash E & 1 & 0 \\ \hline 1 & 0.02 & 0.018 \\ 0 & 0.08 & 0 \end{array}$$

Finally, the normalizing constant is obtained by summation, which gives 0.118, so that

$$p_{M,E|C,L}(m,e|1,1) = \begin{array}{c|cc} M\backslash E & 1 & 0 \\ \hline 1 & 0.169 & 0.153 \\ 0 & 0.678 & 0 \end{array}$$

Marginalizing gives

$$p_{M|C,L}(1|1,1) = 0.322$$

$$p_{E|C,L}(1|1,1) = 0.847.$$

This is an example of *explaining away*; here, $p_{M|C}(1|1) = 0.58$, but $p_{M|C,L}(1|1,1) = 0.322$. The observation that there is a light failure reduces the chance of a mechanical problem with the computer hardware. □

2.7   Object oriented Bayesian networks

Object oriented Bayesian networks are discussed in [54]. In an object oriented Bayesian network (OOBN), a *node* represents an *object* which is a collection of random variables (attributes) rather than a single random variable. The attributes are contained within the object.

The idea is that if there are several different types of object, which share common features, then the *same* potentials and parameters may be used to represent the common features. In this way, there are fewer parameters and the conditional probability potentials related to the common features may be updated using *all* the information available.

This is now illustrated by an example taken from [54]. Old MacDonald (OMD) has a farm where he keeps two milk cows and two meat cows. OMD wants to model his stock using OOBN classes. The *object*, or *class*, is a *cow*. A *generic cow* contains only those features common to both (Figure 2.18). Since the mother of the cow and the food that the cow eats both influence how much milk and meat the cow produces, OMD wants *mother* and *food* to be *input nodes*. OMD wants *milk*, the daily output of milk and *meat*, the amount of meat on a cow, as *output nodes*. Nodes in an instantiation that are neither input nor output nodes are termed *normal nodes*.

The values taken by an instantiation of the 'milk' and 'meat' variables are the quantities of milk and meat respectively produced by the animal. The *generic cow* contains the input and normal nodes that both the milk and meat cows have in common. There may be other nodes peculiar to both. For example, suppose OMD is told by an expert that music influences the state of mind of a milk cow, which in turn influences its metabolism, while 'music' has no influence on a meat cow, but that the weather *does* influence the state of mind of a meat cow (while not that of a milk cow), hence its metabolism and hence the quantities of milk and meat produced. Following all the advice from the expert, the milk cows and meat cows may be represented as illustrated in the DAGs shown in Figures 2.19 and 2.20 respectively.

Here 'cow' is a *class* and 'milk cow' and 'meat cow' are *subclasses*. A class S is a subclass of a class C if the set of nodes of C is a subset of the set of nodes of S. This ensures that an instantiation of S may be used anywhere in the OOBN instead of an instantiation of C.

The idea is that while milk cows and meat cows may be quite different, they nevertheless have some features in common, so that OMD may use the information from the entire herd to update the CPPs (conditional probability potentials) common to both.

Each node in the subclass inherits the conditional probability potentials of the class, unless the parent sets differ.

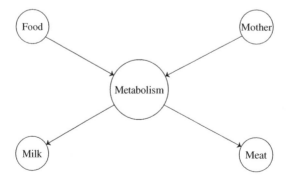

Figure 2.18 A generic cow.

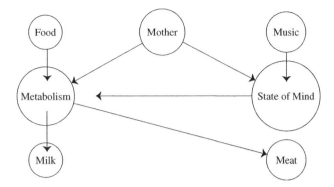

Figure 2.19   A milk cow.

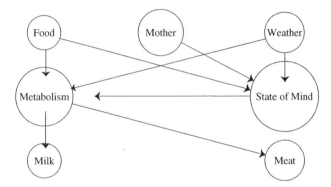

Figure 2.20   A meat cow.

The structure of a class hierarchy will be a *tree*, or a *forest* (recall that a forest is a collection of disjoint trees).

In the example of OMD's cattle, suppose that the mothers of two of the cows are known: Daisy is the mother of one of the meat cows while Matilda is the mother of one of the milk cows, but the other two mothers are unknown. For the two where the mother is unknown, this may be accommodated by introducing a third category, 'mother unknown', for the situation where the identity of the mother is missing from the data.

The *Object Oriented Assumption* is that the CPPs (conditional probability potentials) of a class are the same, wherever that class appears. In the example of OMD's farm, the *generic cow* potentials are the same for all four cows, the *milk cow* potentials are the same for each milk cow and the *meat cow* potentials are the same for each meat cow.

Let X_1 = food, X_2 = mother, X_3 = Music, X_4 = state of mind, X_5 = metabolism, X_6 = milk, X_7 = meat. X_8 = weather. Then the CPPs that need to be specified for the generic cow are p_{X_1}, p_{X_2}, $p_{X_5|X_1,X_2}$, $p_{X_7|X_5}$ and $p_{X_6|X_5}$.

The potentials $p_{X_7|X_5}$, $p_{X_6|X_5}$, p_{X_1} and p_{X_2} remain the same for the *milk* cows and *meat* cows. The potentials connected with food and mother p_{X_1} and p_{X_2} will clearly be irrelevant in the analysis once these variables are fixed in a particular instantiation.

To specify the *milk cow* class, the potentials p_{X_3} and $p_{X_4|X_2,X_3}$ are needed, and the potential $p_{X_5|X_1,X_2}$ from the generic cow has to be replaced with $p_{X_5|X_1,X_2,X_4}$.

To specify the *meat cow* class, the potentials p_{X_8}, $p_{X_4|X_2,X_8}$, $p_{X_5|X_1,X_4,X_8}$ are needed, together with the potentials for the generic cow, except for $p_{X_5|X_1,X_2}$.

2.8   *d*-Separation and conditional independence

The following result shows that, for a given DAG $\mathcal{G}$, d-separation characterizes those conditional independence statements that it is possible to infer from that particular DAG.

Theorem 2.2 (*d*-Separation Implies Conditional Independence) *Let* $\mathcal{G} = (V, E)$ *be a directed acyclic graph and let* p *be a probability distribution that factorizes along* $\mathcal{G}$. *Then for any three disjoint subsets* $A, B, C \subset V$, *it holds that* $A \perp B|C$ *(A and B are independent given C) if A and B are* d-*separated by C.*

Conditions under which a converse to this theorem hold are discussed later in this section, and presented in Theorem 2.3.

Proof of Theorem 2.2 Let $V = \{X_1, \ldots, X_d\}$ denote the set of variables, let $A \subset V$, $B \subset V$ and $C \subset V$ be three disjoint sets of variables and let $\tilde{A}$, $\tilde{B}$, $\tilde{C}$, $\tilde{V}$ denote the indexing sets of the variables in A, B, C and V respectively. For any set $S \subseteq V$ of variables, let $\tilde{S}$ denote the set of indices. Suppose that $A \perp B\|_{\mathcal{G}}C$. Let A, B and C denote also the random vectors $\underline{X}_A$, $\underline{X}_B$ and $\underline{X}_C$ of variables in A, B and C respectively and let $\mathcal{X}_A$, $\mathcal{X}_B$ and $\mathcal{X}_C$ denote their respective state spaces. It is required to show that for all $\underline{a} \in \mathcal{X}_A$, $\underline{b} \in \mathcal{X}_B$ and $\underline{c} \in \mathcal{X}_C$,

$$p_{A,B|C}(\underline{a}, \underline{b}|\underline{c}) = p_{A|C}(\underline{a}|\underline{c})p_{B|C}(\underline{b}|\underline{c}).$$

Let $D = V \backslash (A \cup B \cup C)$. Let

$$E_1 = \{Y \in V | \text{there is a } C\text{-active trail from } A \text{ to } Y\}$$

$$E_2 = \{Y \in V | \text{there is a } C\text{-active trail from } B \text{ to } Y\}$$

$$D_1 = D \cap E_1 \cap E_2, \quad D_2 = D \cap E_1 \cap E_2^c, \quad D_3 = D \cap E_2 \cap E_1^c, \quad D_4 = D \cap (D_1^c \cup D_2^c \cup D_3^c).$$

Since all the nodes of D are uninstantiated and there is no active trail from A to B, it follows that any nodes in D_1 are *colliders* with ancestors (Definition 2.9) in both A and B, together with all the descendants (Definition 2.9) of these colliders. Neither the colliders with ancestors in both A and B nor their descendants are instantiated (that is, belong to C); neither do the nodes in D_1 have descendants that belong to either A or B, otherwise it is from the definitions that there would be an active trail between A and B.

From characterization 5) of Theorem 2.1, it is required to show that there are two functions F and G such that

$$p_{A,B,C}(\underline{a}, \underline{b}, \underline{c}) = F(\underline{a}, \underline{c})G(\underline{b}, \underline{c}).$$

Let $p(X_j|\Pi_j)$ denote the conditional probability function of X_j given the parent variables Π_j. Then

$$p(X_1, \ldots, X_n) = \prod_{j \in \tilde{A}} p(X_j|\Pi_j) \prod_{j \in \tilde{B}} p(X_j|\Pi_j) \prod_{j \in \tilde{C}} p(X_j|\Pi_j)$$

$$\times \prod_{j \in \tilde{D}_1} p(X_j|\Pi_j) \prod_{j \in \tilde{D}_2} p(X_j|\Pi_j) \prod_{j \in \tilde{D}_3} p(X_j|\Pi_j) \prod_{j \in \tilde{D}_4} p(X_j|\Pi_j).$$

Any descendant of a variable in D_1 is also in D_1. Marginalizing over the variables in D_1 does not involve the parent variables of A, B or C, nor does it involve the variables in D_2 or D_3 or their ancestors. Furthermore, the parents of variables in D_4 are either in D_4 or in C.

Now, using ϕ to denote the empty set, let $C_2 = \{X \in C \mid \Pi(X) \cap D_2 \neq \phi\}$, $C_3 = \{X \in C \mid \Pi(X) \cap D_3 \neq \phi\}$ and $C_4 = C \cap C_2^c \cap C_3^c$. Then $C_2 \cap C_3 = \phi$, the empty set, otherwise there would be a collider node in C that would result in an active trail from A to B. It is also clear that $\Pi(C_4) \subseteq C \cup D_4$, where $\Pi(C_4)$ denotes the parent variables of the variables in C_4; that is, $\Pi(C_4) = \{Y|(Y, X) \in E, X \in C_4\}$. The sets C_2, C_3, C_4 are disjoint. Using $\mathcal{X}_S$ to denote the state space of the random vector formed from the variables in a set S, it follows that

$$p(A, B, C) = \left(\sum_{\mathcal{X}_{D_1}} \sum_{\mathcal{X}_{D_4}} \prod_{j \in \tilde{D}_1} p(X_j|\Pi_j) \prod_{j \in \tilde{D}_4} p(X_j|\Pi_j) \prod_{j \in \tilde{C}_4} p(X_j|\Pi_j) \right)$$

$$\times \left(\sum_{\mathcal{X}_{D_2}} \prod_{j \in \tilde{A}} p(X_j|\Pi_j) \prod_{j \in \tilde{C}_2} p(X_j|\Pi_j) \prod_{j \in \tilde{D}_2} p(X_j|\Pi_j) \right)$$

$$\times \left(\sum_{\mathcal{X}_{D_3}} \prod_{j \in \tilde{B}} p(X_j|\Pi_j) \prod_{j \in \tilde{C}_3} p(X_j|\Pi_j) \prod_{j \in \tilde{D}_3} p(X_j|\Pi_j) \right)$$

$$= (\psi_1(C)\psi_2(A, C))\psi_3(B, C)$$

where the definitions of ψ_1, ψ_2 and ψ_3 are clear from the context. This factorization clearly satisfies the required criteria. It follows that d-separation implies conditional independence.   □

2.9   Markov models and Bayesian networks

This section introduces the *local directed Markov condition*, a necessary and sufficient condition so that a probability function p over a set of variables V can be factorized along a graph $\mathcal{G}$.

Definition 2.28 (Local Directed Markov Condition, Locally $\mathcal{G}$-Markovian) *Let* $V = \{X_1, \ldots, X_d\}$ *be a set of discrete random variables. A probability function p over the random vector $\underline{X} = (X_1, \ldots, X_d)$ satisfies the* local directed Markov condition *with respect to a DAG $\mathcal{G} = (V, E)$ or, equivalently, is said to be* locally $\mathcal{G}$-Markovian *if and only if for each $j \in \{1, \ldots, d\}$, X_j is conditionally independent, given Π_j (the set of parents of X_j) of all the variables in the set $V \setminus (V_j \cup \Pi_j)$, where V_j is the set of all descendants of X_j. That is,*

$$V_j = \{Y \in V | \text{there is a directed path from } X_j \text{ to } Y\}. \tag{2.12}$$

That is,

$$X_j \perp V \setminus (V_j \cup \Pi_j) | \Pi_j.$$

The terminology *Markov model* corresponding to a directed acyclic graph $\mathcal{G} = (V, E)$, defined below, was introduced into the literature and may be found in [55].

Definition 2.29 (Markov Model) *Let $V = \{X_1, \ldots, X_d\}$ denote a set of variables and let $\mathcal{G} = (V, E)$ be a directed acyclic graph. Let $\mathcal{V}$ denote the entire set of subsets of V. Let p be a probability function for the random vector $\underline{X} = (X_1, \ldots, X_d)$. Let*

$$\mathcal{I}(p) = \{(X, Y, S) \in V \times V \times \mathcal{V} | X, Y \notin S, \quad X \perp Y | S\}.$$

Note that $\phi \in \mathcal{V}$ and $X \perp Y | \phi$ means that $X \perp Y$.

The Markov model $\mathcal{M}_\mathcal{G}$ *determined by a directed acyclic graph $\mathcal{G} = (V, E)$ is the set of conditional independence statements*

$$\mathcal{M}_\mathcal{G} = \{\mathcal{I} | \mathcal{I} = \mathcal{I}(p) \quad \text{for some p that is locally $\mathcal{G}$- Markovian}\}.$$

That is, the Markov model *is the set of all sets $\mathcal{I}$ of conditional independence relations corresponding to locally $\mathcal{G}$-Markovian distributions. A distribution p is said to belong to the Markov model of $\mathcal{G}$, $p \in \mathcal{M}_\mathcal{G}$, if and only if $\mathcal{I}(p) \in \mathcal{M}_\mathcal{G}$.*

Proposition 2.1 *Let $\mathcal{I}(p)$ denote the entire set of conditional independence statements satisfied by a probability function p for a random vector $\underline{X} = (X_1, \ldots, X_d)$. Then $\mathcal{I}(p) \in \mathcal{M}_\mathcal{G}$ if and only if p factorizes along $\mathcal{G}$.*

Proof of Proposition 2.1 Firstly, if $\mathcal{I}(p) \in \mathcal{M}_\mathcal{G}$ then, by definition, p is locally $\mathcal{G}$-Markovian; that is, for each $j \in \{1, \ldots, d\}$, $X_j \perp V \setminus (V_j \cup \Pi_j)$ where V_j is defined in Equation (2.12). Let $\pi_j(x_1, \ldots, x_{j-1})$ denote the instantiation of Π_j when $\underline{X}$ is instantiated as $(x_1, \ldots, x_d)$. By characterization 1) of Theorem 2.1, for all $j = 1, \ldots, d$ and any π_j such that $p_{\Pi_j}(\pi_j) > 0$,

$$p_{X_j | X_1, \ldots, X_{j-1}}(x_j | x_1, \ldots, x_{j-1}) = p_{X_j | \Pi_j}(x_j | \pi_j)$$

with $p_{X_j | X_1, \ldots, X_{j-1}}(x_j | x_1, \ldots, x_{j-1}) = p_{X_j}(x_j)$ if $\Pi_j = \phi$. It follows directly that

$$p_{X_1, \ldots, X_d} = \prod_{j=1}^{d} p_{X_j | \Pi_j}$$

and hence, by definition, that p factorizes along $\mathcal{G}$.

Secondly, suppose that p factorizes along a graph $\mathcal{G} = (V, E)$. Then it is clear (for example by using the Bayes ball algorithm) that

$$X_j \perp V\backslash(V_j \cup \Pi_j)\|_{\mathcal{G}}\Pi_j$$

where V_j is the set of variables defined by Equation (2.12). If Π_j is instantiated, then any trail from X_j to a variable in $V\backslash(V_j \cup \Pi_j)$ has to pass through a node in Π_j, which will be either a chain or fork connection. It follows from Theorem 2.2 that

$$X_j \perp V\backslash(V_j \cup \Pi_j)|\Pi_j,$$

from which it follows that p is locally $\mathcal{G}$-Markovian.   □

2.10   *I*-maps and Markov equivalence

A particular directed acyclic graph arises from a particular ordering of the variables. This ordering may arise from considerations of cause to effect, but often this is not the case. Faced with a probability distribution $p_{X_1,\dots,X_d}(x_1, \dots, x_d)$, it is often useful to find a factorization along a directed acyclic graph that expresses the structural dependencies between the variables, even if there is no clear *causal* relationship between the variables. This problem was considered in [56] and the references therein.

As in Definition 2.29, let $\mathcal{V}$ denote the set of all subsets of V. The collection of triples

$$\mathcal{M} = \{(X, Y, S) \in V \times V \times \mathcal{V} \mid X \perp Y\|_{\mathcal{G}}S\}$$

(using the notation of Definition 2.17) represents the entire set of conditional independence statements that it is possible to infer from the DAG, but this collection does not necessarily represent the complete set of independence statements that hold for a collection of variables under a given probability distribution. When it does, it is known as a *perfect I-map*.

Definition 2.30 (Perfect *I*-Map, Faithful) *A DAG $\mathcal{G} = (V, E)$ over a set of variables V is known as a* perfect *I*-map *for a probability function p over V if for any three disjoint subsets of variables A, B and C,*

$$A \perp B|C \Leftrightarrow A \perp B \|_{\mathcal{G}} C,$$

using the notation introduced in Definition 2.18, Equation (2.11). If $\mathcal{G}$ is a perfect I-map for p, then $\mathcal{G}$ is said to be faithful *to p.*

If a collection of independence statements is in view, rather than a probability distribution p, the term *consistency* is used to mean exactly the same thing.

Definition 2.31 (Consistency) *Let $V = \{X_1, \dots, X_d\}$ be a collection of random variables and let $\mathcal{M}$ denote the entire collection of conditional independence statements: that is, let $\mathcal{V}$ denote the set of all subsets of V. Then for all $(X_i, X_j, S) \in V \times V \times \mathcal{V}$: $X_i, X_j \notin S$,*

$$(X_i, X_j, S) \in \mathcal{M} \Leftrightarrow X_i \perp X_j|S.$$

A *directed acyclic graph* $\mathcal{G} = (V, E)$ is *consistent with a set of conditional independence statements* $\mathcal{M}$ *if and only if*

$$(X_i, X_j, S) \in \mathcal{M} \Leftrightarrow X_i \perp X_j \|_{\mathcal{G}} S.$$

Let p denote a probability distribution over a set of variables $V = \{X_1, \ldots, X_d\}$ and let $\mathcal{M}_p$ denote the set of conditional independence statements associated with p. That is, for each $(X, Y, S) \in V \times V \times V$, let $(X, Y, S) \in \mathcal{M}_p \Leftrightarrow X \perp Y|S$. Then a DAG $\mathcal{G} = (V, E)$ is consistent with $\mathcal{M}_p$ if and only if $\mathcal{G}$ is *faithful* to p, if and only if $\mathcal{G}$ is a *perfect I-map* of p.

A set of variables $(X_1, \ldots, X_d)$, may be ordered in $d!$ ways. Each permutation σ of $1, \ldots, d$ gives an ordering $(X_{\sigma(1)}, \ldots, X_{\sigma(d)})$. Suppose that an ordering σ of the variables is given and that, for each variable $X_{\sigma(j)}$, a *minimal* set of σ-predecessors $\Pi_j^{(\sigma)}$ is identified that renders $X_{\sigma(j)}$ independent of all the other σ-predecessors. A *σ-predecessor* for variable $X_{\sigma(j)}$ is a variable $X_{\sigma(i)}$ such that $i < j$. Let $\Sigma_j = \{X_{\sigma(1)}, \ldots, X_{\sigma(j)}\}$ A minimal set is a set $U_j \subseteq \Sigma_j$ such that for any $Y \in U_j$, then $X_{\sigma(j)}$ will not be independent of $(\Sigma_j \backslash U_j) \cup \{Y\}$ given $U_j \backslash \{Y\}$. A direct link is then assigned from every variable in $\Pi_j^{(\sigma)}$ to $X_{\sigma(j)}$. The resulting DAG is minimal, in the sense that no edge can be deleted if the DAG is to represent the probability distribution.

The input for this construction consists of a list L of d conditional independence statements, one for each variable, all of the form $\{X_{\sigma(j)}\} \perp U_j^\sigma | \Pi_j^\sigma$, where U_j^σ is the list of σ-predecessors of $X_{\sigma(j)}$, without Π_j^σ. This is equivalent to the statement that the variable $X_{\sigma(j)}$ is d-separated from the set of variables U_j^σ when the set of variables Π_j^σ is instantiated.

For a given collection of variables $V = \{X_1, \ldots, X_d\}$, there may be several *different* DAGs, each representing the *same* independence structure. Two DAGs which represent exactly the same independence structure are said to be I-equivalent.

Definition 2.32 (*I-sub-map, I-map, I-equivalence, Markov Equivalence*) *Let $\mathcal{G}_1$ and $\mathcal{G}_2$ be two DAGs over the same variables. The DAG $\mathcal{G}_1$ is said to be an I-sub-map of $\mathcal{G}_2$ if any pair of variables d-separated by a set in $\mathcal{G}_1$ are also d-separated by the same set in $\mathcal{G}_2$. They are said to be I-equivalent if $\mathcal{G}_1$ is an I-sub-map of $\mathcal{G}_2$ and $\mathcal{G}_2$ is an I-sub-map of $\mathcal{G}_1$.*

I-equivalence is also known as Markov equivalence.

Example 2.11 In the following example on three variables, all three factorizations give the same independence structure. Consider a probability distribution p_{X_1, X_2, X_3} with factorization

$$p_{X_1, X_2, X_3} = p_{X_1} p_{X_2|X_1} p_{X_3|X_2}.$$

It follows that

$$p_{X_1, X_2, X_3} = p_{X_2} p_{X_1|X_2} p_{X_3|X_1, X_2} = p_{X_2} p_{X_1|X_2} p_{X_3|X_2},$$

using Theorem 2.1, since $X_1 \perp X_3|X_2$. Also,

$$p_{X_1, X_2, X_3} = p_{X_3} p_{X_2|X_3} p_{X_1|X_2, X_3} = p_{X_3} p_{X_2|X_3} p_{X_1|X_2},$$

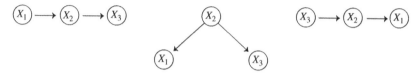

Figure 2.21   Three DAGs, each with the same independence structure.

since $X_1 \perp X_3|X_2$. For the first and last of these, X_2 is a chain node, while in the second of these X_2 is a fork node. The conditional independence structure associated with chains and forks is the same. The three corresponding DAGs are given in Figure 2.21.   □

   In general, the factorizations resulting from different orderings of the variables will not necessarily give *I*-equivalent maps. This is illustrated by the following example on four variables.

Example 2.12   Consider a probability distribution over four variables, which may be factorized as

$$p_{X_1,X_2,X_3,X_4} = p_{X_1}p_{X_2}p_{X_3|X_1,X_2}p_{X_4|X_3}.$$

If the distribution is factorized using the ordering (X_1, X_4, X_3, X_2), proceeding as outlined above, the following factorization results:

$$p_{X_1,X_2,X_3,X_4} = p_{X_1}p_{X_4|X_1}p_{X_3|X_1,X_4}p_{X_2|X_1,X_3,X_4} = p_{X_1}p_{X_4|X_1}p_{X_3|X_1,X_4}p_{X_2|X_1,X_3}$$

since $X_1 \not\perp X_4$ and $X_2 \not\perp X_1|(X_3, X_4)$. The corresponding DAG gives less information on conditional independence; it is not possible, with this ordering of the variables, to conclude that X_4 is conditionally independent of X_1, given X_3. The two corresponding DAGs are shown in Figure 2.22.   □

   While *d*-separated variables are conditionally independent conditioned on the separating set, it does not hold that conditionally independent variables are necessarily *d*-separated; a necessary and sufficient condition is that the DAG is a perfect *I*-map. This result is stated in the following theorem.

Theorem 2.3 *Recall Definition 2.18 and the notation introduced in Equation (2.11). Let p be a probability function for a random vector* $\underline{X} = (X_1, \ldots, X_d)$, *factorized along a DAG* $\mathcal{G}$. *Then p and* $\mathcal{G}$ *are* faithful *to each other if and only if for any three disjoint sets of variables A, B and C*

$$A \perp B \parallel_{\mathcal{G}} C \Leftrightarrow p_{A,B|C} = p_{A|C}p_{B|C}.$$

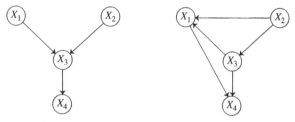

Figure 2.22   DAGs with different independence structures, arising from different factorizations of the same distribution.

In other words, p and $\mathcal{G}$ are faithful to each other if and only if for any three disjoint sets of variables A, B and C,

$$A \perp B \parallel_{\mathcal{G}} C \Leftrightarrow A \perp B|C.$$

Proof of Theorem 2.3 $A \perp B\parallel_{\mathcal{G}} C \Longrightarrow A \perp B|C$ is the result of Theorem 2.2. If p and $\mathcal{G}$ are faithful and $A \perp B|C$ then (by definition of faithfulness) $A \perp B\parallel_{\mathcal{G}} C$. This d-separation follows straight from the definition of faithfulness and is, indeed, the purpose of the definition. □

A DAG that is *faithful* to a probability distribution p may be considered as a true representation of the distribution, in the sense that there are no artificial dependencies introduced by the graph. That is, any statement inferred from the graph that two sets of variables A and B are d-connected when a set S is instantiated implies that the sets of variables A and B are not independent, given S.

2.10.1   The trek and a distribution without a faithful graph

Theorem 2.3 raises an important question; given a probability distribution p_X over a random vector $\underline{X} = (X_1, \ldots, X_d)$, is it always possible to find a directed acyclic graph that is faithful to the independence relations of p? The answer is no, as the following basic example on four variables illustrates.

Definition 2.33 (Trek) *Let $\mathcal{G} = (V, E)$ be a directed acyclic graph. A trek is a sub-graph with four variables (X_1, X_2, X_3, X_4) where there are exactly four directed edges; $X_1 \to X_2$, $X_2 \to X_4$, $X_1 \to X_3$, $X_3 \to X_4$. The sub-graph is illustrated in Figure 2.23.*

The d-separation statements for the trek are: $X_1 \perp X_4\parallel_{\mathcal{G}}\{X_2, X_3\}$ and $X_2 \perp X_3\parallel_{\mathcal{G}}X_1$. This is the entire list; $X_1 \not\perp X_4\parallel_{\mathcal{G}}\phi$, $X_2 \not\perp X_3\parallel_{\mathcal{G}}\phi$, $X_2 \not\perp X_3\parallel_{\mathcal{G}}\{X_4\}$, $X_2 \not\perp X_3\parallel_{\mathcal{G}}\{X_1, X_4\}$. These relations may be seen using the Bayes ball. There are exactly three directed acyclic graphs that share the same d-separation properties as the trek; the other two are shown in Figure 2.24.

Now consider a distribution over (X_1, X_2, X_3, X_4) such that the entire list of conditional independence statements is $X_1 \perp X_4|\{X_2, X_3\}$, $X_1 \perp X_4$, $X_2 \perp X_3|X_1$. Such a distribution is given by taking a distribution that factorizes along the trek;

$$p_{X_1, X_2, X_3, X_4} = p_{X_1}p_{X_2|X_1}p_{X_3|X_1}p_{X_4|X_2, X_3},$$

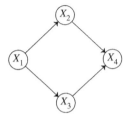

Figure 2.23   A trek.

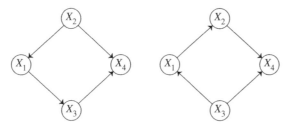

Figure 2.24   Graphs with the same *d*-separation properties as Figure 2.23.

where X_1, X_2, X_3, X_4 are each binary variables, taking values 1 or 0 and where

$$p_{X_2|X_1}(1|0) = 1 - p_{X_3|X_1}(1|1) = a,$$

$$p_{X_2|X_1}(1|1) = 1 - p_{X_3|X_1}(1|0) = b,$$

$$p_{X_4|X_2,X_3}(1|11) = p_{X_4|X_2,X_3}(1|00) = c, \qquad p_{X_4|X_2,X_3}(1|01) = p_{X_4|X_2,X_3}(1|10) = d.$$

Then

$$p_{X_4|X_1}(1|1) = c\,((1-a)b + a(1-b)) + d\,(ab + (1-a)(1-b))$$

while

$$p_{X_4|X_1}(1|0) = c\,(a(1-b) + (1-a)b) + d\,((1-a)(1-b) + ab)$$

so that $X_4 \perp X_1$. Take $a \neq \frac{1}{2}$, $b \neq \frac{1}{2}$, $c \neq \frac{1}{2}$, $d \neq \frac{1}{2}$, $a \neq b$, $c \neq d$. Then, with these conditions, $X_4 \not\perp X_3|S$ for any $S \subseteq \{X_1, X_2\}$ and $X_4 \not\perp X_3|S$ for any $S \subseteq \{X_1, X_2\}$. Furthermore, $X_3 \not\perp X_1|S$ for any $S \subseteq \{X_2, X_4\}$ and $X_2 \not\perp X_1|S$ for any $S \subseteq \{X_3, X_4\}$. The independence relation $X_3 \perp X_2|X_1$ holds and clearly $X_1 \perp X_4|\{X_2, X_3\}$ holds.

There does not exist a directed acyclic graph that expresses all the conditional independence relations that hold for this distribution and no others. If $X_1 \perp X_4$, then there can be no edge between X_1 and X_4. If $X_2 \perp X_3|X_1$ then there can be no edge between X_2 and X_3. Furthermore, any trail between X_1 and X_4 must contain a collider connection. The lack of *d*-separation implies that there are edges between X_1 and X_2, X_1 and X_3, X_2 and X_4, X_3 and X_4. This implies that the nodes X_2 and X_3 are collider nodes. This contradicts the requirement that $X_1 \perp X_4|\{X_2, X_3\}$, since the instantiation of a collider connection opens the communication.

Therefore, there is no directed acyclic graph that is faithful to the distribution described above when $a \neq \frac{1}{2}$, $b \neq \frac{1}{2}$, $c \neq \frac{1}{2}$, $d \neq \frac{1}{2}$, $a \neq b$, $c \neq d$. □

Notes   Perhaps the earliest work that uses directed graphs to represent possible dependencies among random variables is that by S. Wright [57]. A recent presentation of the path analysis methods due to Wright is found in Shipley [58] An early article that considered the notion of a factorization of a probability distribution along a directed acyclic graph representing causal dependencies is that by H. Kiiveri, T.P. Speed and J.B. Carlin [59], where a Markov property for Bayesian networks was defined. This was developed by J. Pearl in [9]; *d*-separation, and the extent to which it characterises independence is discussed by J. Pearl and T. Verma in [60] and by J. Pearl, D. Geiger and T. Verma in [56]. The Bayes ball is taken from R.D. Schachter [50]. See, for example, F. Markowetz and R. Spang [7] for applications of Bayesian networks to cellular models. The application to root cause analysis was discussed in [51]. The counter example showing that it is necessary that the probability function factorizes along a directed *acyclic* graph is found in N. Vorobev [53]. Object oriented Bayesian networks are discussed in [54]. The results for identifying independence in Bayesian networks are taken from D. Geiger, T. Verma and J. Pearl [61] and the results on Markov equivalence are taken from T. Verma and J. Pearl [62]. The web page in [63] contains links to an introduction to Bayesian nets and the Bayes Net Toolbox MATLAB software.

2.11   Exercises: Conditional independence and *d*-separation

The 'HUGIN' software (the educational package) may be used for the analysis of Bayesian networks. A description of the package, and information about how to obtain it, may be found at http://www.hugin.com/Products_Services/Products/Academic/Educational/

1. Let (X, Y, W, Z) be discrete random variables, each with a finite state space. Let $\mathcal{X}$ denote the state space of (X, Y, W, Z).

   (a) Prove that if $X \perp (Y, W)|Z$ then $X \perp Y|Z$ and $X \perp W|Z$.

   (b) Prove that if $X \perp Y|Z$ and $X \perp W|(Y, Z)$ then $X \perp (W, Y)|Z$.

   (c) Assume that $p_{X,Y,W,Z}(x, y, w, z) > 0$ for each $(x, y, w, z) \in \mathcal{X}$. Prove that if $X \perp Y \mid (Z, W)$ and $X \perp W \mid (Z, Y)$, then $X \perp Z \mid (Y, W)$.

2. Let A, B and C be binary random random variables, each of which takes values in $\{0, 1\}$. Suppose that the joint probability function for (A, B, C) is given by

$$p_{A,B,C}\,(0, 0, 0) = 0.028 \quad p_{A,B,C}\,(0, 0, 1) = 0.042$$

$$p_{A,B,C}\,(0, 1, 0) = 0.00003 \quad p_{A,B,C}\,(0, 1, 1) = 0.02997$$

$$p_{A,B,C}\,(1, 0, 0) = 0.072 \quad p_{A,B,C}\,(1, 0, 1) = 0.108$$

$$p_{A,B,C}\,(1, 1, 0) = 0.00072 \quad p_{A,B,C}\,(1, 1, 1) = 0.71928$$

Show that the probability function $p_{A,B,C}$ admits a factorization according to the DAG given in Figure 2.25.

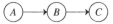

Figure 2.25   Chain connection.

3. Suppose that the following table gives the values for the *joint* probability function $p_{A,B}$:

$$p_{A,B}(.,.) =$$

	b_1	b_2	b_3
a_1	0.02	0.03	0.15
a_2	0.10	0.00	0.30
a_3	0.05	0.15	0.20

Compute p_A, p_B, $p_{A|B}$, $p_{B|A}$.

4. This is the classic example used to illustrate 'explaining away'. A former prime minister and Labour Party leader is now a business consultant. He is in his office in Stockholm, when he receives the news that the burglar alarm in his country mansion has gone off. Convinced that a burglar has broken in, he starts to drive home. But, on his way, he hears on the radio that there has been a minor earth tremor in the

area. Since an earth tremor can set off a burglar alarm, he therefore returns to his office.

(a) Construct the Bayesian network associated with the situation.

(b) Suppose that the variables are listed as R for the radio broadcast (y/n), A for the alarm (y/n), B for the burglary (y/n) and E for the earthquake (y/n), where y stands for 'yes' and n stands for 'no'. Suppose that the conditional probability tables associated with the Bayesian network are

$$p_{R|E} = \begin{array}{c|cc} R\backslash E & n & y \\ \hline n & 0.99 & 0.05 \\ y & 0.01 & 0.95 \end{array}$$

$$p_{A|B,E}(y|.,.) = \begin{array}{c|cc} E\backslash B & n & y \\ \hline n & 0.03 & 0.95 \\ y & 0.95 & 0.98 \end{array}$$

$$p_B = \begin{array}{cc} n & y \\ \hline 0.99 & 0.01 \end{array}$$

$$p_E = \begin{array}{cc} n & y \\ \hline 0.001 & 0.999 \end{array}$$

Find

i. $p_{B|A}(y|y)$, $p_{B|A}(y|n)$ and

ii. $p_{B|A,R}(y|y, y)$.

5. (a) Consider the network given in Figure 2.26, where variables B and J have been instantiated. Which variables are d-separated from A? Which variables are d-separated from F? Explain.

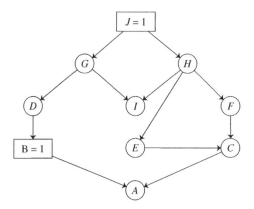

Figure 2.26   Network.

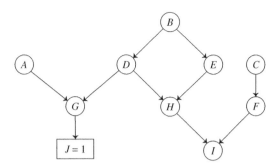

Figure 2.27   Network.

(b) Consider the network given in Figure 2.27, where variable J has been instantiated. Which variables are d-separated from A? Which variables are d-separated
from F? Explain.

6. Let A be a variable in a DAG. Prove that if all the variables in the *Markov blanket* of
A are instantiated, then A is d-separated from the remaining uninstantiated variables.

7. Recall the definition of I-sub-maps and I-equivalence given in Definition 2.32.

(a) Find those graphs from the four given in Figure 2.28 that are I-equivalent to
each other.

(b) Suppose that the probability distributions p, q and r over the variables A, B and
C may be factorized as $p_A p_{B|A} p_{C|A}$, $q_B q_{A|B} q_{C|A}$ and $r_B r_C r_{A|B,C}$. Do p and q
have the same independence structure? Do p and r have the same independence
structure?

8. Consider the first DAG in Figure 2.28 (a fork connection centred at A) and suppose
that $p_{A,B,C}$ may be factorized along the DAG.

(a) Prove that
$$p_{B|A,C} = p_{B|A}.$$

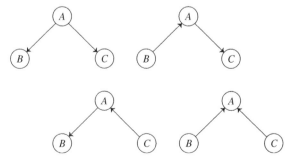

Figure 2.28   Which are I equivalent?

(b) Suppose that $p_{A,B,C} = p_A p_{B|A} p_{C|A}$, with probability tables given by

$$p_A = \begin{array}{cc} a_1 & a_2 \\ \hline 0.3 & 0.7 \end{array}$$

$$p_{B|A} = \begin{array}{c|cc} A\backslash B & b_1 & b_2 \\ \hline a_1 & 0.1 & 0.9 \\ a_2 & 0.8 & 0.2 \end{array}$$

$$p_{C|A} = \begin{array}{c|cc} A\backslash C & c_1 & c_2 \\ \hline a_1 & 0.6 & 0.4 \\ a_2 & 0.4 & 0.6 \end{array}$$

Using appropriate operations of tables, compute

$$p_{A,B}, \ p_B, \ p_{B,C}, \ p_{C|B} \quad \text{and} \quad p_{A,B,C}.$$

9. Let T denote a test result (positive or negative) for the event A (whether or not a driver has too much alcohol in the blood). Suppose the conditional probabilities $p_{T|A}$ are given in the following table (where y = 'yes' and n = 'no'):

$$p_{T|A}(.|.) = \begin{array}{c|cc} A\backslash T & y & n \\ \hline y & 0.95 & 0.05 \\ n & 0.005 & 0.995 \end{array}$$

(a) The police may stop a motorist and perform a blood test on suspicion that the motorist is driving under the influence of alcohol. Experience suggests that 15% of drivers under suspicion do, in fact, drive with too much alcohol. Compute the table $p_{A|T}$. A driver is taken and the blood test is positive. What is the probability that the driver has too much alcohol?

(b) One week, the policy changes so that the police stop drivers randomly and carry out the same test. It is estimated that one in 2000 drivers stopped at random have too much alcohol in their blood. Compute the table $p_{A|T}$. A driver is stopped and gives a positive test result. What is the probability that he is driving under the influence of alcohol?

10. **Conditional independence** Consider three random variables (X, Y, Z) with joint probability function $p_{X,Y,Z}$. Prove that $X \perp Y|Z$ if and only if

$$p_{X,Y,Z}(x, y, z)p_{X,Y,Z}(x', y', z) = p_{X,Y,Z}(x', y, z)p_{X,Y,Z}(x, y', z) \quad \forall(x, y, x', y', z).$$

This observation is due to B. Sturmfels and has important consequences for the application of techniques from algebraic geometry to the study of Bayesian networks (see [64]). The quantity

$$\Xi(x, y; x', y'; z) := p_{X,Y,Z}(x, y, z)p_{X,Y,Z}(x', y', z) - p_{X,Y,Z}(x', y, z)p(x, y', z)$$

is known as the *cross product difference* and is identically zero if and only if $X \perp Y|Z$.

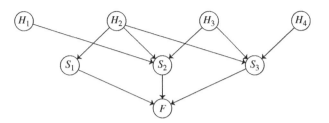

Figure 2.29   Root Cause Analysis.

11. **Root Cause Analysis** Are the variables F and (H_2, H_3) d-connected when (S_1, S_3) are instantiated in the directed acyclic graph in Figure 2.29?

12. **Transitive DAGs** A DAG $\mathcal{G} = (V, E)$ is defined as a *transitive DAG* if it satisfies the following additional condition:

$$(X_i, X_j) \in E \text{ and } (X_j, X_r) \in E \implies (X_i, X_r) \in E.$$

Let an (X_ν) denote the set of ancestors of X_ν. Check that a DAG $\mathcal{G}$ is transitive if and only if for every ν

$$\Pi(X_\nu) = \text{an}(X_\nu).$$

13. **Transitive DAGs** Assume that $\mathcal{G} = (V, E)$ is a transitive DAG following the definition in Exercise 12 and that p a probability distribution over the variables in V in $\mathcal{G} = (V, E)$. Show that p satisfies local directed Markov condition with respect to $\mathcal{G}$ (Definition 2.28) if and only if for all X and Y in V, $X \perp Y \mid X \cap Y$ or, equivalently $X \setminus Y \perp Y \setminus X \mid X \cap Y$. This is known as the *lattice conditional independence property* (LCI). In other words, the LCI is characterized by a Markov model on a transitive DAG. This result is found in [55].

14. The notation $\underline{X}_A$ is used to denote the random (row) vector of all variables in set A. Let $V = \{X_1, \ldots, X_d\}$ be the d variables of a Bayesian network and assume that $\underline{X}_{V \setminus \{X_i\}} = \underline{w}$. That is, all the variables except X_i are instantiated. Assume that X_i is a binary variable, taking values 0 or 1. Consider the odds

$$O_p\left(\{X_i = 1\} \mid \{\underline{X}_{V \setminus \{X_i\}} = \underline{w}\}\right),$$

and show that this depends only on the variables in the Markov blanket (Definition 2.20) of X_i.

3

Evidence, sufficiency and Monte Carlo methods

Let $\underline{X} = (X_1, \ldots, X_d)$ denote a random vector, with finite state space $\mathcal{X} = \times_{j=1}^{d} \mathcal{X}_j$, where $\mathcal{X}_j = (x_j^{(1)}, \ldots, x_j^{(k_j)})$ is the state space for random variable X_j. This chapter considers the various kinds of evidence available: *hard evidence*, *soft evidence* and *virtual evidence*. The definitions used will be the following:

Definition 3.1 (Hard Evidence, Soft Evidence, Virtual Evidence) *The following definitions will be used:*

- A hard finding *is an* instantiation, $\{X_i = x_i^{(l)}\}$ *for a particular value of* $i \in \{1, \ldots, d\}$ *and a particular value of* $l \in \{1, \ldots, k_i\}$. *This specifies that variable* X_i *is in state* $x_i^{(l)}$. *It is expressed as a* $k_1 \times \ldots \times k_d$ *potential e where*

$$e(x_1^{(p_1)}, \ldots, x_d^{(p_d)}) = \begin{cases} 1 & p_i = l \\ 0 & p_i \neq l. \end{cases}$$

   That is, the entries corresponding to configurations containing the instantiation are 1 *and the entries corresponding to all other configurations are* 0.

- Hard evidence *is a collection of hard findings. It is given by a collection of potentials* $\mathbf{e} = (e_1, \ldots, e_m)$ *where* e_m *is a hard finding on one of the variables.*

- A soft finding *on a variable* X_j *specifies the probability distribution of the variable* X_j. *That is, the potential* $p_{X_j | \Pi_j}$ *is replaced by a potential* $p_{X_j}^*$ *with domain* $\mathcal{X}_j$.

Bayesian Networks: An Introduction   T. Koski, J. Noble
© 2009 John Wiley & Sons, Ltd

- Soft evidence *is a collection of soft findings.*

- A virtual finding *on variable* X_j *is a collection of likelihood ratios* $\{L(x_j^{(m)}), m = \{1, \ldots, m\}\}$ *such that the updated conditional probability potential for* $X_j | \Pi_j = \pi_j^{(n)}$ *is, for* $m = 1, \ldots, k_j$,

$$p_{X_j|\Pi_j}^*(x_j^{(m)}|\pi_j^{(n)}) = \frac{1}{\sum_{q=1}^{k_j} p_{X_j|\Pi_j}(x_j^{(q)}|\pi_j^{(n)})L(x_j^{(q)})} p_{X_j|\Pi_j}(x_j^{(m)}|\pi_j^{(n)})L(x_j^{(m)}).$$
(3.1)

- Virtual evidence *is a collection of virtual findings.*

Soft evidence and virtual evidence are different. When soft evidence is received on a variable, the links between the variable and its parents are severed; if soft evidence is received on variable X_j, then the potential $p_{X_j|\Pi_j}$ is replaced by a new potential $p_{X_j}^*$. When virtual evidence is received, the links are preserved. Virtual evidence may be modelled by adding a new node to the network.

3.1 Hard evidence

Consider a situation where the domain variables are $\{X, Y\}$, $\mathcal{X}_X = \{x_1, x_2, x_3\}$ and $\mathcal{X}_Y = \{y_1, y_2, y_3\}$. The following is an example of a finding e that Y is in state y_1.

	X\Y	y_1	y_2	y_3
	x_1	1	0	0
$e \leftrightarrow$	x_2	1	0	0
	x_3	1	0	0

Entering hard evidence Let $p_{\underline{X}}$ denote a joint probability function over a random vector $\underline{X}$ and let e be a finding corresponding to a piece of evidence. The evidence is *entered* into $p_{\underline{X}}$ by multiplying the potentials together. Let $p_{\underline{X};e}$ denote the result;

$$p_{\underline{X};e} = p_{\underline{X}}.e$$
(3.2)

Suppose that several hard findings are received, expressed in potentials $(e_1, \ldots, e_k)$. Then the hard evidence $\mathbf{e} = (e_1, \ldots, e_k)$ is entered by

$$p_{\underline{X};\mathbf{e}} = p_{\underline{X}}. \prod_{j=1}^{k} e_j,$$

where the multiplication is in the sense of potentials, having been extended to the domain $\mathcal{X}$.

Hard evidence renders certain states impossible, giving value 0 for the corresponding configurations in the potential and leaves the other configurations unaltered. The potential

$\prod_{j=1}^{k} e_j$ (where the domains have been extended to $\mathcal{X}$) may be considered as a hard evidence potential with domain $\mathcal{X}$. Clearly, multiplication of hard evidence potentials yields another hard evidence potential.

The potential $p_{\underline{X};e}$ is not a probability function, in the sense that the entries do not add up to 1. To compute the conditional probability potential, conditioned on the evidence received, it is necessary to compute the probability of the evidence. This is given by

$$p(\mathbf{e}) = \sum_{\underline{x} \in \mathcal{X}} p_{\underline{X};e}(\underline{x})$$

and the conditional probability potential is given by

$$p_{\underline{X}|e} = \frac{p_{\underline{X};e}}{p(\mathbf{e})}, \tag{3.3}$$

where the division is taken in the sense of potentials (Definition 2.24).

Example 3.1   Consider a joint probability function

$$p_{X,Y} = \quad \begin{array}{c|ccc} X\backslash Y & y_1 & y_2 & y_3 \\ \hline x_1 & 0.05 & 0.10 & 0.05 \\ x_2 & 0.15 & 0.00 & 0.25 \\ x_3 & 0.10 & 0.20 & 0.10 \end{array}$$

and suppose that evidence is received that $\{Y = y_1\}$. The *evidence potential e* is

$$e = \quad \begin{array}{c|ccc} X\backslash Y & y_1 & y_2 & y_3 \\ \hline x_1 & 1 & 0 & 0 \\ x_2 & 1 & 0 & 0 \\ x_3 & 1 & 0 & 0 \end{array}$$

Entering the evidence gives

$$p_{X,Y;e} = p_{X,Y}.e = \quad \begin{array}{c|ccc} X\backslash Y & y_1 & y_2 & y_3 \\ \hline x_1 & 0.05 & 0 & 0 \\ x_2 & 0.15 & 0 & 0 \\ x_3 & 0.10 & 0 & 0 \end{array}$$

This potential is not a probability function; the numbers do not add up to 1, but the conditional probability potential may be computed by first computing the *probability of the evidence*;

$$p(e) = \sum_{x,y} p_{X,Y;e}(x, y) = 0.05 + 0.15 + 0.10 = 0.30.$$

The *conditional probability* of (X, Y) given the evidence is then computed using the rules for 'division of potentials' (Definition 2.24):

$$p_{X,Y|e} = \frac{p_{X,Y;e}}{p(e)} =$$

X\Y	y_1	y_2	y_3
x_1	$\frac{1}{6}$	0	0
x_2	$\frac{1}{2}$	0	0
x_3	$\frac{1}{3}$	0	0

Example 3.2   Consider an example where X, Y and Z each take on values 0 or 1 and the joint probability function of (X, Y, Z) may be factorized according to the DAG in Figure 3.1. Z is the parent, X and Y are the children.

The factorization of the joint probability function corresponding to the DAG is

$$p_{X,Y,Z} = p_{X|Z} p_{Y|Z} p_Z.$$

Suppose that a hard finding is received, that $\{Y = 0\}$. The evidence potential is

Z\Y	0	1
$e =$ 0	1	0
1	1	0

Now,

$$p_{X,Y,Z;e} = p_{X|Z} p_{Y|Z} p_Z e$$

so that, by the multiplication rules for potentials,

Z\Y	0	1
$p_{Y|Z} p_Z e \leftrightarrow$ 0	$p_{Y|Z}(0|0) p_Z(0)$	0
1	$p_{Y|Z}(0|1) p_Z(1)$	0

By multiplication rules for potentials with different domains,

X\Z\Y	0	1
$p_{X|Z} p_{Y|Z} p_Z e \leftrightarrow$ 0	$(a, 0)$	$(b, 0)$
1	$(c, 0)$	$(d, 0)$

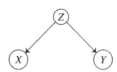

Figure 3.1   Graph for Example 3.2.

where

$$a = p_{X|Z}(0|0) p_{Y|Z}(0|0) p_Z(0),$$

$$b = p_{X|Z}(1|0) p_{Y|Z}(0|0) p_Z(0),$$

$$c = p_{X|Z}(0|1) p_{Y|Z}(0|1) p_Z(1),$$

$$d = p_{X|Z}(1|1) p_{Y|Z}(0|1) p_Z(1).$$

By disregarding the zero elements in the potential, 'entering the evidence' has resulted in a potential defined over the domain $\mathcal{X}_X \times \mathcal{X}_Y$

$$p_{X,Y,Z;e}(., 0, .) = p_{X|Z}(.|.) p_{Y|Z}(0|.) p_Z(.).$$

Let Π_v denote the set of parent variables of X_v and let π_v denote the function such that $\pi_v(x_1, \ldots, x_n) = (x_{j_1}, \ldots, x_{j_v})$ if and only if $\Pi_v = \{X_{j_1}, \ldots, X_{j_v}\}$, for $1 \le j_1 < \ldots < j_v \le n$. Let $\underline{x} = (x_1, \ldots, x_d)$. The general rule for hard evidence is, for a collection of hard evidence potentials $\mathbf{e} = (e_1, \ldots, e_m)$,

$$p_{X_1,\ldots,X_d;e}(x_1, \ldots, x_d) = \prod_{v=1}^{n} p_{X_v|\Pi_v}(x_v|\pi_v(\underline{x})) \prod_{i=1}^{m} e_i.$$

3.2   Soft evidence and virtual evidence

Soft evidence and virtual evidence (Definition 3.1) are both types of evidence that affect the probabilities of a state, but which do not enable any claim to be made that the probability of a state is zero. A virtual finding on variable X_j affects the probability potential $p_{X_j|\Pi_j}$, without affecting the *subsequent* conditional probability potentials. The virtual finding may be considered as an additional variable V (denoting 'virtual finding'), that is added into the network, with a single arrow from V, to the variable affected by the soft evidence. This is the situation in which Pearl's method of virtual evidence is applicable, depending on how the evidence is formulated.

Example 3.3   Consider a DAG on five variables, X_1, X_2, X_3, X_4 and X_5, given in Figure 3.2. Suppose that a piece of virtual evidence is received on the variable X_3. This evidence may be modelled by a variable V, that is inserted to the DAG giving the DAG in Figure 3.3. The variable X_3 is in a certain state, which affects the soft evidence that is observed.

It is clear that $(X_1, X_2, X_4, X_5) \perp V \|_{\mathcal{G}} X_3$. In particular, the decomposition along the DAG gives $p(V|X_1, X_2, X_3, X_4, X_5) = p(V|X_3)$ and $p(X_1, X_2, X_4, X_5|X_3, V) = p(X_1, X_2, X_4, X_5|X_3)$.

In general, consider a set of variables $V = \{X_1, \ldots, X_d\}$, where the joint probability distribution is factorized as

$$p_{X_1,\ldots,X_d} = \prod_{j=1}^{d} p_{X_j|\Pi_j}.$$

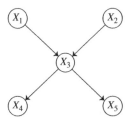

Figure 3.2 Before virtual evidence is added.

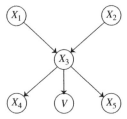

Figure 3.3 After the virtual evidence node is added.

Suppose that virtual evidence is received on variable X_j. This may be expressed as a variable V and, by d-separation properties, the updated distribution $p_{X_1,...,X_d,V}$ has a factorization

$$p_{X_1,...,X_d,V} = \left(\prod_{k \neq j} p_{X_k|\Pi_k} \right) p_{X_j|\Pi_j \cup \{V\}} p_V. \qquad (3.4)$$

The variable V is a 'dummy variable', in the sense that its state space and distribution do not need to be defined; the virtual evidence is interpreted as a particular instantiation $\{V = v\}$ for this variable and this is the only information that is needed. From Equation (3.4),

$$p_{X_1,...,X_d|V}(., ..., .|v) = \left(\prod_{k \neq j} p_{X_k|\Pi_k} \right) p_{X_j|\Pi_j \cup \{V\}}(.|., v).$$

From Equation (3.1),

$$\frac{L(x_j^{(m)})}{\sum_{i=1}^{k_j} L(x_j^{(i)}) p_{X_j|\Pi_j}(x_j^{(i)}|\pi_j^{(n)})} = p_{X_j|\Pi_j \cup \{V\}}(x_j^{(i)}|\pi_j^{(n)}, v) \qquad m = 1, ..., k_j.$$

This is the setting where Pearl's update method may be applied.

3.2.1 Jeffrey's rule

The definition of the Jeffrey's update was given in Equation (1.9) and is restated here. Let p denote the original probability and let q denote the probability obtained after the update is applied. Let $G_1, ..., G_r$ denote the set of mutually exclusive and exhaustive events for which the updated probability $q(G_i)$ is prescribed. Set $\lambda_i = q(G_i)$ and $\mu_i = p(G_i)$. For any set $A \subset \mathcal{X}$ and any $x \in \mathcal{X}$, let $\mathbf{1}_A$ denote the indicator function;

$$\mathbf{1}_A(x) = \begin{cases} 1 & x \in A \\ 0 & x \notin A. \end{cases}$$

Then Jeffrey's rule may be rewritten in the following way: for all $x \in \mathcal{X}$,

$$q(\{x\}) = \sum_{j=1}^{r} \frac{\lambda_j}{\mu_j} p(\{x\}) \mathbf{1}_{G_j}(x).$$

The following example is taken from [65] and discussed in [66].

Example 3.4   A piece of cloth is to be sold on the market. The colour C is either green (c_g), blue (c_b) or violet (c_v). Tomorrow, the piece of cloth will either be sold (s) or not (s^c); this is denoted by the variable S. Experience gives the following probability distribution over C,S

$$p_{C,S} = \begin{array}{c|ccc} S\backslash C & c_g & c_b & c_v \\ \hline s & 0.12 & 0.12 & 0.32 \\ s^c & 0.18 & 0.18 & 0.08 \end{array}$$

The marginal distribution over C is

$$p_C = \begin{array}{ccc} c_g & c_b & c_v \\ \hline 0.3 & 0.3 & 0.4 \end{array}.$$

The piece of cloth is inspected by candle light. Since it cannot be seen perfectly, this only gives *soft evidence*. From the inspection by candle light, the probability over C is assessed as:

$$q_C = \begin{array}{ccc} c_g & c_b & c_v \\ \hline 0.7 & 0.25 & 0.05 \end{array}.$$

This is a situation where Jeffrey's rule may be used to update the probability. For example,

$$q_{S,C}(s, c_g) = \frac{\lambda_g}{\mu_g} p(s, c_g) = \frac{0.7}{0.3} \times 0.12 = 0.28.$$

Updating the whole distribution in this way gives

$$q_{C,S} = \begin{array}{c|ccc} S\backslash C & c_g & c_b & c_v \\ \hline s & 0.28 & 0.10 & 0.04 \\ s^c & 0.42 & 0.15 & 0.01 \end{array}$$

3.2.2   Pearl's method of virtual evidence

Pearl's method deals with virtual evidence. If virtual evidence is received on a variable in a Bayesian network, it may be treated by adding an additional node V. If virtual evidence is received on variable X_3 in the network in Figure 3.2, the network may be extended to the network in Figure 3.3 as before. Pearl's method of virtual evidence was introduced in Section 1.4.3 and considers the situation a piece of soft evidence $\{V = v\}$, which affects a set of mutually exclusive and exhaustive events $G_1, \ldots, G_r$, comes in the form of an odds ratio. The evidence is not specified as a set of new probabilities, but rather for each G_j, $j = 1, \ldots, r$, the ratio $\lambda_j = \frac{q(\{V=v\}|G_j)}{q(\{V=v\}|G_1)}$, $j = 2, \ldots, r$ is given, with $\lambda_1 = 1$. The parameter λ_j therefore represents the likelihood ratio of the event G_j to the event G_1 given the evidence E. Again, using the notation $\mathbf{1}_{G_j}(x) = 1$ for $x \in G_j$ and 0 for $x \notin G_j$, Pearl's method of virtual evidence defines the update $q(.)$ as

$$q(\{x\}) = p(\{x\}) \sum_{j=1}^{r} \frac{\lambda_j}{\sum_{k=1}^{r} p(G_k)\lambda_k} \mathbf{1}_{G_j}(x), \qquad x \in \mathcal{X}.$$

Example 3.5 (Burglary) The following example is taken from [9] and discussed in [66]. It has become a classic example. A variant is included as Exercise 4 in Chapter 2.

On any given day, there is a burglary at any given house with probability 10^{-4}. If there is a burglary, then the alarm will go off with probability 0.95. One day, Professor Noddy receives a call from his neighbour Jemima, saying that she may have heard Professor Noddy's burglar alarm going off. Professor Noddy decides that there is an 80% chance that Jemima did hear the alarm going off.

Let A denote the event that the alarm goes off, B the event that a burglary takes place and let E denote the evidence of the telephone call from Jemima. According to Pearl's method, this evidence can be interpreted as

$$\lambda = \frac{p(E|A)}{p(E|A^c)} = 4.$$

An application of Pearl's virtual evidence rule gives the updated probability that a burglary has taken place as $q(B) = 3.85 \times 10^{-4}$.

3.3   Queries in probabilistic inference

The following are examples of *queries* for a Bayesian network with the variables in U:

Probability updating namely, if evidence **e** is given on some variables, find the posterior probability potentials for the rest of the variables.

Most probable configuration namely, if evidence **e** is given on the variables in a set U, find the most probable values of the rest of the variables.

Maximum aposterior (map) hypothesis namely, if evidence **e** is given on some variables in a set U, find a hypothesis h over a subset of variables which maximizes the probability $p(h|\mathbf{e})$.

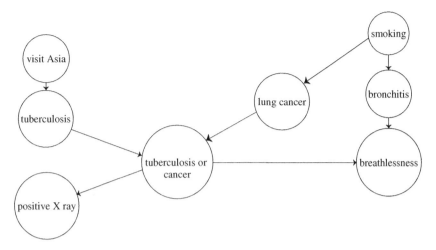

Figure 3.4   The Chest Clinic Problem.

3.3.1   The chest clinic problem

The following is an example of a query. It is known as 'The Chest Clinic Problem' and is due to A.P. Dawid. It may be found in [67]. You are ill, short of breath, and you want to know what is wrong with you. The corresponding DAG is given in Figure 3.4.

Chapters 4 and 10 develop a method for the first query listed, updating the probability distribution when evidence is received. As pointed out by A.P. Dawid, in [68], this method may equally well be applied to the problem of 'maximum aposterior hypothesis', with only minor modifications.

3.4   Bucket elimination

Most of this text is concerned with finding effective algorithmic solutions to probability updating, most probable configuration and maximum aposterior hypothesis. The first technique considered is an algorithm known as *bucket elimination*. It was introduced by R. Dechter in [69].

Consider the DAG in Figure 3.5.

Here

$$p_{A,B,C,D,F,G} = p_{G|F}\, p_{F|C,B}\, p_{D|B,A}\, p_{B|A}\, p_{C|A}\, p_A.$$

Assume there is a *finding* $G = 1$, and the updated probability

$$p_{A|e} = p_{A|G}(.|1)$$

is required. To compute this, both $p_{A;e}$ and $p(e)$ are needed. Since $e = \{G = 1\}$,

$$p_{A;e} = p_{A,G}(.,1) = \sum_{B,C,D,F} p_{G|F}(1|.)\, p_{F|C,B}\, p_{D|B,A}\, p_{C|A}\, p_{B|A}\, p_A.$$

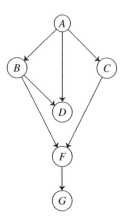

Figure 3.5 Illustration for bucket elimination.

This may be manipulated symbolically using $x(y + z) = xy + xz$. This is the *associative law*, which may seem mathematically trivial, but efficient algorithms will exploit it as much as possible. The left hand side requires two computations ($y + z$, followed by a multiplication by x) while the right hand side requires three (xy, then xz, then add them together).

$$p_{A;e} = p_A \sum_C p_{C|A} \sum_B p_{B|A} \sum_F p_{F|C,B} p_{G|F}(1|.) \sum_D p_{D|B,A}.$$

Note that $\sum_D p_{D|B,A} = 1$, so that

$$p_{A;e} = p_A \sum_C p_{C|A} \sum_B p_{B|A} \sum_F p_{F|C,B} p_{G|F}(1|.).$$

Set

$$\lambda_{C,B;G=1}(.,.) \stackrel{def}{=} \sum_F p_{F|C,B} p_{G|F}(1|.),$$

so that

$$p_{A;e} = p_A \sum_C p_{C|A} \sum_B p_{B|A} \lambda_{C,B;G=1}(.,.)$$

and set

$$\lambda_{C,A;G=1}(.,.) \stackrel{def}{=} \sum_B p_{B|A} \lambda_{C,B;G=1}(.,.),$$

so that

$$p_{A;e} = p_A \sum_C p_{C|A} \lambda_{C,A;G=1}(.,.).$$

Set

$$\lambda_{A;G=1}(.) \stackrel{def}{=} \sum_C p_{C|A} \lambda_{C,A;G=1}(.,.)$$

so that finally

$$p_{A;e} = p_A \lambda_{A;G=1}(.).$$

Here all the variables except for A have been eliminated in a certain order. Let

$$U = \{A, B, C, D, F, G\},$$

then

$$p_{A;e} = \sum_{U \setminus \{A,G\}} p_{A,B,C,D,F,|G}(., ., ., ., .|1).$$

The preceding can be described as follows. First, the conditional probability potentials are put into *buckets* relative to the order to be used in elimination of variables. The first bucket (bucket$_G$) contains all the potentials that have G in the domain. Bucket bucket$_D$ contains all those, from the remaining potentials that contain D in the domain and so on.

The buckets are therefore:

$$\text{bucket}_G = p_{G|F}(1|.),$$

$$\text{bucket}_D = p_{D|B,A},$$

$$\text{bucket}_F = p_{F|B,C},$$

$$\text{bucket}_B = p_{B|A},$$

$$\text{bucket}_C = p_{C|A},$$

$$\text{bucket}_A = p_A.$$

It is shown diagrammatically as follows. Each arrow corresponds to marginalization and multiplication.

bucket$_G$ $p_{G|F}(1|.)$

bucket$_D$ $p_{D|B,A}$ $\searrow$ $\lambda(.) = p_{G|F}(1|.)$

bucket$_F$ $p_{F|B,C}$ $\searrow \downarrow$ $\lambda(.) = p_{G|F}(1|.)$

bucket$_B$ $p_{B|A}$ $\searrow \downarrow$ $\lambda_{C,B;G=1}(., .)$

bucket$_C$ $p_{C|A}$ $\searrow \downarrow$ $\lambda_{C,A;G=1}(., .)$

bucket$_A$ p_A $\searrow \downarrow$ $\lambda_{A;G=1}(.).$

The algorithm is summarized as follows: start with a set of potentials V. Whenever a variable X is to be removed, all potentials from V with X in the domain are taken, and removed from V. Their product is calculated and the potential obtained by summing over the X variable is computed from this product. The resulting potential λ is then added to V. Repeat with the next variable.

3.5   Bayesian sufficient statistics and prediction sufficiency

The conditional independence structure is central to the analysis of a Bayesian network. Sufficient statistics, a central concept in statistics, are random variables that help to establish conditional independence. Suppose that $\mathbf{X}$ is an $n \times d$ random matrix, representing n independent copies of a random (row) vector $\underline{X}$ and $\underline{\Theta}$ is a random vector. A function t of $\mathbf{X}$ such that $\mathbf{X}$ and $\underline{\Theta}$ are independent given $t(\mathbf{X})$ is a *Bayesian sufficient statistic* for $\underline{\Theta}$. The notation is suggestive; an instantiation of the random matrix $\mathbf{X}$ represents the results of n independent replications of an experiment, where d attributes are measured, while $\underline{\Theta}$ is the random vector associated with the parameters. In this text, a random vector that models the outcome of an experiment is, for the most part, discrete and the random vector associated with the parameters is generally taken to be continuous. In the analysis that follows, it will be assumed that $\mathbf{X}$ is discrete and $\underline{\Theta}$ is continuous, although the results are general and the proofs are easily modified to deal with the other cases.

3.5.1   Bayesian sufficient statistics

Let $\mathbf{X}$ be an $n \times d$ random matrix, where each row is an independent copy of a discrete random vector and let $\underline{\Theta}$ be a continuous random vector representing the unknown parameters, from the Bayesian point of view. Suppose that, conditioned on $\underline{\Theta} = \underline{\theta}$, $\mathbf{X}$ has conditional probability function $p_{\mathbf{X}|\underline{\Theta}}(.|\underline{\theta})$. Suppose that $\underline{\Theta}$ has prior density $\pi_{\underline{\Theta}}(\underline{\theta})$ and suppose that t is a function or a *statistic* of $\mathbf{X}$,

$$t = t(\mathbf{X}) .$$

Definition 3.2 (Statistic) *A statistic is the result of applying a function, or statistical algorithm, to a set of data. More formally, a statistic is a function of a random sample (taken in the sense of random variables), where the function itself is independent of the distribution of the sampling distribution. The term 'statistic' is used both for the function and the value of the function on a given sample.*

Definition 3.3 (Bayesian Sufficiency) *A statistic T defined as $T = t(\mathbf{X})$ such that for every prior $\pi_{\underline{\Theta}}$ within the space of prior distributions under consideration, there is a function ϕ such that*

$$\pi_{\underline{\Theta}|X}(\underline{\theta}|\mathbf{x}) = \frac{p_{X|\underline{\Theta}}(\mathbf{x}|\theta)\pi_{\underline{\Theta}}(\theta)}{p_X(\mathbf{x})} = \phi(\underline{\theta}, t(\mathbf{x})) \tag{3.5}$$

is called a Bayesian sufficient statistic for $\underline{\Theta}$.

This definition states that for learning about $\underline{\Theta}$ based on $\mathbf{X}$, the statistic T contains all the relevant information, since the posterior distribution depends on $\mathbf{X}$ only through T.

The following result shows that conditional independence of $\mathbf{X}$ and $\underline{\Theta}$ given $t(\mathbf{X})$ implies Bayesian sufficiency. If the families of probability measures have finite dimensional parameter spaces, then the converse is also true. If there are an infinite number of parameters, counter examples may be obtained to the converse statement.

Proposition 3.1 *Let t denote a function and let $T = t(\mathbf{X})$. If*

$$X \perp \Theta | T, \tag{3.6}$$

then $T = t(\mathbf{X})$ is a Bayesian sufficient statistic for Θ.

Before proving this proposition, the following lemma is required.

Lemma 3.1 *Let $\mathbf{X}$ be a discrete random matrix, $T = t(\mathbf{X})$ where t is a function and let Θ be a continuous random vector. Let $\mathbf{X} \perp \Theta | T$, then*

$$p_{X|T}(\mathbf{x}|t) = p_{X|T,\Theta}(\mathbf{x}|t, \theta). \tag{3.7}$$

Proof of Lemma 3.1 Let $\underline{\theta} \subset \tilde{\Theta}$ and set $A_{\underline{\theta},\epsilon} = \{\underline{\psi} \in \tilde{\Theta} | |\underline{\psi} - \underline{\theta}| < \epsilon\}$. Recall that, in this text, if there are d parameters, then $\tilde{\Theta} \subseteq \mathbf{R}^d$, where $\tilde{\Theta}$ is the parameter space. The distance $|\underline{\psi} - \underline{\theta}|$ is defined as the Euclidean distance:

$$|\underline{\psi} - \underline{\theta}| := \sqrt{\sum_{j=1}^{d}(\psi_j - \theta_j)^2}.$$

Since $\mathbf{X} \perp \Theta | T$, it follows that

$$
\begin{aligned}
p(\{\mathbf{X} = \mathbf{x}\} | \{T = t\}, \{\Theta \in A_{\underline{\theta},\epsilon}\}) &= \frac{p(\{\mathbf{X} = \mathbf{x}\}, \{\Theta \in A_{\underline{\theta},\epsilon}\} | \{T = t\})}{p(\{\Theta \in A_{\underline{\theta},\epsilon}\} | \{T = t\})} \\
&= \frac{p(\{\mathbf{X} = \mathbf{x}\} | \{T = t\}) p(\{\Theta \in A_{\underline{\theta},\epsilon}\} | \{T = t\})}{p(\{\Theta \in A_{\underline{\theta},\epsilon}\} | \{T = t\})} \\
&= p(\{\mathbf{X} = \mathbf{x}\} | \{T = t\}).
\end{aligned}
$$

Since this holds for all $\epsilon > 0$, the result follows by letting $\epsilon \to 0$. □

Proof of Proposition 3.1 In the set up considered here, $\mathbf{X}$ is considered to be a discrete random $n \times d$ matrix (representing n replications of an experiment where d attributes are recorded), while Θ is considered to be a continuous random vector. As usual, let $T = t(\mathbf{X})$.

An application of Bayes' rule gives

$$\pi_{\Theta|X,T}(\theta|\mathbf{x}, t) = \frac{p_{X,T|\Theta}(\mathbf{x}, t|\theta)\pi_{\Theta}(\theta)}{p_{X,T}(\mathbf{x}, t)} = \frac{p_{X|T,\Theta}(\mathbf{x}|t, \theta)p_{T|\Theta}(t|\theta)\pi_{\Theta}(\theta)}{p_{X,T}(\mathbf{x}, t)}, \tag{3.8}$$

and Equation (3.8), together with an application of Equation (3.7) gives

$$
\begin{aligned}
\pi_{\Theta|X,T}(\theta|\mathbf{x}, t) &= \frac{p_{X|T}(\mathbf{x}|t)p_{T|\Theta}(t|\theta)\pi_{\Theta}(\theta)}{p_{X|T}(\mathbf{x}|t)p_T(t)} \\
&= \frac{p_{T|\Theta}(t|\theta)\pi_{\Theta}(\theta)}{p_T(t)} = \pi_{\Theta|T}(\theta|t).
\end{aligned} \tag{3.9}
$$

The proposition is proved by setting $\phi(\underline{\theta}, t(\mathbf{x})) = \pi_{\Theta|T}(\theta|t(\mathbf{x}))$. □

Example 3.6 (Tossing a Thumb-tack) In the thumb-tack experiment described in Section 1.8, there is a single parameter, θ. In this paragraph, a Bayesian sufficient statistic is derived for θ, for a suitable class of prior distributions. Let π_Θ denote the prior density function for θ, and let Θ denote the random variable with this density function. In this case, $\mathbf{X}$ is a $n \times 1$ matrix, a column vector, which will be written as $\underline{X} = (X_1, \ldots, X_n)^t$, a sequence of n independent Bernoulli trials, each with probability θ of success (that is $X_j \sim Be(\theta)$, $j = 1, \ldots, n$ and $X_i \perp X_j | \Theta$ for $i \neq j$). The sequence of outcomes will be denoted by the vector

$$\underline{x} = (x_1, \ldots, x_n)^t.$$

That is, for each $j = 1, \ldots, n$, $x_j = 1$ or 0. The statistic t is a function of n variables, defined as

$$t(\underline{x}) = \sum_{j=1}^{n} x_j.$$

That is, when t is applied to a sequence of n 0s and 1s, it returns the number of 1s in the sequence. Here, $T = t(\underline{X}) = \sum_{j=1}^{n} X_j$ and therefore T has a binomial distribution with the parameters n and θ, since it is the sum of independent Bernoulli trials. The probability function of T is given by

$$p_{T|\Theta}(k|\theta) = \begin{cases} \begin{pmatrix} n \\ k \end{pmatrix} \theta^k (1-\theta)^{n-k} & k = 0, 1, \ldots, n \\ \\ 0 & \text{other } k. \end{cases}$$

Since t is a function of $\underline{x}$, it follows that

$$p_{\underline{X},T|\Theta}(\underline{x}, k|\theta) = \begin{cases} \theta^k (1-\theta)^{n-k} & k = 0, 1, \ldots, n \\ 0 & \text{other } k \end{cases}$$

from which

$$p_{\underline{X}|T,\Theta}(\underline{x}|k, \theta) = \frac{p_{\underline{X},T|\Theta}(\underline{x}, k|\theta)}{p_{T|\Theta}(k|\theta)} = \frac{1}{\begin{pmatrix} n \\ k \end{pmatrix}}.$$

The right hand side does not depend on θ, from which Equation (3.7) holds and hence Equation (3.6) follows. Therefore, if $\underline{x} = (x_1, \ldots, x_n)$ are n independent Bernoulli trials, each with parameter θ, the function t such that $t(\underline{x}) = \sum_{l=1}^{n} x_l$ is a Bayesian sufficient statistic for the parameter θ. In the thumb-tack example, given in Section 1.8, the posterior distribution, based on a uniform prior is an explicit function of the data $\underline{x}$ only through the function $t(\underline{x})$.

Now consider a random vector $\underline{X}$ and suppose now that t is a generic sufficient statistic. Since t is a function of $\underline{X}$ (i.e. $t = t(\underline{X})$), it follows, using the rules of conditional probability and Equation (3.7), that

$$p_{\underline{X}|\Theta}(\underline{x}|\theta) = p_{\underline{X},T|\Theta}(\underline{x}, t(\underline{x})|\theta) = p_{\underline{X}|T,\Theta}(\underline{x}|t(\underline{x}), \theta) p_{T|\Theta}(t(\underline{x})|\theta)$$

$$= p_{\underline{X}|T}(\underline{x}|t(\underline{x})) p_{T|\Theta}(t(\underline{x})|\theta).$$

In other words, there is a *factorization* of the form

$$p_{\underline{X}|\Theta}(\underline{x}|\theta) = g(t(\underline{x}), \theta)h(\underline{x}), \qquad (3.10)$$

where

$$h(\underline{x}) = p_{\underline{X}|T}(\underline{x}|t(\underline{x})) = p_{\underline{X}|t(\underline{X})}(\underline{x}|t(\underline{x})).$$

In statistical literature, $t(\underline{X})$ is often *defined* to be a sufficient statistic if there is a factorization of the type given by Equation (3.10). Equation (3.10) is in fact a characterization of sufficiency in the sense that the likelihood function for θ depends on data only through t; the aspects of data that do not influence the value of t are not needed for inference about θ, as long as $p_{\underline{X}|\Theta}(\underline{x}|\theta)$ is the object of study. In the example above, and in many other cases, this offers a *data reduction*. That is, for any n, a sample of size n can be reduced to a quantity of fixed dimension.

3.5.2 Prediction sufficiency

Let $\underline{X}$ be a discrete random vector, $\underline{Y}$ a discrete random variable or vector, t a function and let $T = t(\underline{X})$. Let Θ be a continuous random variable or vector. Suppose $\underline{X}, \underline{Y}, \Theta$ and T satisfy

$$\underline{X} \perp (\underline{Y}, \Theta) \,|\, T. \qquad (3.11)$$

That is, once $t(\underline{X})$ is given, there is no additional statistical information in $\underline{X}$ about $\underline{Y}$ or Θ. The problem is to predict $\underline{Y}$ statistically using a function of $\underline{X}$. The following proposition is found in [70, 71].

Proposition 3.2

$$\underline{X} \perp (\underline{Y}, \Theta) \,|\, T \Leftrightarrow \begin{cases} \underline{X} \perp \Theta|T \\ \underline{X} \perp \underline{Y}|(\Theta, T). \end{cases} \qquad (3.12)$$

Proof of Proposition 3.2 Showing that $\underline{X} \perp (\underline{Y}, \Theta) \,|\, T \Longrightarrow \underline{X} \perp \Theta|T$ is a simple marginalization. It is required to show that for any sets A and B and any value t such that $p(\{T = t\}) \neq 0$,

$$p(\{\underline{X} \in A\}, \{\Theta \in B\}|\{T = t\}) = p(\{\underline{X} \in A\}|\{T = t\})p(\{\Theta \in B\}|\{T = t\}).$$

Assuming that $\underline{X} \perp (\underline{Y}, \Theta) \,|\, T$ holds, then

$$p(\{\underline{X} \in A\}, \{\Theta \in B\}|\{T = t\}) = \sum_{\underline{y}} p(\{\underline{X} \in A\}, \{\underline{Y} = \underline{y}\}, \{\Theta \in B\}|\{T = t\})$$

$$= \sum_{\underline{y}} p(\{\underline{X} \in A\}|\{T = t\})p(\{\underline{Y} = \underline{y}\}, \{\Theta \in B\}|\{T = t\})$$

$$= p(\{\underline{X} \in A\}|\{T = t\}) \sum_{\underline{y}} p(\{\underline{Y} = \underline{y}\}, \{\Theta \in B\}|\{T = t\})$$

$$= p(\{\underline{X} \in A\}|\{T = t\})p(\{\Theta \in B\}|\{T = t\})$$

and hence $\underline{X} \perp \Theta|T$.

Next, it is proved that $X \perp (Y, \Theta) |T \implies X \perp Y|(\Theta, T)$. Let $A_{\theta,\epsilon} = \{\underline{z} \in \tilde{\Theta} || \underline{z} - \underline{\theta}| < \epsilon\}$. It is required to show that

$$p_{X,Y|\Theta,T}(\underline{x}, \underline{y}|\theta, t) = p_{X|\Theta,T}(\underline{x}|\theta, t)p_{Y|\Theta,T}(\underline{y}|\theta, t).$$

Assuming that $X \perp (Y, \Theta) |T$,

$$p_{X,Y|\Theta,T}(\underline{x}, \underline{y}|\theta, t) = \lim_{\epsilon \to 0} \frac{p(\{X = \underline{x}\}, \{Y = \underline{y}\}, \{\Theta \in A_{\theta,\epsilon}\}|\{T = t\})}{p(\{\Theta \in A_{\theta,\epsilon}\})}$$

$$= \lim_{\epsilon \to 0} \frac{p(\{X = \underline{x}\}|\{T = t\})p(\{Y = \underline{y}\}, \{\Theta \in A_{\theta,\epsilon}\}|\{T = t\})}{p(\{\Theta \in A_{\theta,\epsilon}\})}$$

$$= p(\{X = \underline{x}\}|\{T = t\}) \left(\lim_{\epsilon \to 0} p(\{Y = \underline{y}\}|\{\Theta \in A_{\theta,\epsilon}\}, \{T = t\}) \right)$$

$$= p(\{X = \underline{x}\}|\{T = t\})p(\{Y = \underline{y}\}|\{\Theta = \theta\}, \{T = t\})$$

$$= p(\{X = \underline{x}\}|\{\Theta = \theta\}, \{T = t\})p(\{Y = \underline{y}\}|\{\Theta = \theta\}, \{T = t\})$$

$$= p_{X|\Theta,T}(\underline{x}|\theta, t),$$

so that

$$p_{X,Y|\Theta,T}(\underline{x}, \underline{y}|\theta, t) = p_{Y|\Theta,T}(\underline{y}|\theta, t).$$

By definition, therefore, $X \perp Y|(\Theta, T)$. It has therefore been shown that $X \perp (Y, \Theta)|T$ implies both $X \perp \Theta|T$ and $X \perp Y|(\Theta, T)$.

Finally, the converse, that $X \perp \Theta|T$ and $X \perp Y|(\Theta, T)$ imply $X \perp (Y, \Theta)|T$, is relatively straightforward. It is required to prove that, under the assumptions, for any sets A, B, C and any value t such that $p(\{T = t\}) \neq 0$,

$$p(\{X \in A\}, \{Y \in B\}, \{\Theta \in C\}|\{T = t\})$$
$$= p(\{X \in A\}|\{T = t\})p(\{Y \in B\}, \{\Theta \in C\}|\{T = t\}).$$

Using $X \perp Y|(\Theta, T)$ for the second equality and $X \perp \Theta|T$ for the third,

$$p(\{X \in A\}, \{Y \in B\}, \{\Theta \in C\}|\{T = t\})$$
$$= p(\{X \in A\}, \{Y \in B\}|\{\Theta \in C\}, \{T = t\})p(\{\Theta \in C\}|\{T = t\})$$
$$= p(\{X \in A\}|\{\Theta \in C\}, \{T = t\})p(\{Y \in B\}|\{\Theta \in C\}, \{T = t\})p(\{\Theta \in C\}|\{T = t\})$$
$$= p(\{X \in A\}|\{T = t\})p(\{Y \in B\}, \{\Theta \in C\}|\{T = t\})$$

so that $X \perp (Y, \Theta)|T$. The proof of Proposition 3.6 is complete.   □

The following result shows that if $X \perp (Y, \Theta)|T$ then, in a sense, (Y, T) is Bayesian sufficient for Θ.

Proposition 3.3 *Let t denote a function and let $T = t(X)$. If X, Y, T, Θ satisfy $X \perp (Y, \Theta)|T$, then*

$$\pi_{\Theta|Y,X,T}(\theta \mid \underline{y}, \underline{x}, t) = \pi_{\Theta|Y,T}(\theta \mid \underline{y}, t). \tag{3.13}$$

Proof of Proposition 3.3 Firstly, $\underline{X} \perp (\underline{Y}, \Theta)|T$ implies (by a simple marginalization over Θ) that $\underline{X} \perp \underline{Y}|T$. The previous result also gives that $\underline{X} \perp (\underline{Y}, \Theta)|T$ implies $\underline{X} \perp \Theta|T$ and $\underline{X} \perp \underline{Y}|(T, \Theta)$. It follows that

$$p_{\underline{X},\underline{Y}|T}(\underline{x}, \underline{y}|t) = \int_{\tilde{\Theta}} p_{\underline{X},\underline{Y}|\Theta,T}(\underline{x}, \underline{y}|\theta, t)\pi_{\Theta|T}(\theta|t)d\theta$$

$$= \int_{\tilde{\Theta}} p_{\underline{X}|\Theta,T}(\underline{x}, \underline{y}|\theta, t) p_{\underline{Y}|\Theta,T}(\underline{y}|\theta, t)\pi_{\Theta|T}(\theta|t)d\theta$$

$$= p_{\underline{X}|T}(\underline{x}|t)\int_{\tilde{\Theta}} p_{\underline{Y}|\Theta,T}(\underline{y}|\theta, t)\pi_{\Theta|T}(\theta|t)d\theta$$

$$= p_{\underline{X}|T}(\underline{x}|t) p_{\underline{Y}|T}(\underline{y}|t).$$

It follows that

$$p_{\underline{Y}|\underline{X},T}(\underline{y}|\underline{x}, t) = \frac{p_{\underline{X},\underline{Y}|T}(\underline{x}, \underline{y}|t)}{p_{\underline{X}|T}(\underline{x}|t)} = p_{\underline{Y}|T}(\underline{y}|t). \tag{3.14}$$

An application of Bayes' rule gives

$$\pi_{\Theta|\underline{Y},\underline{X},T}(\theta|\underline{y}, \underline{x}, t) = \frac{p_{\underline{Y},\underline{X},T|\Theta}(\underline{y}, \underline{x}, t|\theta)\pi_\Theta(\theta)}{p_{\underline{Y},\underline{X},T}(\underline{y}, \underline{x}, t)} = \frac{p_{\underline{Y},\underline{X}|T,\Theta}(\underline{y}, \underline{x}|t, \theta) p_{T|\Theta}(t|\theta)\pi_\Theta(\theta)}{p_{\underline{Y},\underline{X},T}(\underline{y}, \underline{x}, t)}$$

$$= \frac{p_{\underline{Y}|T,\Theta}(\underline{y}|t, \theta) p_{\underline{X}|T,\Theta}(\underline{x}|t, \theta) p_{T|\Theta}(t|\theta)\pi_\Theta(\theta)}{p_{\underline{Y},\underline{X},T}(\underline{y}, \underline{x}, t)},$$

where the conditional independence $\underline{X} \perp \underline{Y}|(\Theta, T)$ was used. Then, since $\underline{X} \perp \Theta|T$, it follows that $p_{\underline{X}|T,\Theta}(\underline{x}|t, \theta) = p_{\underline{X}|T}(\underline{x}|t)$ and hence, using the Equation (3.14), that

$$\pi_{\Theta|\underline{Y},\underline{X},T}(\theta|\underline{y}, \underline{x}, t) = \frac{p_{\underline{Y}|T,\Theta}(\underline{y}|t, \theta) p_{\underline{X}|T}(\underline{x}|t) p_{T|\Theta}(t|\theta)\pi_\Theta(\theta)}{p_{\underline{Y}|\underline{X},T}(\underline{y} \mid \underline{x}, t) p_{\underline{X}|T}(\underline{x} \mid t) p_T(t)}$$

$$= \frac{p_{\underline{Y}|T,\Theta}(\underline{y}|t, \theta) p_{T|\Theta}(t|\theta)\pi_\Theta(\theta)}{p_{\underline{Y}|T}(\underline{y} \mid t) p_T(t)}.$$

It follows that

$$\pi_{\Theta|\underline{Y},\underline{X},T}(\theta|\underline{y}, \underline{x}, t) = \frac{p_{\underline{Y},T|\Theta}(\underline{y}, t \mid \theta)\pi_\Theta(\theta)}{p_{\underline{Y}|T}(\underline{y}, t)} = \pi_{\Theta|\underline{Y},T}(\theta|\underline{y}, t),$$

as claimed. □

3.5.3 Prediction sufficiency for a Bayesian network

Let $\mathcal{G} = (V, E)$ denote a DAG with $V = \{X_1, \ldots, X_d\}$, where the nodes are numbered, for convenience such that for each j,

$$\Pi_j \subseteq \{X_1, \ldots, X_{j-1}\},$$

where Π_j (as usual) denotes the parent set for X_j. Such a numbering is always possible by Lemma 2.1.

Using a fully Bayesian approach to the problem, the parameter vector θ is considered as an observation on a random vector Θ and for each $j = 1 \ldots, d$ the parameter vector θ_j an observation on a random vector Θ_j.

Definition 3.4 (Parameter Modularity) *A set of parameters Θ for a Bayesian network satisfies* parameter modularity *if it may be decomposed into d distinct parameter sets $\Theta_1, \ldots, \Theta_d$ such that for $j = 1, \ldots, d$, the parameters in vector Θ_j are directly linked only to node X_j.*

This definition was introduced in [72], and is a necessary condition for the sensitivity analysis discussed in Chapter 7.

Under the assumption of parameter modularity, the DAG may be expanded by adding the parameter nodes as parent variables in the graph, and directed links from each node in the set Θ_j to the node X_j giving an extended graph that is directed and acyclic, where $p_{X_1, \ldots, X_d | \Theta}$ has the decomposition

$$p_{X_1, \ldots, X_d | \Theta} = \prod_{j=1}^{d} p_{X_j | \Theta_j, \Pi_j}. \tag{3.15}$$

Furthermore, under the assumption of modularity, $\Theta_1, \ldots, \Theta_d$ are independent random vectors and the joint prior distribution is a product of individual priors; $\pi_\Theta = \prod_{j=1}^{d} \pi_{\Theta_j}$.

Following Proposition 3.2, the following notation is useful:

$$\tilde{X}_j := \left((X_1, \Theta_1), \ldots, (X_{j-1}, \Theta_{j-1}) \right), \qquad j = 1, \ldots, d$$

and, for $j = 1, \ldots, d$, t_j is used to denote the function such that

$$t_j(\tilde{X}_j) = \Pi_j.$$

It follows directly from (3.15) that

$$\tilde{X}_j \perp \left(X_j, \Theta_j \right) | \Pi_j.$$

In other words, the parent set Π_j is a *prediction sufficient statistic* for (X_j, Θ_j) in the sense that there is no further information in $\left((X_1, \Theta_1), \ldots, (X_{j-1}, \Theta_{j-1}) \right)$ relevant to uncertainty about either Θ_j or X_j.

In a Bayesian network where the parameters satisfy the modularity assumption (Definition 3.4), (Π_j, X_j) are a Bayesian sufficient statistic for Θ_j. The modularity assumption is clearly satisfied when Equation (3.15) holds.

3.6   Time variables

This section considers *Markov models*.

Figure 3.6 Illustration of a Markov model.

Definition 3.5 (Markov Chain) *A sequence* $(X_n)_{n \geq 1}$ *is a Markov Chain if for all n, N,*

$$(X_{n+N}, \ldots, X_{n+1}) \perp (X_{n-1}, \ldots, X_1)|X_n.$$

In other words, the past is conditionally independent of the future given the present.

A Markov chain may be represented by the DAG given in Figure 3.6. The index n, for $n = 1, 2, 3, 4, 5$, may be considered as a time variable. For a Markov chain, the conditional probability potentials $p_{X_{n+1}|X_n}$ are known as *transition probabilities*. The Markov chain is said to be *time homogeneous* if $p_{X_{n+1}|X_n}$ does not depend on n. For a time homogeneous Markov chain on a binary space,

$$p_{X_{n+1}|X_n} = \begin{array}{c|cc} X_{n+1}\backslash X_n & 0 & 1 \\ \hline 0 & p & 1-q \\ 1 & 1-p & q \end{array}$$

where $0 \leq p \leq 1$ and $0 \leq q \leq 1$. The *joint probability* may be decomposed as

$$p_{X_1, X_2, X_3, X_4, X_5} = p_{X_1} p_{X_2|X_1} p_{X_3|X_2} p_{X_4|X_3} p_{X_5|X_4},$$

which is a product of transition probability potentials.

Consider the Bayesian network described by the graph in Figure 3.7 and suppose that the variables $(X_1, X_2, X_3, \ldots)$ cannot be observed, but that the values taken by $(Y_1, Y_2, Y_3, \ldots)$ can be observed. For example, X_n may be the state of an infection on day n and Y_n the test result on day n.

The model supposes that the past and future are independent given the current state and therefore the sequence of variables $(X_j)_{j \geq 1}$ in Figure 3.7 form a Markov chain. Since only the variables $(Y_n)_{n \geq 1}$ may be observed, the sequence $(X_n)_{n \geq 1}$ is referred to as a *hidden Markov chain*. From the graph,

$$P_{(X_1, Y_1), \ldots, (X_5, Y_5)} = p_{Y_5|X_5} p_{X_5|X_4} p_{Y_4|X_4} p_{X_4|X_3} p_{Y_3|X_3} p_{X_3|X_2} p_{Y_2|X_2} p_{X_2|X_1} p_{Y_1|X_1} p_{X_1}.$$

It also follows from the graph that

$$Y_n \perp Y_{n-1}|X_n \quad \text{and} \quad Y_{n+1} \perp Y_n|X_n.$$

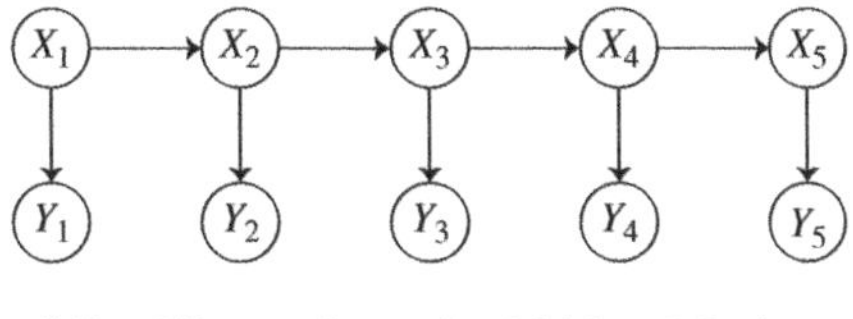

Figure 3.7 Illustration of a hidden Markov model.

In other words, the observable variables are conditionally independent given the state sequence X_n.

Note that the observed variables $Y_1, \ldots, Y_n$ may be considered in the same way as the virtual evidence node V in Figure 3.3.

Queries for Markov models   The queries for *hidden Markov models* formulated above are given the following names:

- filtering: Find $p_{X_n|Y_1,\ldots,Y_n}$.

- prediction: Find $p_{X_{n+1}|Y_1,\ldots,Y_n}$.

The maximum aposterior hypothesis problem is that of finding the most probable path; i.e. finding the path $X_1, \ldots, X_5$ which maximizes

$$p_{X_1,\ldots,X_5|Y_1,\ldots,Y_5}.$$

Often, it is natural to consider discrete time variables when one considers a Markov model. Associated with each time, one may have several variables, hidden or observable, with causal relations between them. These variables, together with the associated DAG, are known as a *time slice*. The DAG for the entire system is then decomposed into the consecutive time slices, together with additional causal links between the time slices to indicate the direct causal relationships. Time slices connected by temporal links constitute *dynamic Bayesian networks*, as discussed by K. Murphy in [73].

3.7   A brief introduction to Markov chain Monte Carlo methods

Since the problem of locating an appropriate graph structure of a Bayesian network is NP-hard (Chickering [74]), exact methods are of limited use and Markov chain Monte Carlo methods are essential for the analysis of larger networks. An McMC method for locating the graph structure is described in Section 6.4 and the necessary preparatory material on Markov chains and Monte Carlo methods is given here.

Since there are many good treatments of Markov chains available (see, for example, [75]), only a sketch is presented here; the necessary results and features of Markov Chains are recapitulated and the key ideas outlined.

For a random variable X taking values in a finite state space $\mathcal{E} = (s^{(1)}, \ldots, s^{(k)})$, the purpose of Markov chain Monte Carlo (McMC) is to find an estimate $\hat{p}_E(.)$ of the distribution of $p_E(.)$ for the random variable E. In the setting where it will be used, $\mathcal{E}$ is the space of all possible graph structures. This may then used to find an optimal 'value'; for example, the s_j that maximizes $\hat{p}_E(s_j)$. The McMC approach to the problem is to develop a *time homogeneous Markov chain* $X = (X_0, X_1, X_2 \ldots)$ (defined below) with state space $\mathcal{E}$ and has *stationary distribution* (defined below) $p_E(.)$.

Definition 3.6 (Time Homogeneous Markov Chain, Stationary Distribution) *A Markov chain $\underline{X}$ with state space $\mathcal{E} = (s_1, \ldots, s_k)$ is defined as a sequence of random variables*

$\underline{X} = (X_0, X_1, X_2, \ldots)$ such that for any $n \geq 1$ and any sequence $(s_0, s_1, \ldots, s_n) \in \mathcal{E}^n$,

$$p_{X_n|X_{n-1},\ldots,X_0}(s_n|s_{n-1}, \ldots, s_1, s_0) = p_{X_n|X_{n-1}}(s_n|s_{n-1}).$$

The Markov chain is time homogeneous if, in addition, there is a $k \times k$ matrix P with entries P_{ij} for all $n \geq 1$ and all $(s^{(i)}, s^{(j)}) \in \mathcal{E}^2$,

$$p_{X_n|X_{n-1}}(s^{(j)}|s^{(i)}) = P_{ij}.$$

The matrix P is known as the one step transition matrix, or transition matrix when the 'one step' is clear.

A stationary distribution for the time homogeneous Markov chain is a row vector $\underline{\pi} = (\pi_1, \ldots, \pi_k)$ such that $\pi_j \geq 0$ for all $j \in \{1, \ldots, k\}$ and $\sum_{j=1}^{k} \pi_j = 1$ and such that if $p_{X_0}(s_j) = \pi_j$ for all $j \in \{1, \ldots, k\}$, then $p_{X_n}(s_j) = \pi_j$ for all $n \geq 1$ and all $j \in \{1, \ldots, k\}$.

It follows directly from the definition of the Markov chain that the future is independent of the past, conditioned on the present. That is,

$$(X_{n+1}, X_{n+2}, \ldots) \perp (X_0, \ldots, X_{n-1})|X_n.$$

Since only time homogeneous Markov chains will be considered, the term 'Markov chain' will be used for 'time homogeneous Markov chain'.

Lemma 3.2 (Stationary Distribution) A distribution $\underline{\pi} = (\pi_1, \ldots, \pi_k)$ (that is, a row vector such that $0 \leq \pi_j \leq 1$ for each $j \in \{1, \ldots, k\}$ and $\sum_{j=1}^{k} \pi_j = 1$) is a stationary distribution for the Markov chain $\underline{X}$ with transition matrix P if and only if

$$\underline{\pi} P = \underline{\pi}.$$

Proof of Lemma 3.2 Firstly, if $p_{X_n}(s_j) = \pi_j$ for all $j \in \{1, \ldots, k\}$ and all $n \geq 0$, then

$$\pi_j = p_{X_1}(s_j) = \sum_{i=1}^{k} p_{X_1|X_0}(s_j|s_i) p_{X_0}(s_i) = \sum_{i=1}^{k} \pi_i P_{ij}$$

so that a stationary distribution satisfies $\underline{\pi} = \underline{\pi} P$. Secondly, if $p_{X_0}(s_j) = \pi_j$ for all $j \in \{1, \ldots, k\}$ and if $\underline{\pi} P = \underline{\pi}$, then

$$p_{X_n}(s_j) = \sum_{i=1}^{k} p_{X_n|X_{n-1}}(s_j|s_i) p_{X_{n-1}}(s_i).$$

Let $\underline{\psi}^{(n)} = (p_{X_n}(s_1), \ldots, p_{X_n}(s_k))$ (the row vector). Then

$$\underline{\psi}^{(n)} = \underline{\psi}^{(n-1)} P = \underline{\psi}^{(0)} P^n,$$

so if $\underline{\psi}^{(0)} = \underline{\pi}$ and $\underline{\pi}$ satisfies $\underline{\pi} P = \underline{\pi}$, then $\underline{\psi}^{(n)} = \underline{\pi}$ for all $n \geq 0$. It follows that $\underline{\pi}$ is a stationary distribution and the result is proved. $\square$

For each $i \in \{1, \ldots k\}$ (where k is finite, but typically can be very large), the row $P_{i.}$ in the transition matrix should try to have as few $j \in \{1, \ldots, k\}$ as possible such that P_{ij} is non-zero. Then, if the MC has been *designed properly* (the requirements will be discussed below), and it can be simulated well enough, the empirical distribution from the first n steps, defined by

$$\hat{p}_E(s_i) = \frac{1}{n} \sum_{j=1}^{n} K_{s_i}(X_j), \qquad s_i \in \mathcal{E},$$

where $K_s(e) = 1$ if $s = e$ and 0 otherwise, will be close to $p_E(.)$ for sufficiently large n.
Designed properly means that the MC satisfies certain conditions, as defined below:

1. It has to be *irreducible*,

2. It has to be *aperiodic*,

3. p_E (the target distribution) is a stationary distribution for the Markov chain.

With these conditions satisfied, it is possible to show that, for sufficiently n, the empirical distribution will be close to the target distribution, no matter what initial value is chosen.

Let $\underline{X} = (X_0, X_1, X_2, \ldots)$ be a Markov chain with state space $\mathcal{E} = (s_1, \ldots, s_k)$ A realization of the process is a sequence of states. The state associated with time 0 is the initial state, and it must be specified. After that the *one step transition matrix P* describes how to select successive states.

Example 3.7   There is (in spring 2009) a coffee shop at every corner of the main square (Stora torget) in the town of Linköping, Ostrogothia Province in Sweden. These are $v_1 = $ Cioccolata, $v_2 = $ Coffee-house by George, $v_3 = $ Linds, and $v_4 = $ Santinis. Suppose that an Ostrogothian is at corner v_j at time n. He tosses a coin. If the coin comes up heads, he will move to corner v_{j+1} at time $n + 1$, otherwise he will move to corner v_{j-1} at time $n + 1$, where $v_5 \equiv v_1$ and $v_0 \equiv v_4$. The states of this process and the transition probabilities may be represented by the circle diagram in Figure 3.8.

Any time homogeneous Markov chain with a finite state space may be represented this way. In this case, the transition matrix P with entries defined by $P_{ij} = p_{X_{n+1}|X_n}(v_j|v_i)$ is given by

$$P = \begin{pmatrix} 0 & \frac{1}{2} & 0 & \frac{1}{2} \\ \frac{1}{2} & 0 & \frac{1}{2} & 0 \\ 0 & \frac{1}{2} & 0 & \frac{1}{2} \\ \frac{1}{2} & 0 & \frac{1}{2} & 0 \end{pmatrix}.$$

Figure 3.8   Diagram for state transitions.

If an initial distribution $\underline{\mu} = (\mu_1, \ldots \mu_k)$ is specified, such that $\mu_j \geq 0$ for $j = 1, \ldots, k$, $\sum_{j=1}^{k} \mu_j = 1$ and such that $p_{X_0}(s_j) = \mu_j$, then the distribution p_{X_1} is determined by

$$p_{X_1}(s_j) = \sum_{i=1}^{k} p_{X_1, X_0}(s_j, s_i) = \sum_{i=1}^{k} p_{X_1 | X_0}(s_j | s_i) p_{X_0}(s_i) = \sum_{i=1}^{k} \mu_i P_{ij}.$$

Let $\underline{\mu}^{(n)} = (p_{X_n}(s_1), \ldots, p_{X_n}(s_k))$, where the probability distribution is described as a row vector with k components. Then it follows that

$$\underline{\mu}^{(1)} = \underline{\mu}^{(0)} P$$

and, in general, it is easy to see that

$$\underline{\mu}^{(n)} = \underline{\mu}^{(0)} P^n.$$

3.7.1   Simulating a Markov chain

This discussion will *assume* that there is access to a source of uniformly distributed random variables. In practice, this involves a pseudo random number generator of reasonable quality. To simulate a time homogeneous Markov process X for n steps, the following are required:

1. An initial distribution.

2. A procedure to determine the next state given the present state, so that the input of a sequence of independent uniform variables of length $n + 1$, $(U_0, U_1, \ldots, U_n)$ produces the initial value and the first n steps of the Markov chain $(X_0, X_1, \ldots, X_n)$. For some algorithms, several independent uniformly distributed variables are required to produce the next step of the chain.

3. It has to be possible verify that the simulation is a plausible trajectory of $\underline{X}$.

Specifying the initial value   Suppose that an initial distribution $\underline{\mu} = (\mu_1, \ldots, \mu_k)$ is specified. Given $u \in [0, 1]$, an observation from a $U(0, 1)$ distributed random variable, an observation from this distribution can be simulated quite easily using a function $\psi : [0, 1] \to \mathcal{E}$, defined as

$$\psi(u) = \begin{cases} s_1 & u \in [0, \mu_1) \\ s_i & u \in [\sum_{j=0}^{i-1} \mu_j, \sum_{j=0}^{i} \mu_j) \quad i = 1, \ldots, k. \end{cases}$$

This function ψ is known as the *initiation function*.

Updating   Suppose $X_n = s_m$, then X_{n+1} may be determined from the function $\phi(s_m, u)$ by

$$\phi(s_m, u) = \begin{cases} s_1 & u \in [0, p_{s_m, s_1}) \\ e_i & u \in [\sum_{j=0}^{i-1} p_{s_m, s_j}, \sum_{j=0}^{i} p_{s_m, s_j}) \quad i = 1, \ldots, k. \end{cases}$$

To simulate the chain now, given a sequence $(u_0, u_1, \ldots, u_n)$ of observations from independent $U(0, 1)$ random variables, set

$$X_0 = \psi(u_0),$$

$$X_{j+1} = \phi(X_j, u_{j+1}), \qquad j = 0, \ldots, n - 1.$$

3.7.2   Irreducibility, aperiodicity and time reversibility

This subsection clarifies the requirements on a Markov chain that is to be used for Monte Carlo simulation. It has to be *irreducible* and *aperiodic* with a finite state space. Many efficient methods (such as the Metropolis-Hastings, considered below) also require that the Markov chain is *time reversible*.

Definition 3.7 (Irreducible Markov Chain) *An* irreducible Markov chain *is a chain in which all the states communicate. Two states s_i and s_j are said to* communicate *if there is an $n_1 < +\infty$ such that $p_{X_{n_1}|X_0}(s_j|s_i) > 0$ and an $n_2 < +\infty$ such that $p_{X_{n_2}|X_0}(s_i|s_j) > 0$.*

Aperiodic Markov chains   Let $P_{ij}(n) := p_{X_n|X_0}(s_j|s_i)$. The matrix $P(n)$ is known as the n *step transition matrix*. The *greatest common divisor* of a finite, or infinite set of positive integers $A = \{a_1, a_2, a_3, \ldots\}$, will be written $GCD(A)$.

Definition 3.8 (Period of a State) *Suppose $\mathcal{E} = \{s_1, \ldots, s_k\}$. For each $j = 1, \ldots, k$, set*

$$A_j = \{n | P_{jj}(n) > 0\}.$$

The period *of state s_j is defined as $d(s_j) = GCD(A_j)$, the greatest common divisor of the set of times that the chain starting at a state s_j can return to state s_j with positive probability.*

Definition 3.9 (Aperiodic Markov Chain) *A state s_j is said to be* aperiodic *if $d(s_j) = 1$, A Markov chain is said to be* aperiodic *if all of its states are aperiodic. Otherwise, the chain is said to be* periodic.

The following theorem describes a key property of aperiodic Markov chains, that is necessary for Monte Carlo simulation.

Theorem 3.1 *Suppose that $\underline{X} = (X_0, X_1, \ldots)$ is a Markov chain with state space $\mathcal{E} = (s_1, \ldots, s_k)$ and one-step transition matrix P. Then there exists an $N < \infty$ such for all that for all $j = 1, \ldots, k$ and all $n \geq N$, $P_{jj}(n) > 0$.*

The proof of this theorem requires the following standard lemma from number theory, which is stated here without proof.

Lemma 3.3 *Let $A = \{a_1, a_2, \ldots\}$ be a set of positive integers such that*

1. $GCD(A) = 1$

2. *For any $a_i, a_j \in A$, $a_i + a_j \in A$.*

Then there exists an integer $N < +\infty$ such that $n \in A$ for all $n \geq N$.

Proof of Lemma 3.3 Omitted: it may be found in any standard introduction to number theory. □

Proof of Theorem 3.1 For each $s_j \in \mathcal{E}$, let $A_j = \{n \geq 1 | P_{jj}(n) > 0\}$, the set of possible return times to the state s_j for the chain E starting from site s_j. Since the chain is aperiodic, $GDC(A_j) = 1$. Suppose that $a_1, a_2 \in A_j$, then

$$P_{jj}(a_1 + a_2) = p_{X_{a_1+a_2}|X_0}(s_j|s_j) \geq p_{X_{a_1},X_{a_1+a_2}|X_0}(s_j,s_j|s_j)$$

$$= p_{X_{a_1+a_2}|X_{a_1},X_0}(s_j|s_j,s_j)p_{X_{a_1}|X_0}(s_j|s_j) = P_{jj}(a_2)P_{jj}(a_1) > 0,$$

so that the conditions of Lemma 3.3 are satisfied. The result follows directly from Lemma 3.3. □

Corollary 3.1 *Suppose that $\underline{X}$ is an irreducible and aperiodic MC with state space $\mathcal{E} = \{s_1, \ldots, s_k\}$ with one-step transition matrix P. Then there exists a finite positive integer M such that $P_{ij}(n) > 0$ for all $n \geq M$ and all $1 \leq i \leq k, 1 \leq j \leq k$.*

Proof of Corollary 3.1 Since $\underline{X}$ is aperiodic, there is an $N < +\infty$ such that $P_{jj}(n) > 0$ for all $j = 1, \ldots, k$ and all $n \geq N$. For any s_i, s_j, there is, by irreducibility, an n_{ij} such that $P_{ij}(n_{ij}) > 0$. Let $M_{ij} = N + n_{ij}$. Then, for all $m \geq M_{ij}$,

$$p_{X_m|X_0}(s_j|s_i) \geq p_{X_m,X_{m-n_{ij}}|X_0}(s_j,s_i|s_i)$$

$$= p_{X_m|X_{m-n_{ij}}}(s_j|s_i)p_{X_{m-n_{ij}}|X_0}(s_i|s_i) = P_{ij}(n_{ij})P_{ii}(m - n_{ij}) > 0.$$

Take $M = \max_{ij} M_{ij}$. □

Stationary distributions   When carrying out a Markov chain Monte Carlo simulation, it is necessary that the distributions of $X(n)$ converge as $n \to +\infty$ to a distribution that is independent of the initial condition and that they converge to the correct target distribution. That is,

$$(p_{X_n}(s_1), \ldots, p_{X_n}(s_k)) \overset{n \to +\infty}{\longrightarrow} \underline{\pi}$$

for any initial distribution μ.

The following results concern existence, uniqueness, and convergence to stationarity for irreducible, aperiodic Markov chains.

Theorem 3.2 (Existence and Uniqueness) *Let $\underline{X}$ be a time homogeneous irreducible aperiodic Markov chain with a finite state space $\mathcal{E} = (s_1, \ldots, s_k)$. There exists a unique stationary distribution $\underline{\pi}$.*

Proof of Theorem 3.2 This is found in any standard introduction to Markov chains and is omitted. □

When applying a Monte Carlo simulation, it is necessary that the process eventually visits all sites in the state space, hence the first hitting time and the expected return times for a site are of interest.

Definition 3.10 (First Hitting Time, Mean Return Time) *The first hitting time of site s_j is defined as*

$$T_j = \min\{n : X_n = s_j\},$$

with $T_j = +\infty$ if the chain never visits site s_j. The expected first hitting time of site s_j for a process started at site s_i is defined as

$$\tau_{ij} = E[T_j | X_0 = s_i].$$

The mean return time *for site s_i is defined as τ_{ii} for $i = 1, \ldots, k$.*

Lemma 3.4 *Let $\underline{X}$ be an irreducible and aperiodic time homogeneous Markov chain with state space $\mathcal{E} = (s_1, \ldots, s_k)$ and one step transition matrix P and let $s_i, s_j \in \mathcal{E}$. Then for all i, j, $p(\{T_j < +\infty\} | \{X_0 = s_i\}) = 1$. Furthermore, $\tau_{ij} < +\infty$. That is, the mean hitting time for any state is finite.*

Proof of Lemma 3.4 By Corollary 3.1, there is an integer $M < +\infty$ such that $P_{ij}(n) > 0$ for all $i, j \in \{1, \ldots, k\}$ and all $n \geq M$. Let $\alpha = \min_{1 \leq i \leq k, \, 1 \leq j \leq k}\{P_{ij}(M)\}$. Note that $\alpha > 0$. For states s_i, s_j,

$$p(\{T_j > M\} | \{X_0 = s_i\}) \leq p(\{X_M \neq s_j\} | \{X_0 = s_i\}) \leq 1 - \alpha.$$

Also,

$$p(\{T_j > 2M\} | \{X_0 = s_i\}) = \sum_{r_1 \neq s_i, \ldots, r_{2M} \neq s_i} p_{X_{2M}, X_{2M-1}, \ldots, X_1 | X_0}(r_{2M}, \ldots, r_1 | s_i)$$

$$\leq \sum_{r_{2M} \neq s_i, r_M \neq s_i} p_{X_{2M}, X_M | X_0}(r_{2M}, r_M | s_i)$$

$$= \sum_{r_{2M} \neq s_i, r_M \neq s_i} p_{X_{2M} | X_M}(r_{2M} | r_M) p_{X_M | X_0}(r_M | s_i)$$

$$\leq (1 - \alpha) \sup_q p(\{X_{2M} \neq s_i\} | \{X_M = s_q\}) \leq (1 - \alpha)^2.$$

By a similar argument, it is clear that for each $l > 1$,

$$p(\{T_j > lM\} | \{X_0 = i\}) \leq (1 - \alpha)^l \xrightarrow{l \to +\infty} 0.$$

It follows that

$$\lim_{N \to +\infty} p(\{T_j < N\} | \{X_0 = s_i\}) = 1.$$

Furthermore,

$$\tau_{ij} = E[T_j | \{X_0 = s_i\}] = \sum_{n=1}^{\infty} p(\{T_j \geq n\} | \{X_0 = s_i\}) \leq \sum_{l=0}^{\infty} \sum_{n=lM}^{(l+1)M-1} p(\{T_j \geq lM\} | \{X_0 = s_i\})$$

$$= M \sum_{l=0}^{\infty} p(\{T_j > lM\} | \{X_0 = s_i\}) \leq M \sum_{l=0}^{\infty} (1 - \alpha)^l = \frac{M}{\alpha} < +\infty.$$

The lemma is proved.   □

Theorem 3.3 (The Markov Chain Convergence Theorem) *Let $\underline{X}$ be an irreducible, aperiodic Markov chain with state space $\mathcal{E} = (s_1, \ldots, s_k)$ and transition matrix P, with arbitrary initial distribution $\underline{\mu} = (\mu_1, \ldots, \mu_k)$. Then for any distribution $\underline{\pi} = (\pi_1, \ldots, \pi_k)$ that is stationary for the transition matrix P,*

$$\lim_{n \to +\infty} \max_{j \in \{1,\ldots,k\}} |\mu_j^{(n)} - \pi_j| = 0,$$

where $\underline{\mu}^{(n)} = \underline{\mu} P^n$ (where $\underline{\mu}, \underline{\pi}$ and $\underline{\mu}^{(n)}$ are, as usual, taken as row vectors).

Proof of Theorem 3.3 The proof uses a coupling argument. Let $\underline{X} = (X_0, X_1, X_2, \ldots)$ and $\underline{Y} = (Y_0, Y_1, Y_2, \ldots)$ denote two independent copies of the Markov chain. Let $\tau = \min\{n \geq 1 : X_n = Y_n\}$, with $\tau = +\infty$ if the chains never meet. The aim is to show that the random time τ satisfies $p(\{\tau < +\infty\}) = 1$. Since the Markov chain is irreducible and aperiodic, there exists an integer $N < +\infty$ by Corollary 3.17 such that $P_{ij}(M) > 0$ for all $i, j \in \{1, 2, \ldots, k\}$. Let $\alpha = \min\{P_{ij}(M), \ i, j \in \{1, \ldots, k\}\} > 0$. It follows that

$$p(\{\tau \leq M\}) = \sum_{j=1}^{M} p(\{X_j = Y_j\})$$

$$\geq p(\{X_M = Y_M\}) \geq p(\{X_M = Y_M = s_1\}) = p(\{X_M = s_1\}) p(\{Y_M = s_1\})$$

$$= \left(\sum_{i=1}^{k} P_{i1}(M) p(\{X_0 = s_i\})\right) \left(\sum_{i=1}^{k} P_{i1}(M) p(\{Y_0 = s_i\})\right)$$

$$\geq \alpha^2.$$

It follows that $p(\{\tau > M\}) \leq 1 - \alpha^2$. Similarly,

$$p(\{\tau > lM\}) = p(\{\tau > M\}) \prod_{j=1}^{l} p(\{\tau > (j+1)M\}|\{\tau > jM\})$$

and a similar argument to the one that shows $p(\{\tau > M\}) \leq 1 - \alpha^2$ also gives that $p(\{\tau > (j+1)M\}|\{\tau > jM\}) \leq 1 - \alpha^2$ for each j. From this, it follows directly that

$$p(\{\tau > lM\}) \leq (1 - \alpha^2)^l$$

and hence that $\lim_{N \to +\infty} p(\tau < N) = 1$, so that $\lim_{N \to +\infty} p(\tau > N) = 0$. It is also straightforward to show from the bounds that $E[\tau] < +\infty$.

For any $j \in \{1, \ldots, k\}$, note that $p(\{Y_n = s_j\}|\{\tau \leq n\}) = p(\{X_n = s_j\}|\{\tau \leq n\})$. It follows that, for each $j = 1, \ldots, k$,

$$|p_{X_n}(s_j) - p_{Y_n}(s_j)| \leq |(p(\{X_n = s_j\}|\{\tau \leq n\}) - p(\{Y_n = s_j\}|\{\tau \leq n\})|p(\{\tau \leq n\})$$

$$+ |p(\{X_n = s_j\}|\{\tau > n\}) - p(\{Y_n = s_j\}|\{\tau > n\})|p(\{\tau > n\})$$

$$\leq p(\{\tau > n\}) \xrightarrow{n \to +\infty} 0.$$

The result follows directly. $\qquad\qquad\square$

Time reversible Markov chains   Several McMC algorithms require the Markov chain to be *time reversible*. This is the case with the Metropolis-Hastings algorithm, discussed in Section 3.7.3.

Definition 3.11 (Time Reversible Markov Chain) *Let $\underline{X} = (X_1, X_2, \ldots)$ be a time homogeneous Markov chain with state space $\mathcal{E} = (s_1, \ldots, s_k)$. Let μ be the entrance law (that is $\mu_j = p_{X_0}(s_j)$) and P be the transition matrix. The Markov chain is* reversible *if there exists a distribution $\underline{\pi}$ (that is $\pi_j \geq 0$ for all $j \in \{1, \ldots, k\}$ and $\sum_{j=1}^{k} \pi_j = 1$) such that*

$$\pi_i P_{ij} = \pi_j P_{ji} \tag{3.16}$$

for all $1 \leq i \leq k$, $1 \leq j \leq k$. A distribution $\underline{\pi}$ with this property is said to be a reversible distribution.

Proposition 3.4 *Let $\underline{X}$ be a Markov chain, with finite state space $\mathcal{E}$, one step transition matrix P and entrance law μ. Let $\underline{\pi}$ be a reversible distribution. Then $\underline{\pi}$ is a stationary distribution for the chain.*

Proof of Proposition 3.4   Since $\sum_{j=1}^{k} P_{ij} = 1$, it follows by Equation (3.16) that

$$\pi_i = \pi_i \sum_{j=1}^{k} P_{ij} = \sum_{j=1}^{k} \pi_i P_{ij} = \sum_{j=1}^{k} \pi_j P_{ji}$$

for all $i \in \{1, \ldots, k\}$. In other words,

$$\underline{\pi} = \underline{\pi} P,$$

where $\underline{\pi}$ is taken as a $1 \times k$ row vector, so that $\underline{\pi}$ is a stationary distribution by Lemma 3.2.   □

3.7.3   The Metropolis-Hastings algorithm

Since direct sampling from a posterior distribution may not be possible, the Metropolis-Hastings algorithm starts by generating candidate draws from a so-called proposal distribution. These draws are then corrected so that they behave asymptotically as random observations from the desired invariant or target distribution.

The MC constructed by the algorithm at each stage is therefore built in two steps: a **proposal** step and an **acceptance** step. These two steps are associated with the proposal and acceptance distributions, respectively.

The Metropolis problem   Let $f = (f_1, \ldots, f_k)$ be an arbitrary probability function, which is the **target distribution**, on a finite state space $\mathcal{E} = (s_1, \ldots, s_k)$. That is,

- $f_j \geq 0$ for $j \in \{1, \ldots, k\}$,
- $\sum_{j=1}^{k} f_j = 1$.

The *Metropolis problem* is to construct a Markov chain with invariant distribution f. The following discussion shows that it is always possible to construct an appropriate transition matrix to solve the Metropolis problem. There are, in fact, infinitely many solutions to the stated problem.

A solution of the Metropolis problem   Let Q be a *symmetric* $k \times k$ matrix, that is

$$Q_{ij} = Q_{ji}, \quad \forall (i, j) \in \{1, \ldots, k\}^2 \tag{3.17}$$

such that $Q_{ij} \geq 0$ for each (i, j) and $\sum_{j=1}^{k} Q_{ij} = 1$ for each $i \in \{1, \ldots, k\}$.

The aim is to construct a Markov chain $\underline{X}$ with the state space $\mathcal{E}$ and the invariant distribution f. Consider the following rules for transition:

For $X_n = s_i$, **propose** a value of $Y_{n+1} = s_j$ where s_j is drawn with probability Q_{ij}, for $j = 1, \ldots, k$, independently of $X_0, \ldots, X_{n-1}$. Then **accept** j with probability

$$\alpha_{ij} = \min \left\{ 1, \frac{f_j}{f_i} \right\}. \tag{3.18}$$

Acceptance means that the chain moves to $X_{n+1} = s_j$. **Reject** the proposed value j with probability

$$1 - \alpha_{i,j}. \tag{3.19}$$

Rejection means that the chain stays at s_i. That is, $X_{n+1} = s_i$. The procedure is implemented in terms of an independent random toss of coin the probability function

$$(p(\text{heads}), p(\text{tails})) = \left(\alpha_{i,j}, 1 - \alpha_{i,j} \right).$$

Heads means acceptance and tails means rejection.

The next task is to compute the transition probabilities P_{ij} of the Markov chain $\underline{X}$. Let $\{\mathbf{ta}\}$ denote the event that the proposed transition is accepted, and let $\{\mathbf{ta}\}^c$ denote the complement. Then, for $i \neq j$,

$$P_{ij} = p_{X_{n+1}|X_n}(s_j|s_i) = p\left(\{Y_{n+1} = s_j\} \cap \mathbf{ta}|\{X_n = s_i\}\right)$$

from the definition of proposal and acceptance. It follows that

$$= P\left(\{Y_{n+1} = j\}|\mathbf{ta} \cap \{X_n = s_i\}\right) P\left(\mathbf{ta}|\{X_n = s_i\}\right) = p\left(\{Y_{n+1} = s_j\}|\{X_n = s_i\}\right) \alpha_{i,j}.$$

since, conditioned on X_n proposal is generated independently of acceptance. In other words,

$$P_{ij} = Q_{ij} \cdot \min \left\{ 1, \frac{f_j}{f_i} \right\} = Q_{ij} \cdot \alpha_{i,j} \qquad i \neq j. \tag{3.20}$$

For $i = j$,

$$P_{ii} = p_{X_{n+1}|X_n}(s_i|s_i)$$

$$= p\left(\{Y_{n+1} = s_i\} \cap \mathbf{ta}|\{X_n = s_i\}\right) + p\left(\{Y_{n+1} \neq s_i\} \cap \mathbf{ta}^c|\{X_n = s_i\}\right)$$

$$= p_{Y_{n+1}|X_n}(s_i|s_i) \, p\left(\mathbf{ta}|\{X_n = s_i\}\right) + p_{Y_{n+1}|X_n}(s_i|s_i) \, p\left(\mathbf{ta}^c|\{X_n = s_i\}\right)$$

$$= p_{Y_{n+1}|X_n}\left(s_i|s_i\right)\alpha_{i,i} + \sum_{j\neq i} p_{Y_{n+1}|X_n}\left(s_j|s_i\right)\left(1-\alpha_{i,j}\right)$$

$$= Q_{ii}\alpha_{i,i} + \sum_{j\neq i} p_{Y_{n+1}|X_n}\left(s_j|s_i\right)\left(1-\alpha_{i,j}\right)$$

$$= Q_{ii} + \sum_{j\neq i} Q_{ij}\left(1-\alpha_{i,j}\right).$$

The matrix P thus defined is a legitimate transition probability matrix; $P_{ij} \geq 0$ for each (i, j) and $\sum_{j=1}^{k} P_{ij} = 1$ for each $k \in \{1, \ldots, k\}$.

It remains to show that f is an invariant distribution for Markov chain $\underline{X}$ with transition matrix P and an arbitrary initial distribution μ.

The following argument shows that f satisfies the reversibility condition $f_i P_{ij} = f_j P_{ji}$ for all (i, j). Consider $i \neq j$ such that

$$f_i < f_j.$$

Then, by Equation (3.20), and the symmetry of Q (which is assumed in the construction),

$$f_i P_{ij} = f_i Q_{ij} \min\left\{1, \frac{f_j}{f_i}\right\}$$

$$= f_i Q_{ij}$$

$$= Q_{ij} \min\left\{1, \frac{f_i}{f_j}\right\} f_j$$

$$= Q_{ji}\alpha_{j,i} f_j$$

$$= P_{ji} f_j.$$

Now consider $i \neq j$ such that

$$f_j < f_i,$$

Then

$$f_j P_{ji} = Q_{ji} \min\left\{1, \frac{f_i}{f_j}\right\} f_j.$$

Continuing in the same way as before gives the result. Hence the reversibility condition (Equation (3.16)) holds for all (i, j), and therefore, by the result of Proposition 3.4, the Metropolis problem is solved.

Example 3.8 Let

$$\underline{f} = \left(\frac{1}{4}, \frac{1}{4}, \frac{1}{6}, \frac{1}{3}\right),$$

and

$$Q = \begin{pmatrix} \frac{1}{6} & \frac{1}{6} & \frac{1}{6} & \frac{1}{2} \\ \frac{1}{6} & \frac{1}{2} & \frac{1}{6} & \frac{1}{6} \\ \frac{1}{6} & \frac{1}{6} & \frac{2}{3} & 0 \\ \frac{1}{2} & \frac{1}{6} & 0 & \frac{1}{3} \end{pmatrix}.$$

Then the acceptance probabilities are

$$\alpha = \begin{pmatrix} 1 & 1 & \frac{2}{3} & 1 \\ 1 & 1 & \frac{2}{3} & 1 \\ 1 & 1 & 1 & 0 \\ \frac{3}{4} & \frac{3}{4} & 0 & 1 \end{pmatrix}.$$

The transition matrix may be computed using MATLAB. It is

$$P = \begin{pmatrix} 0.2222 & 0.1667 & 0.1111 & 0.5 \\ 0.1667 & 0.5556 & 0.1111 & 0.1667 \\ 0.1667 & 0.1667 & 0.667 & 0 \\ 0.3750 & 0.125 & 0 & 0.5 \end{pmatrix}.$$

3.7.4 The one-dimensional discrete Metropolis algorithm

The solution of the Metropolis problem, as established above, can be used to simulate a Markov chain $\underline{X}$ with preassigned stationary distribution f. The pertinent simulation algorithm is known as the **Metropolis algorithm**.

Definition 3.12 (Metropolis Algorithm) *Let* Q *be a symmetric transition probability matrix. Given that* $X_n = s_i$,

1. *Generate:* $Y_{n+1} = s_j$ *with probability* Q_{ij} *for* $j \in \{1, \ldots, k\}$.

2. *Take*

$$X_{n+1} = \begin{cases} Y_{n+1} & \text{with probability } \alpha_{i,j} \\ s_i & \text{with probability } 1 - \alpha_{i,j} \end{cases}$$

   where

$$\alpha_{i,j} = \min\left\{1, \frac{f_j}{f_i}\right\}.$$

The distributions $Q_{i,\cdot}$ *are called the* proposal *distributions.*

In W.K. Hastings [76], the algorithm of Metropolis was generalized by relaxing the requirement that the matrix of proposal distributions **Q** be symmetric. The more general simulation algorithm is known as the **Metropolis-Hastings algorithm**.

Definition 3.13 (Metropolis-Hastings Algorithm) *Let* Q *be a transition probability matrix. Given that* $X_n = s_i$

1. *Generate* $Y_{n+1} = s_j$ *with probability* Q_{ij}, $j \in \{1, \ldots, k\}$.

2. *Take*

$$X_{n+1} = \begin{cases} Y_{n+1} & \text{with probability} \quad \alpha_{i,j}^H \\ s_i & \text{with probability} \quad 1 - \alpha_{i,j}^H, \end{cases}$$

where

$$\alpha_{i,j}^H = \min\left\{1, \frac{f_j Q_{ji}}{f_i Q_{ij}}\right\}. \tag{3.21}$$

The distributions $Q_{i.}$ for $i \in \{1, \ldots, k\}$ are called the proposal *distributions.*

Notes   The 'Chest Clinic Problem' example is found in A.P. Dawid [68] and in S. Lauritzen [77]. Conditional independence and sufficiency is discussed in [70] and [71]. McMC is a major computational workhorse in modern Bayesian inference. A thorough introduction and treatment of McMC and its applications to Bayesian inference is found in [26]. The software WinBUGS and BUGS for Bayesian statistical inference [78] on a number of statistical models using McMC has a source code which is analogous to a Bayesian network, and represents an application of Bayesian networks to computer software. Simulation methods like McMC can, of course, be used for many other purposes than computing expectations or probabilities by empirical averages; an example is found in Pĕna [49]. The work in [79] presents a non-reversible McMC technique which has turned out to be useful in, for example, learning of structures, as discussed in Chapter 6. A study of sufficiency in machine learning is found in [80].

3.8 Exercises: Evidence, sufficiency and Monte Carlo methods

1. **Jeffrey's Rule** In a certain country, people use only two car models, Volvo and Saab, which come in two colours, red and blue. The sales statistics suggest $p(\text{Volvo}) = p(\text{Saab}) = 1/2$. Furthermore, $p(\text{red}|\text{Volvo}) = 0.7$ and $p(\text{red}|\text{Saab}) = 0.2$. You are on holiday in this region and you are standing outside a large underground garage, which you may not enter. The attendant of the garage communicates his impression that 40% of the cars in the garage are red. What is the probability that the first car leaving the garage is a Volvo?

2. **Pearl's Method** Let A denote an event that gives uncertain information (or virtual/soft evidence) about the partition (that is a collection of mutually exclusive and exhaustive events) $\{G_j\}_{j=1}^n$. Suppose that A satisfies

$$p\left(A \mid G_j, B\right) = p\left(A \mid G_j\right), \quad j = 1, 2, \ldots, n$$

for *every* event B. This is an assumption of conditional independence; the event A is independent of all other events given the partition G_j. Set $\lambda_j = p(A|G_j)$ and show that

$$p\left(C \mid A\right) = \frac{\sum_{j=1}^n \lambda_j p\left(C \cap G_j\right)}{\sum_{j=1}^n \lambda_j p\left(G_j\right)}.$$

Check that $p(.|A)$ satisfies the definition of the Pearl update (Definition 1.8).

The following two exercises are about converting soft evidence in a format where Jeffrey's rule is applicable, to a format where Pearl's rule is applicable and vice versa. They are taken from [81].

3. Let p denote a probability distribution before evidence is obtained and suppose that a piece of evidence gives uncertain information about the partition (that is, the collection of mutually exclusive and exhaustive events) $\{G_j\}_{j=1}^n$. Suppose that this evidence is specified by the posterior probabilities

$$p^*\left(G_j\right) = q_j, \quad j = 1, 2, \ldots, n$$

and by

$$\lambda_j = \frac{q_j}{p\left(G_j\right)}, \quad j = 1, 2, \ldots, n.$$

For any event C, compute the probability $p\left(C \mid A\right)$ using Pearl's method of virtual evidence and show that this gives the same result as Jeffrey's rule of update.
Hint: Use the formula in the preceding exercise.

4. Suppose that p is a probability distribution before any new information has been received and that the virtual evidence A gives uncertain information about the

partition (that is, the collection of mutually exclusive and exhaustive events) $\{G_j\}_{j=1}^n$. Suppose that this evidence is specified by likelihood ratios

$$\frac{p(A \mid G_j)}{p(A \mid G_1)} = \lambda_j, \quad j = 1, 2, \ldots, n.$$

Assume that

$$p^*(G_j) = q_j = p(G_j \mid A), \quad j = 1, 2, \ldots, n.$$

For any event C, compute the probability $p^*(C)$ using Jeffrey's rule and show that this gives the same result as Pearl's method of virtual evidence.

5. Let X_1, X_2, X_3 be three binary random variables, each taking values in $\{0, 1\}$, such that

$$p_{X_1, X_2, X_3}(x_1, x_2, x_3) = \frac{1}{8},$$

for $(x_1, x_2, x_3) \in \{0, 1\}^3$.

Now let V be an additional binary random variable and let $E = \{V = 1\}$. Here V stands for virtual information. Suppose that the conditional probability function of V given X_3 satisfies

$$p_{V \mid X_3}(1 \mid 1) = \lambda p_{V \mid X_3}(1 \mid 0).$$

Let G_1 and G_2 be the two events

$$G_1 = \{(x_1, x_2, x_3) \in \{0, 1\}^3 \mid x_3 = 0\}$$

and

$$G_2 = \{(x_1, x_2, x_3) \in \{0, 1\}^3 \mid x_3 = 1\}.$$

The events G_1 and G_2 are mutually exclusive and exhaustive. Use Pearl's method of virtual evidence to obtain the updated probability distribution

$$\tilde{p}_{X_1, X_2, X_3}(x_1, x_2, x_3) = p_{X_1, X_2, X_3 \mid V}(x_1, x_2, x_3 \mid 1) \qquad (x_1, x_2, x_3) \in \{0, 1\}^3.$$

Comment Virtual evidence is an event that is not accommodated by the statistical model. In this case, the model is described by the probability function p_{X_1, X_2, X_3} and it does not accommodate V, in the sense that the event $\{V = 1\}$ is not an event in the event algebra generated by the model. Information of this type (that is, in terms of events not in the event algebra generated by the model) can be expected to occur in practice, since a model cannot be expected to consider all scenarios and all possible sources of information. When virtual evidence arrives, some assessment has to be made as to how it should be incorporated. Here, the additional modelling assumption has been added that the likelihood of $\{V = 1\}$ depends only on X_3. That is, $\{V = 1\}$ is conditionally independent of X_1, X_2 given X_3. The statistical model for V is not complete, since only the ratio of the likelihoods has been specified. This adds to the modelling capacities of Bayesian networks, as virtual nodes may be added, as shown in Figure 3.3.

6. Let (X_1, X_2) be discrete random variables, with joint probability function

$$p_{X_1, X_2}(x_1, x_2 | \theta_1, \theta_2)$$

$$= \begin{cases} \frac{n!}{x_1! x_2! (n - x_1 - x_2)!} \theta_1^{x_1} \theta_2^{x_2} (1 - \theta_1 - \theta_2)^{n - x_1 - x_2} & x_1 \geq 0, x_2 \geq 0, x_1 + x_2 \leq n \\ 0 & \text{otherwise,} \end{cases}$$

where $\theta_1 \geq 0$, $\theta_2 \geq 0$ and $\theta_1 + \theta_2 \leq 1$. Suppose that the prior distribution over (θ_1, θ_2) is taken from the class

$$\pi(\theta_1, \theta_2) = \frac{\Gamma(\alpha_1 + \alpha_2 + \alpha_3)}{\prod_{j=1}^{3} \Gamma(\alpha_j)} \theta_1^{\alpha_1 - 1} \theta_2^{\alpha_2 - 1} (1 - \theta_1 - \theta_2)^{\alpha_3} \quad \theta_j \geq 0,$$

$$j = 1, 2, \quad \theta_1 + \theta_2 \leq 1$$

where $\alpha_j > 0$, $j = 1, 2, 3$.
Let $\psi = \theta_1 + \theta_2$. Show that $T = X_1 + X_2$ is a sufficient statistic for ψ.

7. Let $X = (X_1, \ldots, X_n)$ be a random sample from $U(\theta_1, \theta_2)$ (uniformly distributed between θ_1 and θ_2). That is, the density function is

$$\pi(x | \theta_1, \theta_2) = \begin{cases} \frac{1}{\theta_2 - \theta_1} & \theta_1 \leq x \leq \theta_2 \\ 0 & \text{otherwise.} \end{cases}$$

Let $T(X) = (\min_j X_j, \max_j X_j)$. Find the distribution of T and show that it is a sufficient statistic for $\theta = (\theta_1, \theta_2)$.

8. **Bucket Elimination** Suppose A, B, C, D, E, F and G are all binary variables (taking values 0 or 1) and that their joint probability may be factorized along the DAG given in Figure 3.9.

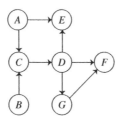

Figure 3.9 Bucket elimination.

Suppose that $p_{F,C}(1|1)$ is to be computed.

(a) Suppose the elimination order E, A, B, G, D is chosen. Compute the sizes of the tables that have to be manipulated at each stage in the computation. Compute the sizes of the tables that have to be manipulated for an elimination order: D, G, B, A, E.

(b) Write a MATLAB code for the problem of finding an efficient bucket elimination to compute $p_{F|C}(1|1)$ where the variables A, B, C, D, E, F, G are all binary variables, the probability distribution $p\,(A, B, C, D, E, F,G)$ may be factorized according to a directed acyclic graph, the input, for each variable A, B, C, D, E, F, G is the set of parents and the programme returns the elimination order that yields the smallest sum of table sizes.

9. **Divorcing**   The following example considers a simple principle known as *divorcing*, for reducing the sizes of the tables required. If the parent set $\Pi(X)$ of a variable X contains variables $A_1, \ldots, A_m$, $B_1, \ldots, B_n$, where $(A_1, \ldots, A_m) \perp (B_1, \ldots, B_n)\|_{\mathcal{G}}X$, then there are circumstances where it may be possible to introduce two *divorcing* variables C_1 and C_2, where $C_1 \perp C_2\|_{\mathcal{G}}X$, C_1 and C_2 are parents of X, $\Pi(C_1) = \{A_1, \ldots, A_m\}$, $\Pi(C_2) = \{B_1, \ldots, B_n\}$ and the direct links from $A_1, \ldots, A_m, B_1, \ldots, B_n$ to X are removed.

Consider a Bayesian network connected with a land search-and-rescue operation. The purpose is to predict the condition of an individual when found. There are a large number of variables that influence this: temperature, wind, precipitation, mental and physical health, age (which influences mental and physical health). But the most important feature about the weather is whether or not the combination of wind, precipitation and temperature can cause a dangerous wind chill, which leads to hypothermia. The various aspects of a person's health may be summarized as his condition when lost, and the extend to which a person's health deteriorates depends on the person's age and condition.

Construct a full Bayesian network containing all the variables and, from this, construct a Bayesian network with appropriate divorcing variables. Assuming all variables are binary, compute the sizes of the tables for the full network and compute the sizes of the tables for the network with the divorcing variables.

10. **Hidden Variables** Consider a situation where H is a hidden variable and observations are made on the variables $(I_j)_{j=1}^n$. Let $p(H)$ denote the prior distribution of H. Show that, given observations on the variables $(I_j)_{j=1}^n$, the posterior is given by

$$p(H|(I_j)_{j=1}^n) = \mu p(H) \prod_{j=1}^n p(I_j|H)$$

where μ is a normalization constant if and only if the variables satisfy the causal relations shown in the diagram in Figure 3.10.

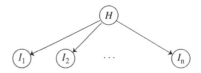

Figure 3.10   Hidden variables example.

11. **Monte Carlo**   Simulate the Markov chain $X = (X_0, X_1, X_2, \ldots)$ with state space $\{1, 2, 3, 4\}$ and one-step transition matrix

$$
P = \begin{pmatrix}
0 & \frac{1}{2} & 0 & \frac{1}{2} \\
\frac{1}{2} & 0 & \frac{1}{2} & 0 \\
0 & \frac{1}{2} & 0 & \frac{1}{2} \\
0 & \frac{1}{2} & 0 & \frac{1}{2}
\end{pmatrix}.
$$

Simulate it for $n = 10$, 100 and 1000 steps, with initial conditions $X_0 = 1$ and $X_0 = 2$. What are $(\hat{p}(1), \hat{p}(2), \hat{p}(3), \hat{p}(4))$ for each simulation? Is this chain aperiodic?

12. **A Metropolis Algorithm for Probabilistic Inference in a Bayesian Network**   Let $\underline{X} = (X_1, \ldots, X_d)$ be a random vector where each variable is binary. That is, $\mathcal{X} = \mathcal{X}_1 \times \ldots \times \mathcal{X}_d$ where $\mathcal{X}_j = \{0, 1\}$ for each $j \in \tilde{V} = \{1, \ldots, d\}$. An instantiation of $\underline{X}$ is therefore a vector $\underline{x} = (x_1, \ldots, x_d)$, where $x_i \in \{0, 1\}$ for $i = 1, \ldots, d$. Let $\tilde{V} = \{1, \ldots, d\}$ and $\tilde{U} = \{i_1, \ldots, i_m\} \subseteq \tilde{V}$. Let $\underline{X}_{\tilde{U}} = (X_{i_1}, \ldots, X_{i_m})$ and assume that $\underline{X}_{\tilde{U}} = \underline{y}$; that is, the variables $\underline{X}_{\tilde{U}}$ are instantiated. as $\underline{y}$. Let $\underline{X}_{\tilde{V}\backslash\tilde{U}}$ denote the uninstantiated variables and let $\underline{x}_{\tilde{V}\backslash\tilde{U}}$ denote a generic element of $\mathcal{X}_{\tilde{V}\backslash\tilde{U}} = \{0, 1\}^t$, where t is the number of uninstantiated variables.

The aim is to compute the probabilities

$$
p_{\underline{X}_{\tilde{V}\backslash\tilde{U}}|\underline{X}_{\tilde{U}}} \left(\underline{x}_{\tilde{V}\backslash\tilde{U}} \mid \underline{y} \right) = \frac{p_{\underline{X}_{\tilde{V}\backslash\tilde{U}}, \underline{X}_{\tilde{U}}} \left(\underline{x}_{\tilde{V}\backslash\tilde{U}}, \underline{y} \right)}{p_{\underline{X}_{\tilde{U}}} \left(\underline{y} \right)}.
$$

As will be shown in the later chapters, there are deterministic algorithms for probabilistic inference, which compute the desired conditional probabilities exactly, but since probabilistic inference in a Bayesian network is NP-hard in the number of variables (G.F. Cooper [82]), these exact algorithms are not expected to run in polynomial time. It therefore makes sense to run an approximate algorithm, if it converges rapidly.

This exercise designs a Markov chain Monte Carlo method for approximate probabilistic inference with $p_{\underline{X}_{\tilde{V}\backslash\tilde{U}}|\underline{X}_{\tilde{U}}} \left(\underline{x}_{\tilde{V}\backslash\tilde{U}} \mid \underline{y} \right)$ as the target distribution, based on the Metropolis algorithm. Firstly, define a random walk $\{Z_n\}_{n\geq 1}$ with values in $\{0, 1\}^t$ as follows.

- Let $\underline{x}_{\tilde{V}\backslash\tilde{U}} = \left(x_{\tilde{V}\backslash\tilde{U},i} \right)_{i=1}^{t}$ and assume that $Z_n = \underline{x}_{\tilde{V}\backslash\tilde{U}}$.

- **Proposal.**   Choose at random and independently of Z_n an integer l in $\{1, \ldots, t\}$ (For, convenience of writing, $\tilde{V}\backslash\tilde{U} = \{1, \ldots, t\}$ has been renumbered), i.e. $q(l) = \frac{1}{t}$. Then define a new vector $\underline{x}^*_{\tilde{V}\backslash\tilde{U}}$ in $\{0, 1\}^t$ by

$$
x^*_{\tilde{V}\backslash\tilde{U},l} = x_{\tilde{V}\backslash\tilde{U},l} \oplus 1
$$

where $1 \oplus 1 = 0 \oplus 0 = 0$, $1 \oplus 0 = 0 \oplus 1 = 1$, and

$$x^*_{\tilde{V}\setminus\tilde{U},i} = x_{\tilde{V}\setminus\tilde{U},i}, \qquad \text{if } i \neq l$$

In other words, $\underline{x}^*_{\tilde{V}\setminus\tilde{U}}$ differs from $\underline{x}_{\tilde{V}\setminus\tilde{U}}$ only in the lth coordinate.

- **Acceptance.** Take

$$Z_{n+1} = \begin{cases} \underline{x}^*_{\tilde{V}\setminus\tilde{U}} & \text{with probability } \alpha_{\underline{x}_{\tilde{V}\setminus\tilde{U}}\cdot\underline{x}^*_{\tilde{V}\setminus\tilde{U}}} \\ \underline{x}_{\tilde{V}\setminus\tilde{U}} & \text{with probability } 1 - \alpha_{\underline{x}_{\tilde{V}\setminus\tilde{U}}\cdot\underline{x}^*_{\tilde{V}\setminus\tilde{U}}} \end{cases}$$

where

$$\alpha_{\underline{x}_{\tilde{V}\setminus\tilde{U}}\cdot\underline{x}^*_{\tilde{V}\setminus\tilde{U}}} = \min\left\{1, \frac{f^*}{f}\right\},$$

where for a potential ϕ,

$$\frac{f^*}{f} = \frac{\phi\left(X_{\tilde{V}\setminus\tilde{U},l} = x^*_{\tilde{V}\setminus\tilde{U},l}, M_l\right)}{\phi\left(X_{\tilde{V}\setminus\tilde{U},l} = x_{\tilde{V}\setminus\tilde{U},l}, M_l\right)}$$

and where, finally, M_l is the Markov blanket of the variable $X_{\tilde{V}\setminus\tilde{U},l}$ in the DAG. Here, the variable $X_{\tilde{V}\setminus\tilde{U},l}$ is the variable in $\underline{X}_{\tilde{V}\setminus\tilde{U}}$ that was picked in the proposal step of the algorithm and ϕ is a function of $X_{\tilde{V}\setminus\tilde{U},l}$ and the variables in M_l. Note that this algorithm requires no operations of marginalization (required by bucket elimination).

- The questions are:

  (a) Show that $\{Z_n\}_{n\geq 1}$ is an irreducible, aperiodic Markov chain in $\{0, 1\}^t$.

  (b) Find the expression for the potential ϕ.

  (c) Show that $p_{\underline{X}_{\tilde{V}\setminus\tilde{U}}|\underline{X}_{\tilde{U}}}\left(\underline{x}_{\tilde{V}\setminus\tilde{U}} \mid \underline{y}\right)$ is the stationary distribution of $\{Z_n\}_{n\geq 1}$.

Hence $\{Z_n\}_{n\geq 1}$ is a sequence of samples from the target distribution, which is $p_{\underline{X}_{\tilde{V}\setminus\tilde{U}}|\underline{X}_{\tilde{U}}}\left(\cdot \mid \underline{y}\right)$, and any value $p_{\underline{X}_{\tilde{V}\setminus\tilde{U}}|\underline{X}_{\tilde{U}}}\left(\underline{x}_{\tilde{V}\setminus\tilde{U}} \mid \underline{y}\right)$ may be computed by counting the number of times $\underline{x}_{\tilde{V}\setminus\tilde{U}}$ occurs in the samples.
This is a Metropolis algorithm based on Pearl [83].
The crucial thing to be investigated, but not requested here, is the run time of the algorithm needed to approximate the target distribution with a given error bound.

13. The purpose of this exercise is to use the software technology from HUGIN Expert A/S to make queries on the Bayesian network specified later, in Chapter 9 (The Wet Pavement and the Sprinkler).
The Wet Pavement and the Sprinkler A simplified version of a standard example of a Bayesian network is given by the DAG in Figure 3.11 and by the conditional probability potentials given below. (In HUGIN, the term 'conditional probability table', abbreviated CPT is used).

$$C = \texttt{Cloudy}, S = \texttt{Sprinkler},$$

$$R = \texttt{Rain}, W = \texttt{WetGrass}.$$

The joint probability distribution is assumed to factorize according to the graph as product of the following conditional probability tables:

$$p_{S|C} = \begin{array}{c|cc} S\backslash C & 0 & 1 \\ \hline 0 & 0.5 & 0.9 \\ 1 & 0.5 & 0.1 \end{array}$$

$$p_{R|C} = \begin{array}{c|cc} R\backslash C & 0 & 1 \\ \hline 0 & 0.8 & 0.2 \\ 1 & 0.2 & 0.8 \end{array}$$

$$p_{W|R,S}(1|.,.) = \begin{array}{c|cc} S\backslash R & 0 & 1 \\ \hline 0 & 0.00 & 0.90 \\ 1 & 0.9 & 0.99 \end{array}$$

The prior distribution is

$$p_C(0) = p_C(1) = 0.5.$$

(a) Find $p_{C|W}(1 \mid 1)$ using HUGIN.

(b) Find $p_{C|S}(1 \mid 1)$ and $p_{R|S}(1 \mid 1)$ using HUGIN.

(c) What is the state of maximum probability for C if $W = 1$? This query can be handled by using the **max propagation** button in the tool bar of the run mode of HUGIN.

Using HUGIN Lite for propagating evidence and making queries Start the programme by clicking on the icon. This opens a window with a command tool bar. You are now in the **edit mode** of HUGIN. You use the tool bar to introduce your DAG in the window.

The package includes a help file (click on Help in the tool bar of the starting window) with a manual that explains how to use the software. The steps needed are outlined below.

- Open the menu under 'Network', click on **Auto propagate** and choose Do not auto propagate (this is more instructive for our aims than auto propagation).

- *Draw your DAG*:

  (a) By clicking with left mouse button in the command tool bar at the button depicting an ellipse with a single boundary choose first (discrete) nodes and place them in the window. Then draw your DAG in the window by connecting the nodes with arrows, press the left mouse button and move from the centre of a parent node to the centre of the child node and release the mouse button.

  (b) You may give names to nodes by double clicking on a node, thus opening an appropriate window.

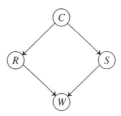

Figure 3.11 Sprinkler, Rain and Wet Pavement.

- *Introduce your CPTs* (CPT–Conditional Probability Table in HUGIN).

  (a) Click on a node in the DAG you created in the edit mode.

  (b) Then use ctrl + left (?) mouse click. A window should now open and you may write the probabilities in the window as text. Note the correct order of probabilities, where a column is a conditional probability distribution.

  (c) Repeat the same for all nodes.

  (d) The CPT windows may be closed/opened by clicking on the table button in the tool bar.

- Click the yellow flash button in the tool bar, and you enter the **run mode** in HUGIN. (You may get back to **edit mode** by clicking the yellow pen button.) The network is now compiled.

- *Introduce evidence*:

  (a) Check the initial marginal distributions on the nodes by using expand node list in the tool bar. You will see the distribution in green coloured staples and in digits.

  (b) Collapse the node list using the button on the right hand side of the expand button.

  (c) Click on a node.

  (d) Click then on the **enter likelihood evidence** button (three green bars) in the tool bar.

  (e) Write in the evidence (here 0 and 1).

  (f) a tag with little red **e** inside a box should appear in the node where evidence is introduced.

  (g) repeat for all nodes where you want evidence to be inserted.

  (h) expand the node list (a button in the tool bar). The nodes with evidence should be shown and have red colour, and the other nodes have a grey colour.

- PROPAGATE (query: posterior probability): click in the tool bar (run mode) on the **sum propagate button** (depicting a Σ under two arrows in each direction). The posterior probabilities of the states will be emerge in green colour and will also be shown as digits in the left hand part of the window (expand this part if necessary).

- PROPAGATE (query: state of maximum probability): with evidence inserted as in the procedure above, click on the **max propagate** button The states with maximum probabilities will be emerge with number 100 assigned in the left hand part of the window (extend this part if necessary).

- Use the print button to dump down the result, i.e. the screen, on a printer.

By using the refresh button (indicated by a loop arrow) you can always restart with the DAG initially specified. In the edit mode (yellow pencil) you may modify the graph and the CPTs.

4

Decomposable graphs and chain graphs

Having discussed the way that information and uncertainty about variables are represented by a Bayesian network, the next task is to develop the graph theory that is used for learning the graphical structure and for constructing a junction tree to enable the probability distribution to be updated efficiently in the light of new information.

The second of these topics will be considered first. Once the graph structure has been learned, there are many ways of updating the probability distribution in light of new information, but attention is focused on the fact that Bayesian networks are 'tree-like', or 'almost trees', in the sense that they exhibit many of the important properties of trees that are useful for computations. This was the basic idea of Pearl, Lauritzen and Speed in the 1980s, as pointed out by S. Arnborg in [84] and [85]. One key idea is that the Bayesian network may be expressed as a *junction tree*, where the nodes are groups of variables. Some selected edges are added in, and the edges made undirected, to form a *decomposable graph*, from which the junction tree is constructed.

Next, the necessary structures for considering classes of Markov equivalent graphs will be considered. Markov equivalent directed acyclic graphs were introduced in Chapter 2 and it is convenient to develop a language to characterize them, in terms of the *essential graph*, which will be introduced in this chapter and used in Chapter 6. The essential graph is a graph that retains all the directed edges that are common to all the Markov equivalent graphs, and removes the direction of the remaining edges, by replacing them with undirected edges. The collection of essential graphs is important when applying numerical methods to learn the graph structure, the topic of Chapter 6; if a Markov chain Monte Carlo algorithm is to be efficient, then it should move between graph structures that are essentially different. A brief outline of the basic ideas of

Markov chains is given in Chapter 3, the essential graph is introduced in this chapter and a suitable Markov chain Monte Carlo algorithm for locating the essential graph is discussed in Chapter 6.

4.1   Definitions and notations

Many of the basic notations and definitions necessary for the text have already been given in Chapter 2, the remainder are given here.

Recall the definition of an *induced sub-graph* (Definition 2.6).

Definition 4.1 (Complete Graph, Complete Subset) *A graph $\mathcal{G}$ is* complete *if every pair of nodes is joined by an undirected edge. That is, for each $(\alpha, \beta) \in V \times V$ with $\alpha \neq \beta$, $(\alpha, \beta) \in E$ and $(\beta, \alpha) \in E$. In other words, $\langle \alpha, \beta \rangle \in U$, where U denotes the set of undirected edges.*

A subset of nodes is called complete *if it induces a complete sub-graph.*

Definition 4.2 (Clique) *A clique is a* complete sub-graph *that is maximal with respect to $\subseteq$. In other words, a clique is not a sub-graph of any other complete graph.*

Definition 4.3 (Simplicial Node) *Recall the definition of family, found in Definition 2.3. For an undirected graph, the family of a node β is $F(\beta) = \{\beta\} \cup N(\beta)$, where $N(\beta)$ denotes the set of neighbours of β. A node β in an undirected graph is called* simplicial *if its family $F(\beta)$ is a clique.*

Definition 4.4 (Connectedness, Strong Components) *Let $\mathcal{G} = (V, E)$ be a simple graph, where $E = U \cup D$. That is, E may contain both directed and undirected edges. Let $\alpha \to \beta$ denote that there is a path (Definition 2.8) from α to β. If there is both $\alpha \to \beta$ and $\beta \to \alpha$ then α and β are said to be* connected. *This is written:*

$$\alpha \leftrightarrow \beta.$$

This is clearly an equivalence relation. The equivalence class *for α is denoted by $[\alpha]$. In other words, $\beta \in [\alpha]$ if and only if $\beta \leftrightarrow \alpha$. These equivalence classes are called* strong components *of $\mathcal{G}$.*

Note that a *graph* is connected if between any two nodes there exists a trail (Definition 2.7), but any two nodes α and β are only said to be connected if there is path from α to β and a path between β and α, where the definition of a 'path' is given in Definition 2.8.

Definition 4.5 (Separator) *A subset $S \subseteq V$ is called an (α, β) separator if every trail from α to β intersects S. Let $A \subseteq V$ and $B \subseteq V$, such that A, B and S are disjoint. The subset S is said to* separate A from B *if it is an (α, β) separator for every $\alpha \in A$ and $\beta \in B$.*

Definition 4.6 (Minimal Separator) *Let $A \subseteq V$, $B \subseteq V$ and $S \subseteq V$ be three disjoint subsets of V. Let S separate A and B. The separator S is said to be a* minimal separator *of A and B if no proper subset of S is itself an (α, β) separator for any $(\alpha, \beta) \in A \times B$.*

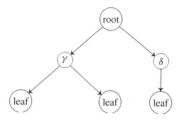

Figure 4.1   Illustration of a rooted tree.

Definition 4.7 (Rooted Tree) *A rooted tree $\mathcal{T}$ is a tree graph with a designated node ρ called the* root. *A* leaf *of a tree is a node that is joined to at most one other node. Figure 4.1 gives an illustration of a rooted tree.*

Definition 4.8 (Diameter) *The* diameter *of a tree is the length of the longest trail between two leaf nodes.*

Definition 4.9 (Moral Graph) *Let $\mathcal{G}$ be a DAG. Then $\mathcal{G}$ is said to be* moralized *if all undirected edges between all pairs of parents of each node which are not already joined are added and then all edges are made undirected.*

An example of a DAG is given in Figure 4.2. The result of moralizing is given in Figure 4.3. The *cliques* of the moral graph are illustrated in Figure 4.4.

Definition 4.10 (Chord) *Let $\mathcal{G} = (V, E)$ be an undirected graph. Let σ be an n cycle in $\mathcal{G}$. A* chord *of this cycle is a pair (α_i, α_j) of* non-consecutive *nodes in σ such that $\alpha_i \sim \alpha_j$ in $\mathcal{G}$.*

Definition 4.11 (Triangulated) *An undirected graph $\mathcal{G} = (V, E)$ is* triangulated *if every one of its cycles of length ≥ 4 possesses a chord.*

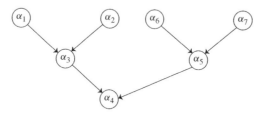

Figure 4.2   A directed acyclic graph.

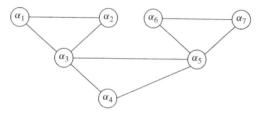

Figure 4.3   The graph in Figure 4.2, after moralizing.

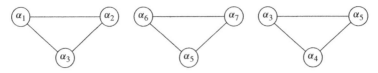

Figure 4.4   Cliques of the graph in Figure 4.3.

Lemma 4.1 *If* $G = (V, E)$ *is triangulated, then the induced graph* G_A *is also triangulated.*

Proof of Lemma 4.1 Consider any cycle of length ≥ 4 in the restricted graph. All the edges connecting these nodes remain. If the cycle possessed a chord in the original graph, the chord remains in the restricted graph.   □

Definition 4.12 (Decomposition) *A triple* (A, B, S) *of disjoint subsets of the node set* V *of an undirected graph is said to form a* decomposition *of* G *or to* decompose G *if*

$$V = A \cup B \cup S$$

and

- *S separates A from B,*
- *S is a* complete *subset of V.*

A, B or S may be the empty set. If both A and B are non-empty, then the decomposition is proper.

Clearly, every graph can be decomposed to its connected components (Definition 2.7). If the graph is undirected, then the connected components are the strong components (Definition 4.4).

Definition 4.13 (Decomposable Graph) *An undirected graph* G *is* decomposable *if either*

1. *it is complete, or*
2. *it possesses a proper decomposition* (A, B, S) *such that both sub graphs* $G_{A \cup S}$ *and* $G_{B \cup S}$ *are decomposable.*

This is a *recursive* definition, which is permissible, since the decomposition (A, B, S) is required to be proper, so that $G_{A \cup S}$ and $G_{B \cup S}$ have fewer nodes than the original graph G.

Example 4.1 A decomposable graph   Consider the graph in Figure 4.5. In the first stage, set $S = \{\alpha_3\}$, with $A = \{\alpha_1, \alpha_2\}$ and $B = \{\alpha_4, \alpha_5, \alpha_6\}$. Then S is a clique and S separates A from B. Then $A \cup S = \{\alpha_1, \alpha_2, \alpha_3\}$ and $G_{A \cup S}$ is a clique. $B \cup S = \{\alpha_3, \alpha_4, \alpha_5, \alpha_6\}$.

The graph $G_{B \cup S}$ is decomposable; take $S_2 = \{\alpha_3, \alpha_5\}$, $A_2 = \{\alpha_4\}$ and $B_2 = \{\alpha_6\}$. Then $G_{A_2 \cup S_2}$ and $G_{B_2 \cup S_2}$ are cliques.   □

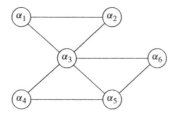

Figure 4.5   Example of a decomposable graph.

4.2   Decomposable graphs and triangulation of graphs

Decomposable graphs provide the basis for one of the key methods for updating a probability distribution described in terms of a Bayesian network. The DAG is moralized and then triangulated using the most efficient triangulation algorithms available. The triangulated graph is then decomposed and organized to form a junction tree, which supports effective algorithms, one of which will be described in Chapter 10.

Theorem 4.1 *Let $G = (V, E)$ be an undirected graph. The following conditions 1), 2) and 3) are equivalent.*

1. *G is decomposable.*

2. *G is triangulated.*

3. *For every pair of nodes $(\alpha, \beta) \in V \times V$, their minimal separator is complete.*

Proof of Theorem 4.1 The proof below follows the lines of Cowell, Dawid, Lauritzen and Spiegelhalter in [67].

Proof of 1) $\Longrightarrow$ 2), Theorem 4.1

Inductive hypothesis: All undirected decomposable graphs with n nodes or less are triangulated. This is true for one node.

Let G be a decomposable graph with $n + 1$ nodes. There are two alternatives:

Either G is complete, in which case it is triangulated,

Or: by the definition of decomposable, there are three disjoint subsets A, B, S such that S is a complete subset, S separates A from B, $V = A \cup B \cup S$ and $G_{A \cup S}$ and $G_{B \cup S}$ are decomposable. The decomposition is proper, hence $G_{A \cup S}$ and $G_{B \cup S}$ have less than or equal to n nodes. Therefore, by the *inductive hypothesis* $G_{A \cup S}$ and $G_{B \cup S}$ are triangulated. Therefore, a cycle of length ≥ 4 without a chord, will be a cycle from A which passes through B. By decomposability, S separates A from B and therefore any such cycle must pass S at least twice. But then this cycle has a chord, since S is a complete subset. $\square$

Proof of 2) $\Longrightarrow$ 3), Theorem 4.1 Assume that $G = (V, E)$ is an undirected, triangulated graph. Let S be a minimal separator for two nodes α and β. Let A denote the set such that $\alpha \in A$ and G_A is the largest connected sub-graph of $G_{V \setminus S}$ such that α is in the node set. Let $B = V \setminus (A \cup S)$. For every node $\gamma \in S$, there is a node $\tau \in A$ such that $\langle \gamma, \tau \rangle \in E$ and there is a node $\sigma \in B$ such that $\langle \gamma, \sigma \rangle \in E$. Otherwise $S \setminus \{\gamma\}$ would be a separator for α and β, contradicting the minimality of S. Hence, for any pair $(\gamma, \delta) \in S \times S$, there exist paths $\gamma, \tau_1, \ldots, \tau_m, \delta$ and $\gamma, \sigma_1, \ldots, \sigma_n, \delta$ where all the nodes $\{\tau_1, \ldots, \tau_m\}$ are in A and all the nodes $\{\sigma_1, \ldots \sigma_n\}$ are in B. Then $\gamma, \tau_1, \ldots, \tau_m, \delta, \sigma_n, \ldots, \sigma_1, \gamma$ is a cycle of length ≥ 4 and therefore has a chord. Assume that $\tau_1, \ldots, \tau_m$ and $\sigma_1, \ldots, \sigma_n$ have been chosen so that the paths are as short as possible (that is, there is no shorter path from γ to δ with all intervening nodes in A and no shorter path from γ to δ with all intervening nodes in B).

The chord cannot be of the form $\langle \tau_i, \tau_j \rangle$ for some (i, j) or $\langle \sigma_k, \sigma_l \rangle$ for any (k, l) because of the minimality of the lengths of the chosen paths. Therefore, γ and δ are adjacent for every pair $(\gamma, \delta) \in S \times S$. It follows that S is a clique. $\qquad \square$

Proof of 3) $\implies$ **2), Theorem 4.1** (This part is unnecessary for the argument, but the direct proof is reasonably straightforward). Assume that for every pair of nodes (α, β), their minimal separator is complete. Let $\alpha, \gamma, \beta, \tau_1, \ldots, \tau_r, \alpha$ be a cycle of length ≥ 4 in $\mathcal{G}$. If $\langle \alpha, \beta \rangle$ is not a chord of the cycle, then denote by S the minimal separator that puts α and β in different components of $\mathcal{G}_{V \backslash S}$. Then S clearly contains γ and τ_j for some $j \in \{1, \ldots, r\}$. By hypothesis, S is a clique, so that $\langle \gamma, \tau_j \rangle \in E$ and $\langle \gamma, \tau_j \rangle$ is a chord of the cycle. Therefore, any cycle of length ≥ 4 has a chord, therefore $\mathcal{G}$ is triangulated.

Proof of 3) $\implies$ **1), Theorem 4.1** If $\mathcal{G}$ is complete, then the result is clear. If $\mathcal{G}$ is not complete, then choose two distinct nodes $(\alpha, \beta) \in V \times V$ that are not adjacent. Let $S \subseteq V \backslash \{\alpha, \beta\}$ denote the minimal separator for the pair (α, β). Let A denote the node set of the maximal connected component of $\mathcal{G}_{V \backslash S}$ and let $B = V \backslash (A \cup S)$. Then (A, B, S) provides a decomposition. This procedure can be repeated on $\mathcal{G}_{A \cup S}$ and $\mathcal{G}_{B \cup S}$, and repeated recursively, hence the graph is decomposable. $\qquad \square$

Definition 4.14 (Perfect Node Elimination Sequence) *Let $V = \{\alpha_1, \ldots, \alpha_d\}$ denote the node set of a graph $\mathcal{G}$. A* perfect node elimination sequence *of a graph $\mathcal{G}$ is an ordering of the node set $\{\alpha_1, \ldots, \alpha_d\}$ such that for each j in $1 \leq j \leq d-1$, α_j is a simplicial node of the sub-graph of $\mathcal{G}$ induced by $\{\alpha_j, \alpha_{j+1}, \ldots, \alpha_d\}$*

Lemma 4.2 *Every triangulated graph $\mathcal{G}$ has a simplicial node. Moreover, if $\mathcal{G}$ is* not *complete, then it has two non-adjacent simplicial nodes.*

Proof of Lemma 4.2 The lemma is trivial if either $\mathcal{G}$ is complete, or else $\mathcal{G}$ has two or three nodes. Assume that $\mathcal{G}$ is not complete. Suppose the result is true for all graphs with fewer nodes than $\mathcal{G}$. Consider two non-adjacent nodes α and β. Let S denote the minimal separator of α and β. Let $\mathcal{G}_A$ denote the largest connected component of $\mathcal{G}_{V \backslash S}$ such that $\alpha \in A$ and let $B = V \backslash (A \cup S)$, so that $\beta \in B$.

By induction, either $\mathcal{G}_{A \cup S}$ is complete, or else it has two non-adjacent simplicial nodes. Since $\mathcal{G}_S$ is complete, it follows that at least one of the two simplicial nodes is in A. Such a node is therefore also simplicial in $\mathcal{G}$, because none of its neighbours is in B.

If $\mathcal{G}_{A \cup S}$ is complete, then any node of A is a simplicial node of $\mathcal{G}$.

In all cases, there is a simplicial node of $\mathcal{G}$ in A. Similarly, there is a simplicial node in B. These two nodes are then non-adjacent simplicial nodes of $\mathcal{G}$. $\qquad \square$

Theorem 4.2 *A graph $\mathcal{G}$ is triangulated if and only if it has a perfect node elimination sequence.*

Proof of Theorem 4.2 Suppose that $\mathcal{G}$ is triangulated. Assume that every triangulated graph with fewer nodes than $\mathcal{G}$ has a perfect elimination sequence. By the previous lemma, $\mathcal{G}$ has a simplicial node α. Removing α returns a triangulated graph. (Consider any cycle of length ≥ 4 with a chord. If the cycle remains after the node is removed, then the chord is not removed). By proceeding inductively, it follows that $\mathcal{G}$ has a perfect elimination sequence.

Conversely, assume that $\mathcal{G}$ has a perfect sequence, say $\{\alpha_1, \ldots, \alpha_d\}$. Consider any cycle of length ≥ 4. Let j be the first index such that α_j is in the cycle. Let $V(C)$ denote the node set of the cycle and let $V_j = \{\alpha_j, \ldots, \alpha_d\}$. Then $V(C) \subseteq V_j$. Since α_j is simplicial in $\mathcal{G}_{V_{j+1}}$, the neighbours of α_j in the cycle are adjacent, hence the cycle has a chord. Therefore $\mathcal{G}$ is triangulated. □

Definition 4.15 (Eliminating a Node) *Let $\mathcal{G} = (V, E)$ be an undirected graph. A node α is eliminated from an undirected graph $\mathcal{G}$ in the following way:*

1. *For all pairs of neighbours (β, γ) of α add a link if $\mathcal{G}$ does not already contain one. The added links are called* fill ins.

2. *Remove α.*

The resulting graph is denoted by $\mathcal{G}^{-\alpha}$.

Example 4.2 Consider the graph in Figure 4.6.

This graph is already triangulated. But suppose one did not notice this and decided to eliminate node α_3 from the graph in Figure 4.6. The resulting graph is given in Figure 4.7.

Definition 4.16 (Elimination Sequence) *An elimination sequence of $\mathcal{G}$ is a linear ordering of its nodes.*

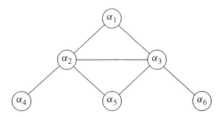

Figure 4.6   Example for Definition 4.5, eliminating a node.

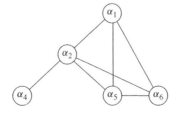

Figure 4.7   Graph 4.6 with α_3 eliminated.

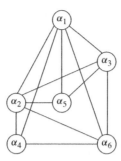

Figure 4.8 $\mathcal{G}^\sigma$. Elimination sequence $(\alpha_3, \alpha_2, \alpha_4, \alpha_1, \alpha_5, \alpha_6)$.

Let σ be an elimination sequence and let Λ denote the fill-ins produced by eliminating a node of $\mathcal{G}$ in the order σ. Denote by $\mathcal{G}^\sigma$ the graph $\mathcal{G}$ extended by Λ.

Example 4.3 Consider the graph in Figure 4.6. Suppose the elimination sequence $\alpha_3, \alpha_2,$ $\alpha_4, \alpha_5, \alpha_6$ is employed. Then the fill-ins, for each stage, will be $\langle \alpha_1, \alpha_6 \rangle, \langle \alpha_1, \alpha_5 \rangle, \langle \alpha_2, \alpha_6 \rangle$ for α_3, then $\langle \alpha_1, \alpha_4 \rangle, \langle \alpha_4, \alpha_6 \rangle$ for α_2. No further fill-ins are required. The graph $\mathcal{G}^\sigma$ is given in Figure 4.8.

Definition 4.17 (Elimination Domains) *Consider an elimination sequence σ; that is a linear ordering of the nodes, such that for any node α, $\sigma(\alpha)$ denotes the number assigned to α. A node β is said to be of higher elimination order than α if $\sigma(\beta) > \sigma(\alpha)$. The elimination domain of a node α is the set of neighbours of α of higher elimination order.*

In $\mathcal{G}^\sigma$, any node α together with its neighbours of higher elimination order form a *complete* subset. The neighbours of α of higher elimination order are denoted by $N_{\sigma(\alpha)}$. The sets $N_{\sigma(\alpha)}$ are the *elimination domains* corresponding to the elimination sequence σ.

An efficient algorithm clearly tries to minimize the number of fill-ins. If possible, one should find an elimination sequence that does not introduce fill-ins.

Proposition 4.1 *All cliques in a $\mathcal{G}^\sigma$ are a $N_{\sigma(\alpha)}$ for some $\alpha \in V$.*

Proof Let C be a clique in $\mathcal{G}^\sigma$ and let α be a variable in C of the lowest elimination order. Then $C = N_{\sigma(\alpha)}$. □

An efficient algorithm ought to find an elimination sequence for the domain graph that yields cliques of minimal total size.

The following proposition is clear.

Proposition 4.2 *Any $\mathcal{G}^\sigma$ is a triangulation of $\mathcal{G}$.* □

It is known (proof omitted) that the algorithms for triangulating a graph are NP - complete; i.e. a graph can be triangulated in a number of steps that is polynomial in the number of nodes.

Recall that a graph is *triangulated* if and only if it has an elimination sequence without fill-ins. This is equivalent to the statement that an undirected graph is triangulated if and only if all nodes can be eliminated by successively eliminating a node α such that the family $F_\alpha = \{\alpha\} \cup N_\alpha$ is complete. From the definition, such a node α is a *simplicial node*.

4.3   Junction trees

The purpose of this section is to define junction trees and to show how to construct them. They provide a key tool for updating a Bayesian network.

Definition 4.18 (Junction Trees) *Let C be a collection of subsets of a finite set V and T be a tree with C as its node set. Then T is said to be a* junction tree *(or* join tree*) if any intersection $C_1 \cap C_2$ of a pair C_1, C_2 of sets in C is contained in every node on the unique path in T between C_1 and C_2. Let G be an undirected graph and C the family of its cliques. If T is a junction tree with C as its node set, then T is known as* junction tree for the graph G.

Theorem 4.3 *There exists a junction tree T of cliques for the graph G if and only if G is decomposable.*

Proof of Theorem 4.3 The proof is by construction; a sequence is established in the following way. Firstly, a *simplicial node* α is chosen; F_α is therefore a clique. The algorithm continues by choosing nodes from F_α that *only have neighbours in* F_α. Let i be the number of nodes in F_α that only have neighbours in F_α. The set of nodes F_α is labelled V_i and the set of those nodes in F_α that have neighbours *not* in F_α is labelled S_i. This set is a *separator*.

Now remove the nodes in F_α that do not have neighbours outside F_α and name the new graph G'. Choose a new node α in the graph G' such that F_α is a clique. Repeat the process, with the index j, where j is the previous index, *plus* the number of nodes in the *current* F_α that only have neighbours in F_α.

When the parts have been established (as indicated in the diagram below), each separator S_i is then connected to a clique V_j with $j > i$ and such that $S_i \subset V_j$. This is always possible, because S_i is a *complete* set and, in the elimination sequence described above, the first point of S_i is eliminated when dealing with a clique of index greater than i.

It is necessary to prove that the structure constructed is a tree and that it has the junction tree property.

Firstly, each clique has at most one parent, so there are not multiple paths. The structure is therefore a tree.

To prove the *junction tree* condition, consider two cliques, V_i and V_j with $i > j$ and let α be a member of both. There is a *unique* path between V_i and V_j.

Because α is not eliminated when dealing with V_i, it is a member of S_i. It is also a member of the parent of V_i, say V_k and is a member of the parent of V_k and, by induction it is also a member of V_j and, of course, all the separators in between.   □

Example 4.4 Consider the directed acyclic graph in Figure 4.9. The corresponding moral graph is given in Figure 4.10.

An appropriate elimination sequence for this moral graph is

$$(\alpha_8, \alpha_7, \alpha_4, \alpha_9, \alpha_2, \alpha_3, \alpha_1, \alpha_5, \alpha_6).$$

There are two fill-ins; these are $\langle \alpha_1, \alpha_5 \rangle$ corresponding to the elimination of α_2 and $\langle \alpha_1, \alpha_6 \rangle$, corresponding to the elimination of α_3. The corresponding triangulated graph is given in Figure 4.11.

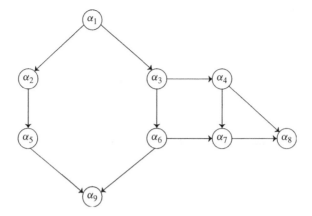

Figure 4.9   A directed acyclic graph for Example 4.3.

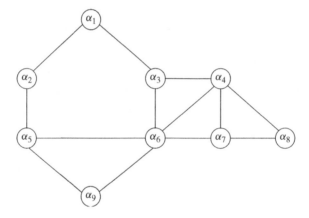

Figure 4.10   Moral graph corresponding to Figure 4.9.

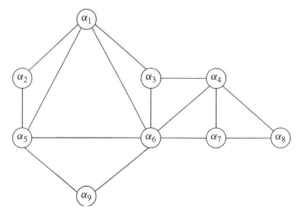

Figure 4.11   The triangulated graph corresponding to Figure 4.10.

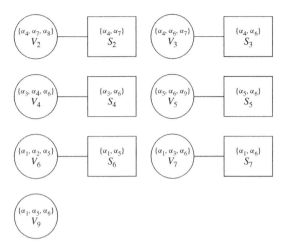

Figure 4.12   The Cliques and Separators from Figure 4.11.

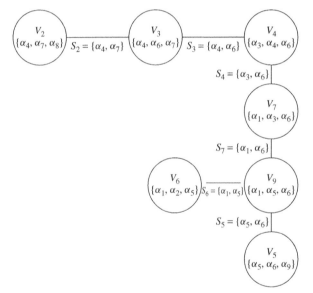

Figure 4.13   A junction tree (or join tree) constructed from the triangulated graph in Figure 4.11.

The junction tree construction may be applied. The cliques and separators, with the labels resulting from the diagram, are shown in Figure 4.12 and put together to form the junction tree, or join tree, shown in Figure 4.13.

4.4   Markov equivalence

Section 2.10 discussed the concept of *Markov equivalence* for directed acyclic graphs. That is, two different directed acyclic graphs over the same set of variables V are said to

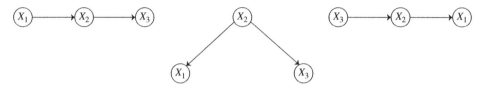

Figure 4.14   Three DAGs, each with the same independence structure.

Figure 4.15   Not Markov equivalent to the graphs in Figure 4.14.

be Markov equivalent if they have exactly the same d-separation properties. The formal definition of Markov equivalence for directed acyclic graphs is given by Definition 2.32. When trying to fit a graphical model to data, the subject of Chapter 6, all the DAGs in a Markov equivalence class will fit the data equally well and efficient algorithms for finding the structure will therefore only examine the different equivalence classes, rather than all the different possible DAGs. If there are direct *causal* dependencies between the variables, additional information about the causal structure is needed to choose the DAG from the equivalence class that has an appropriate causal interpretation.

Figure 4.14 shows three directed acyclic graphs, each with the same independence structure; for any probability distribution that factorizes according to these DAGs, $X_1 \perp X_3 | X_2$. Markov equivalence is considered in Exercise 7. In that exercise, it is shown that a distribution factorized along the DAG given in Figure 4.15 does *not* have the same independence structure as those in Figure 4.14.

One key result in this section is Theorem 4.4, which states that two features of the directed acyclic graph are necessary and sufficient to determine its Markov structure; its *immoralities* and its *skeleton*. These are defined below.

Definition 4.19 (Immorality) *Let $\mathcal{G} = (V, E)$ be a graph. Let $E = D \cup U$, where D contains directed edges, U contains undirected edges and $D \cap U = \phi$. An immorality in a graph is a triple (α, β, γ) such that $(\alpha, \beta) \in D$ and $(\gamma, \beta) \in D$, but $(\alpha, \gamma) \notin D$, $(\gamma, \alpha) \notin D$ and $\langle \alpha, \gamma \rangle \notin U$.*

Definition 4.20 (Skeleton)   *The skeleton of a graph $\mathcal{G} = (V, E)$ is the graph obtained by making the graph undirected. That is, the skeleton of $\mathcal{G}$ is the graph $\tilde{\mathcal{G}} = (V, \tilde{E})$ where $\langle \alpha, \beta \rangle \in \tilde{E} \Leftrightarrow (\alpha, \beta) \in D$ or $(\beta, \alpha) \in D$ or $\langle \alpha, \beta \rangle \in U$.*

Theorem 4.4 states simply that two DAGs are Markov equivalent if and only if they have the same skeleton and the same immoralities. The key to establishing this criteria will be to consider the active trails (Definition 2.16) in the graph. The following two definitions are also required.

Definition 4.21 (*S*-active node) *Let $G = (V, E)$ be a directed acyclic graph and let $S \subset V$. Recall the definition of a trail (Definition 2.5) and the definition of an active trail (Definition 2.16). A node $\alpha \in V$ is said to be S-active if either $\alpha \in S$ or there is a directed path from the node α to a node $\beta \in S$.*

Definition 4.22 (Minimal *S*-active trail) *Let $G = (V, E)$ be a directed acyclic graph and let $S \subset V$. An S-active trail τ in G between two nodes α and β is said to be a* minimal *S-active trail if it satisfies the following two properties:*

1. *If k is the number of nodes in the trail, the first node is α and the kth node is β, then there does not exist an S-active trail between α and β with fewer than k nodes and*

2. *There does not exist a* different *S-active trail ρ between α and β with exactly k nodes such that for all $1 < j < k$ either $\rho_j = \tau_j$ or ρ_j is a descendant of τ_j.*

Theorem 4.4 was proved by P. Verma and J. Pearl; Corollary 3.2 in [62].

Theorem 4.4 *Two DAGs are Markov equivalent if and only if they have the same* skeleton *and the same* immoralities.

The proof of Theorem 4.4 follows directly from Lemma 4.3.

Lemma 4.3 *Let $G_1 = (V, E_1)$ and $G_2 = (V, E_2)$ be two directed acyclic graphs with the same skeletons and the same immoralities. Then for all $S \subset V$, a trail is S-active trail in G_1 if and only if it is S-active in G_2.*

Proof of Lemma 4.3 Recall the notation from Definition 2.3: $\alpha \sim \beta$ denotes that two nodes $(\alpha, \beta) \in V \times V$ are neighbours. That is, either $(\alpha, \beta) \in E$ or $(\beta, \alpha) \in E$. Since G_1 and G_2 have the same skeletons, any trail τ in G_1 is also a trail in G_2. Let $S \subset V$. Assume that τ is an S-active trail in G_1. It is now proved, by induction on the number of collider nodes along the path, that τ is also an S-active trail in G_2. By definition, a single node will be considered an S-active trail, for any $S \subset V$. The proof is in three parts: Let τ be a *minimal* S-active trail in G_1. Then

1. If τ contains no colliders in G_1, then it is S-active in G_2.

2. If τ contains at least one collider connection centred at node τ_j, then τ is S-active in G_2 if and only if τ_j is S-active in G_2.

3. If τ contains at least one collider centred at node τ_j, then τ_j is an S-active node in G_2.

Part 1:   If τ is an S-active trail in G_1 and does not contain any collider connections in G_1, then none of the nodes on τ are in S. This can be seen by considering the Bayes ball algorithm, which characterizes d-separation. It follows that the path is S-active in G_2 if and only if it does not contain a collider connection in G_2.

Let τ be a minimal S-active trail in $\mathcal{G}_1$ with k nodes and no collider connections in $\mathcal{G}_1$. Suppose that a node τ_i is a collider node in $\mathcal{G}_2$, so that τ_{i-1} and τ_{i+1} are parents of τ_i in $\mathcal{G}_2$. Then, so that no new immoralities are introduced, it follows that $\tau_{i-1} \sim \tau_{i+1}$. Since τ_i is either a chain or a fork in $\mathcal{G}_1$, it follows that in $\mathcal{G}_1$, the connections between nodes $\tau_{i-2}, \tau_{i-1}, \tau_i, \tau_{i+1}, \tau_{i+2}$ take one of the forms shown in Figure 4.16 when τ_i a chain node or those in Figure 4.17 when τ_i a fork node.

It is clear from Figure 4.16 and 4.17 that the trail of length $k - 1$ in $\mathcal{G}_1$, obtained by removing τ_i and using the direct link from τ_{i-1} to τ_{i+1} is also an S-active trail in $\mathcal{G}_1$, contradicting the assumption that τ was a minimal S-active trail. Hence τ_i is a chain node or a fork node in $\mathcal{G}_2$.

It follows that there are no collider connections along the trail τ taken in $\mathcal{G}_2$ and hence, since it does not contain any nodes that are in S, it is an S- active trail in $\mathcal{G}_2$.

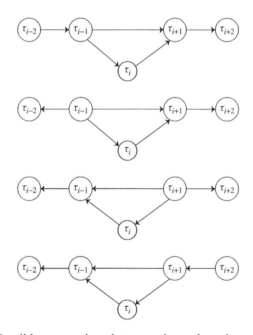

Figure 4.16   Possible connections between the nodes when τ_i is chain node.

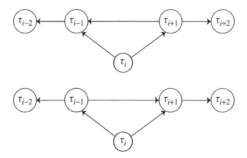

Figure 4.17   Possible connections between the nodes if τ_i is a fork node.

Part 2:   Assume that any minimal S-active trail in $\mathcal{G}_1$ containing n collider connections is also S-active in $\mathcal{G}_2$. This is true for $n = 0$ by part 1. Let τ be a trail with k nodes that is minimal S-active in $\mathcal{G}_1$ and with $n + 1$ collider connections in $\mathcal{G}_1$. Consider one of the collider connections centred at τ_j, with parents τ_{j-1} and τ_{j+1}. Let $\tilde{\tau}^{(0,j-1)} = (\tau_0, \tau_1, \ldots, \tau_{j-2}, \tau_{j-1})$ and let $\tilde{\tau}^{(j+1,k)} = (\tau_{j+1}, \ldots, \tau_k)$. Both $\tilde{\tau}^{(0,j-1)}$ and $\tilde{\tau}^{(j+1,k)}$ are minimal S-active in $\mathcal{G}_1$ and they both have a number of collider connections less than or equal to n. By the inductive hypothesis, they are therefore both S-active in $\mathcal{G}_2$.

Because the trail τ is minimal S-active in $\mathcal{G}_1$, it follows that $\tau_{j-1} \not\sim \tau_{j+1}$. This is because both τ_{j-1} and τ_{j+1} are S- active nodes in $\mathcal{G}_1$ (they have a common descendant in S to make the trail active), and neither is in S (neither is the centre of a collider along τ) it follows that if $\tau_{j-1} \sim \tau_{j+1}$, then the trail on $k - 1$ nodes obtained by removing the node τ_j would be S-active in $\mathcal{G}_1$, for the following reason: any chain or fork $(\tau_{j-2}, \tau_{j-1}, \tau_{j+1})$ or $(\tau_{j-1}, \tau_{j+1}, \tau_{j+2})$ would be active because both τ_{j-1} and τ_{j+1} are uninstantiated. Any collider $(\tau_{j-2}, \tau_{j-1}, \tau_{j+1})$ or $(\tau_{j-1}, \tau_{j+1}, \tau_{j+2})$ would be active because both τ_{j-1} and τ_{j+1} have a descendant in S. It follows that $\tau_{j-1} \not\sim \tau_{j+1}$. This holds in both $\mathcal{G}_1$ and $\mathcal{G}_2$, since the skeletons are the same.

Since $\tilde{\tau}^{(1,i-1)}$ and $\tilde{\tau}^{(i+1,k)}$ are both active, and $\tau_{j-1} \to \tau_j \leftarrow \tau_{j+1}$ is a collider, the trail τ is active if and only if τ_j is an active node. That is, it is either in S or has a descendant in S.

Part 3:   Let τ be a minimal S-active trail in $\mathcal{G}_1$ and let $\tau_j \in \tau$ be a collider node in $\mathcal{G}_1$. Since the trail τ is a minimal S-active trail in $\mathcal{G}_1$, it follows either that $\tau_j \in S$ or τ_j, considered in $\mathcal{G}_1$, has a descendant in S. That is, considered in $\mathcal{G}_1$, there is a directed *path* from τ_j to a node $w \in S$. Let ρ denote the shortest such path. If $\tau_j \in S$, then the length of the path is 0 and τ_j is also an S-active node in $\mathcal{G}_2$.

Assume there is a directed edge from τ_j to $w \in S$ in $\mathcal{G}_1$. If there are links from τ_{j-1} to w or τ_{j+1} to w, then these links are $\tau_{j-1} \to w$ or $\tau_{j+1} \to w$ respectively, otherwise the DAG would have cycles. If both are present, then the trail τ violates the second assumption of the minimality requirement. This is seen by considering the trail formed by taking w instead of τ_j in τ. It follows that either $\tau_{j-1} \not\sim w$ or $\tau_{j+1} \not\sim w$ or neither of the edges are present. Without loss of generality, assume $\tau_{j-1} \not\sim w$ (since the argument proceeds in the same way if $\tau_{j+1} \not\sim w$). The diagram in Figure 4.18 may be useful.

Since neither τ_{j-1} nor w are parents of τ_j in $\mathcal{G}_1$, they cannot *both* be parents of τ_j in $\mathcal{G}_2$, since both graphs have the same immoralities. Furthermore, $\tau_{j-1} \not\sim \tau_{j+1}$ (since they are both uninstantiated, and, in $\mathcal{G}_1$ both have a common descendant in S, so that if $\tau_{j-1} \sim \tau_{j+1}$ then the trail with τ_j removed would be active whether the connections at τ_{j-1} and τ_{j+1} are chain, fork or collider, contradicting the minimality assumption). Since both graphs have the same immoralities and $\tau_{j-1} \not\sim \tau_{j+1}$, it follows that $(\tau_{j-1}, \tau_j, \tau_{j+1})$ is an immorality in both $\mathcal{G}_1$ and $\mathcal{G}_2$ and hence that τ_{j-1} is a parent of τ_j in $\mathcal{G}_2$. Therefore, τ_j is a parent of w in $\mathcal{G}_2$ and is therefore w is an S- active node in $\mathcal{G}_2$.

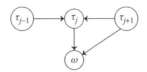

Figure 4.18   Illustration where τ_j is an uninstantiated collider node.

Assume that for the shortest directed path ρ from τ_j to w in $\mathcal{G}_1$, the first l links have the same directed edges in $\mathcal{G}_2$. Now suppose that the shortest directed path is ρ, where $\tau_j = \rho_0, \ldots, \rho_{l+p} = w$ and consider the links $\rho_l \sim \rho_{l+1}$ and $\rho_{l+1} \sim \rho_{l+2}$. If $\rho_l \sim \rho_{l+2}$, then in $\mathcal{G}_1$, the directed edge $\rho_l \to \rho_{l+2}$ is present, otherwise there is a cycle. If the directed edge $\rho_l \to \rho_{l+2}$ is present in $\mathcal{G}_1$, then the path ρ is not minimal. Therefore, $\rho_l \not\sim \rho_{l+2}$. This holds in both $\mathcal{G}_1$ and $\mathcal{G}_2$, because both graphs have the same skeletons. By a similar argument, $\rho_{l-1} \not\sim \rho_{l+1}$. (there would be a cycle in $\mathcal{G}_1$ if (ρ_{l+1}, ρ_{l-1}) were present; ρ would not be minimal in $\mathcal{G}_1$ if (ρ_{l-1}, ρ_{l+1}) were present. Since the skeletons are the same, $\rho_{l-1} \not\sim \rho_{l+1}$ in either $\mathcal{G}_1$ or $\mathcal{G}_2$). Since $\rho_l \not\sim \rho_{l+2}$, it follows that ρ_l and ρ_{l+2} are not both parents of ρ_{l+1} in $\mathcal{G}_2$; otherwise $\mathcal{G}_2$ would contain an immorality not present in $\mathcal{G}_1$. Similarly, since $\rho_{l-1} \not\sim \rho_{l+1}$, the edge $\rho_l \to \rho_{l+1}$ is present in $\mathcal{G}_2$, otherwise $\mathcal{G}_2$ would have either the immorality $(\rho_{l-1}, \rho_l, \rho_{l+1})$, since the edge (ρ_{l-1}, ρ_l) is present in $\mathcal{G}_2$ by assumption. It follows that the directed edges (ρ_l, ρ_{l+1}) and (ρ_{l+1}, ρ_{l+2}) are both present in $\mathcal{G}_2$. By induction, therefore, the whole directed path ρ is also present in $\mathcal{G}_2$ and hence τ_j is an S-active in both $\mathcal{G}_1$ and $\mathcal{G}_2$.   □

Proof of Theorem 4.4 This follows directly: let $\mathcal{G}_1$ and $\mathcal{G}_2$ denote two DAGs with the same skeleton and the same immoralities. For any set S and any two nodes X_i and X_j, it follows from the lemma, together with the definition of d-separation, that

$$X_i \perp X_j \|_{\mathcal{G}_1} S \Leftrightarrow X_i \perp X_j \|_{\mathcal{G}_2} S; \tag{4.1}$$

if there is an S-active trail between the two variables in one of the graphs, then there is an S-active trail between the two variables in the other. If there is no S-active trail between the two variables in one of the graphs then there is no S-active trail between the two variables in the other. By definition, two variables are d-separated by a set of variables S if and only if there is no S-active trail between the two variables. Two graphs are Markov equivalent, or I-equivalent (Definition 2.32), if and only if Equation (4.1) holds for all $(X_i, X_j, S) \in V \times V \times V$.   □

4.5 Markov equivalence, the essential graph and chain graphs

Consider the DAG in Figure 4.19. Using the characterization given by Theorem 4.4, the DAG in Figure 4.19 is equivalent to the DAGs in Figure 4.20.

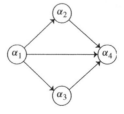

Figure 4.19 A DAG on four nodes.

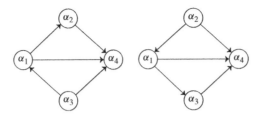

Figure 4.20 The equivalent DAGs.

For the DAG in Figure 4.19, all the DAGs with the same skeleton can be enumerated, and it is clear that those in Figure 4.20 are the only two that satisfy the criteria. To find the DAGs equivalent to the one in Figure 4.19, the immorality $(\alpha_2, \alpha_4, \alpha_3)$ has to be preserved and no new immoralities may be added. The directed edges (α_1, α_4), (α_2, α_4) and (α_3, α_4) are therefore *essential*, the directed edges (α_2, α_4) and (α_3, α_4) to form the immorality, the directed edge (α_1, α_4) because the connection $(\alpha_2, \alpha_1, \alpha_3)$ is either a fork or chain, forcing (α_1, α_4) to prevent a cycle. These three directed edges will be present in any equivalent DAG. The other three edges may be oriented in 2^3 different ways, but only five of these lead to DAGs (the other graphs contain cycles) and of these five, only the three shown in Figures 4.19 and 4.20 have the same immoralities.

A useful starting point for locating all the DAGs that are Markov equivalent to a given DAG is to locate the *essential graph*, given in the following definition.

Definition 4.23 (Essential Graph) *Let $\mathcal{G}$ be a directed acyclic graph. The* essential graph *$\mathcal{G}^*$ associated with $\mathcal{G}$ is the graph with the same skeleton as $\mathcal{G}$, but where an edge is directed in $\mathcal{G}^*$ if and only if it occurs as a directed edge with the same orientation in every DAG that is Markov equivalent to $\mathcal{G}$. The* directed edges *of $\mathcal{G}^*$ are the* essential edges *of $\mathcal{G}$.*

Once the *essential graph* is located, if it has k undirected edges, then there are 2^k directed graphs to be checked for possible Markov equivalence to $\mathcal{G}$. They are equivalent if they are acyclic, with the same immoralities. The essential graph contains both directed and undirected edges, and is an example of a *chain graph*. The following material gives the definition of a chain graph and deals mainly with the properties that will be used when considering the essential graph, with some extensions.

Definition 4.24 (Chain Graph) *A chain graph is a graph $\mathcal{G} = (V, E)$ containing both directed and undirected edges, where the node set V can be partitioned into n disjoint subsets $V = V_1 \cup \ldots \cup V_n$ such that*

1. *$\mathcal{G}_{V_j}$ is an undirected graph for all $j = 1, \ldots, n$*

2. *For any $i \neq j$, and any $\alpha \in V_i$, $\beta \in V_j$, there is no cycle in $\mathcal{G} = (V, E)$ (Definition 2.10) containing both α and β.*

The chain graph consists of components where the edges are undirected, which are connected by directed edges. The components with undirected edges are known as *chain components*, which are defined below.

Definition 4.25 (Chain Component) *Let* $G = (V, E)$ *be a chain graph. Let* $\hat{G} = (V, \hat{E})$ *denote the graph obtained by removing all the directed edges from* E. *Each connected component of* $\tilde{G}$ *is known as a* chain component.

The *chain components* (V_j, F_j), $j = 1, \ldots, n$ of G therefore satisfy the following conditions:

1. $V_j \subseteq V$ and F_j is the edge set obtained by retaining all *undirected* edges $\langle \alpha, \beta \rangle \in E$ such that $\alpha \in V_j$ and $\beta \in V_j$.

2. There is no undirected edge in E from any node in $V \backslash V_j$ to any node in V_j.

Theorem 4.5 states any essential graph is necessarily a chain graph and presents the additional features required to ensure that a chain graph is an essential graph corresponding to a directed acyclic graph. One of the features is that the *directed* edges in a chain graph have to be *strongly protected*.

Definition 4.26 (Strongly Protected) *Let* $G = (V, E)$ *be a chain graph, where* $E = D \cup U$. *A directed edge* $(\alpha, \beta) \in D$ *is said to be* strongly protected *if it occurs in at least one of the configurations in Figure 4.21.*

The following theorem, stated here without proof, characterizes essential graphs. The statement has been included, because it is a vital step in the Monte Carlo algorithm presented later, in Section 6.4 for locating the graph structure.

Theorem 4.5 *Let* $G = (V, E)$ *be a graph, where* $E = D \cup U$. *There exists a directed acyclic graph* G^* *for which* G *is the corresponding essential graph if and only if* G *satisfies the following conditions:*

1. G *is a chain graph,*

2. *Each chain component of* G *is triangulated,*

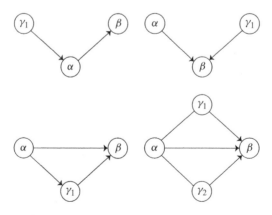

Figure 4.21   The directed edge (α, β) is strongly protected (Definition 4.26).

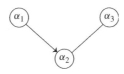

Figure 4.22 A forbidden sub-graph on $\{\alpha_1, \alpha_2, \alpha_3\}$ in an essential graph.

3. *The configuration shown in Figure 4.22 does not occur in any induced sub-graph of a three variable set $\{\alpha_1, \alpha_2, \alpha_3\} \subset V$ (that is, a directed edge (α_1, α_2), an undirected edge $\langle\alpha_2, \alpha_1\rangle$ and no edge between α_1 and α_3), and*

4. *Every directed edge $(\alpha_1, \alpha_2) \in D$ is strongly protected in $\mathcal{G}$.*

Proof of Theorem 4.5 Omitted.   □

It is useful to extend the definition of *faithfulness* to the essential graph, since all the DAGs with the same immoralities and skeleton as the essential graph preserve the same independence structure. The following discussion extends the definition of faithfulness to the chain graph which, in view of Theorem 4.5 covers the essential graph.

Recall definition of parents, directed and undirected neighbours, from Definition 2.3. Since the chain graph contains both directed and undirected edges, the notion of the 'parents' of a node has to be extended to 'ancestral boundary', which will play a similar role. It is simply the collection of parents together with the undirected neighbours.

Definition 4.27 (Ancestral Boundary) *The* ancestral boundary *of a node α in a graph $\mathcal{G}$ is defined as*

$$\Lambda_\mathcal{G}(\alpha) = \Pi(\alpha) \cup N_{(u)}(\alpha) \qquad (4.2)$$

and the ancestral boundary of a subset $A \subseteq V$ is defined as

$$\Lambda_\mathcal{G}(A) = \Pi(A) \cup N_{(u)}(A). \qquad (4.3)$$

The closure *of a subset $A \subseteq V$ in $\mathcal{G}$ is defined as*

$$\Phi_\mathcal{G}(A) = A \cup \Lambda_\mathcal{G}(A). \qquad (4.4)$$

Crucial to the extension of the definition of faithfulness is the definition of *ancestral set*. For a directed acyclic graph, the minimal ancestral set of a node α is simply the set of all nodes β such that there is a directed path from β to α.

Definition 4.28 (Ancestral Set, Minimal Ancestral Set) *A subset $A \subseteq V$ is said to be* ancestral *if $\Lambda_\mathcal{G}(\alpha) \subseteq A$ for each $\alpha \in A$, where $\Lambda_\mathcal{G}(\alpha)$ is defined in Equation (4.2). For any subset $A \subseteq V$, the* minimal ancestral set *is defined as the smallest ancestral set containing A and is denoted $An(A)$.*

For a chain graph, in many situations, a *complex*, defined below, plays the same role as an immorality in a directed acyclic graph.

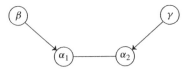

Figure 4.23 A simple chain graph; here $S = \{\alpha_1, \alpha_2\}$ and (β, γ, S) is a minimal complex.

Definition 4.29 (Complex, Minimal Complex) *Recall that (β, α) denotes the directed edge $\beta \rightarrow \alpha$. Let $\mathcal{G} = (V, E)$ be a chain graph. Let $\alpha, \beta \in V$ and let $S \subseteq V$. The triple (α, β, S) is known as a* complex *in $\mathcal{G}$ if S is a connected subset of a chain component C of $\mathcal{G}$ and α, β are two non-adjacent nodes in $\Lambda_{\mathcal{G}}(S) \cap \Lambda_{\mathcal{G}}(C)$. The triple (α, β, S) is known as a* minimal complex *in $\mathcal{G}$ if there is no strict subset $S' \subset S$ such that (α, β, S') forms a complex in $\mathcal{G}$.*

It follows directly from the definition that if S contains a single node, then the complex (α, β, S) is an immorality, where the node in S is a collider.

An example of a simple chain graph, where $S = \{\alpha_1, \alpha_2\}$ and (β, γ, S) is a minimal complex, is given in Figure 4.23.

The following concepts and results have been established. For directed acyclic graphs, Markov equivalence is defined in Definition 2.32. That is, two DAGs are *Markov equivalent* if they have the same d-separation properties. Theorem 4.4 states that two DAGs are Markov equivalent if and only if they have the same skeleton and the same immoralities. The essential graph (Definition 4.23) is the partially directed graph that retains the same skeleton as a given directed acyclic graph and where the set of directed edges is simply the set of all those directed edges that are common to all the DAGs that are Markov equivalent; all the other edges are undirected. The resulting essential graph is a chain graph.

The notion of local Markov property, the moral graph, and global Markov property (faithfulness) are now extended to chain graphs.

The definitions of *descendants* and *ancestors* are given in Definition 2.9; the definition of *ancestral boundary* is given in Definition 4.27. The definition of the local Markov condition (given in Definition 2.28) may be defined for a chain graph:

Definition 4.30 (Local Markov Condition for a Chain Graph) *Let $\mathcal{G} = (V, E)$ be a chain graph. A probability distribution p over a set of variables $V = \{X_1, \ldots, X_d\}$ is said to be* local $\mathcal{G}$-Markovian, *or satisfies the local Markov condition if for each $j \in \{1, \ldots, d\}$*

$$X_j \perp (V \backslash (D(X_j) \cup \Phi_{\mathcal{G}}(\{X_j\})) | \Lambda_{\mathcal{G}}(\{X_j\}),$$

where $\Phi_{\mathcal{G}}$ and $\Lambda_{\mathcal{G}}$ are defined by Equations (4.4) and (4.3) respectively.

In other words, p satisfies the local Markov condition for a graph $\mathcal{G} = (V, E)$ if for each j, X_j is conditionally independent, conditioned on the ancestral boundary (which is the generalization to chain graphs of the set of parents) of all the other variables that are not either descendants or belonging to the ancestral boundary. It is left as an exercise (Exercise 9) to see that this reduced to the local directed Markov condition of Definition 2.28 when the $\mathcal{G}$ is a directed acyclic graph.

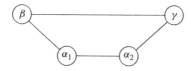

Figure 4.24   The moralized graph for the chain graph in Figure 4.23.

The definition of *faithfulness* (Definition 2.30) may also be extended to chain graphs. The following definitions are necessary to make the extension.

Definition 4.31 (Moral Graph for a Chain Graph) *Let $\mathcal{G} = (V, E)$ be a chain graph. The moral graph $\mathcal{G}^m$ is defined as the graph (V, E^m), where E^m contains the edge $\langle \beta, \gamma \rangle$ if and only if either*

1. *$(\beta, \gamma) \in E$ or $(\gamma, \beta) \in E$ or $\langle \alpha, \beta \rangle \in E$, or*

2. *There exist nodes α_1 and α_2 (not necessarily distinct) in the same chain component such that $(\beta, \alpha_1) \in E$ and $(\gamma, \alpha_2) \in E$.*

For example, the moral graph for the chain graph of Figure 4.23 is shown in Figure 4.24.

The global Markov property (defined below) generalizes the definition of faithfulness to probability distributions represented by chain graphs.

Definition 4.32 (Global Markov Property) *Recall the definition $An(C)$, the minimal ancestral set of a subset $C \subseteq V$, given in Definition 4.28 and recall the Definition 4.5, where a* separator *is defined. Let $\mathcal{G} = (V, E)$ be a chain graph, where $V = \{X_1, \ldots, X_d\}$ and let p be a probability function over the variable set V, then p is said to be* globally $\mathcal{G}$-Markovian *if for any sets $A, B, S \subset V$ such that S separates A and B in $\mathcal{G}^m_{An(A \cup B \cup S)}$ (the moralized graph restricted to the set $An(A \cup B \cup S)$), it holds that $A \perp B | S$.*

It is left as an exercise (Exercise 10) to show that this definition is equivalent to the definition of faithfulness when $\mathcal{G} = (V, E)$ is a directed acyclic graph.

The definition of Markov equivalence (Definition 2.32) for directed acyclic graphs, may be extended to chain graphs.

Definition 4.33 (Markov Equivalence for Chain Graphs) *Two chain graphs $\mathcal{G}_1 = (V, E_1)$ and $\mathcal{G}_2 = (V, E_2)$ are Markov equivalent if for any three sets A, B and S, S separates A from B in $\mathcal{G}^m_1$ if and only if S separates A from B in $\mathcal{G}^m_2$.*

Definition 4.34 (Graphical Equivalence for Chain Graphs) *Two chain graphs are said to be graphically equivalent if they have the same skeleton and the same minimal complexes.*

The following basic result about Markov equivalence of chain graphs was first proved by M. Frydenberg [86] in a special case (Theorem 5.6 in that article) and then by Madigan, Andersson and Perlman [87] for the general case.

Theorem 4.6 *Two chain graphs $\mathcal{G}_1 = (V, E_1)$ and $\mathcal{G}_2 = (V, E_2)$ are Markov equivalent if and only if they have the same skeleton and the same minimal complexes. That is, $\mathcal{G}_1$ and $\mathcal{G}_2$ are Markov equivalent if and only if they are graphically equivalent.*

This result extends Theorem 4.4. The proof is omitted.

Notes   This chapter presents the basic graph theory necessary for the treatment of Bayesian networks covered in this text. The theory of decomposable graphs is standard material from algorithmic graph theory [88]. Some specialized additional material on the algorithms for Bayesian networks is found in Chapter 4 of [67]. A rigorous and up-to-date research treatise covering chain graphs is [89] (M. Studený).

4.6   Exercises: Decomposable graphs and chain graphs

1. Determine the simplicial nodes of the graph in Figure 4.25. Is the graph triangulated?

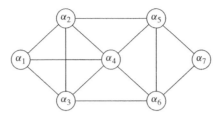

Figure 4.25   Graph for Example 1.

2. Determine the simplicial nodes for the graph in Figure 4.26. Is the graph triangulated? If not, state a single edge which, if added, would triangulate the graph.

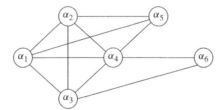

Figure 4.26   Graph for Exercises 2 and 3.

3. Consider the graph in Figure 4.26. Determine the cliques and construct a junction tree for the graph.

4. Consider the graph in Figure 4.27. Is it decomposable? Justify your answer.

5. Consider the Bayesian network shown in Figure 4.28.

   (a) Moralize the graph.

   (b) By finding an appropriate elimination sequence, show that the moral graph triangulated.

   (c) Find a junction tree.

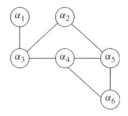

Figure 4.27   Graph for Exercise 4.

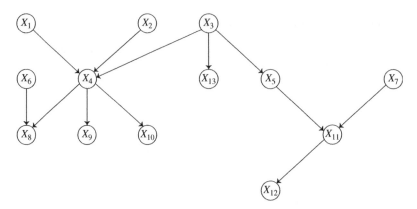

Figure 4.28   Bayesian network for Exercise 5.

6. A directed acyclic graph is *singly connected* if the graph obtained by dropping the directions of the links is a tree. For example, the Bayesian network in Figure 4.28 is singly connected.

   (a) Prove that the moral graph of a singly connected graph is triangulated.

   (b) Prove that the separators in a junction tree for a singly connected graph consist of exactly one node.

7. Let $\mathcal{G} = (V, E)$ be a triangulated graph, let $A \subset V$ and let $\mathcal{G}_A = (A, E_A)$, where $E_A = E \cap A \times A$ is the graph restricted to A. Prove (formally) that $\mathcal{G}_A$ is triangulated.

8. Consider the Bayesian network given in Figure 4.29.

   (a) Find the moral graph $\mathcal{G}^{(m)}$

   (b) Find a triangulation of the moral graph, adding in as few edges as possible.

   (c) Find a junction tree for the triangulated graph in (b).

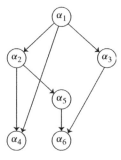

Figure 4.29   Graph for Exercise 8.

9. Prove that Definition 4.30 is equivalent to Definition 2.28 if $\mathcal{G} = (V, E)$ is a directed acyclic graph.

10. Let $G = (V, E)$ be a directed acyclic graph. Prove that if a probability distribution over the variables V is *Globally G-Markovian* (Definition 4.32), then p and G are *faithful* (Definition 2.32).

11. Let T be a junction tree of cliques constructed from an undirected triangulated graph $G = (V, E)$. Let α be a node in G. Show that if all nodes *not* containing α are removed from T, then the remaining tree is connected.

5

Learning the conditional probability potentials

In most applications of artificial learning, a machine is expected, using a database of instantiations, to 'learn' (in other words estimate) both the *structure* of the DAG (in other words, which directed edges should be used and which should be omitted) and, once the structure of the DAG has been established, the conditional probability potentials (CPPs). Let $V = \{X_1, \ldots, X_d\}$ then, provided the graph structure has been established, the probability potentials are $\{(p_{X_v|\Pi_v})_{v=1}^d\}$ where Π_v (as usual) denotes the parent set of variable X_v.

Chapter 5 considers the second of these problems; how to learn the CPPs *given* a DAG structure, while Chapter 6 considers the problem of learning a suitable DAG structure from a set of instantiations.

In the analysis given below, it is assumed that the prior distribution over the conditional probability potentials may be expressed as a product of Dirichlet distributions. The analysis in [90] illustrates that under rather broad modelling assumptions, the Dirichlet prior is inevitable.

5.1 Initial illustration: maximum likelihood estimate for a fork connection

Consider a network with three variables, $U = (X, Y, Z)$, where the distribution factorizes as $p_U = p_{X|Z} p_{Y|Z} p_Z$. Suppose that the variables X, Y and Z are binary, and suppose there are four *complete* and independent instantiations of U; namely, $(u^{(i)})_{i=1,2,3,4}$, given by

$$\begin{array}{cccc} & x & y & z \\ u^{(1)} & 0 & 1 & 0 \\ u^{(2)} & 1 & 1 & 0 \\ u^{(3)} & 0 & 0 & 1 \\ u^{(4)} & 0 & 1 & 0 \end{array}$$

Definition 5.1 (Complete Instantiation) *An instantiation of a set of variables* $U = (X_1, \dots, X_d)$ *is said to be* complete *if there are* no missing values.

The probability of the four instantiations listed above from independent observations is

$$p_U(u^{(1)}) p_U(u^{(2)}) p_U(u^{(3)}) p_U(u^{(4)}),$$

where

$$p_U(u^{(i)}) = p_{X|Z}(x^{(i)}|z^{(i)}) p_{Y|Z}(y^{(i)}|z^{(i)}) p_Z(z^{(i)}).$$

The probabilities of success, conditioned on the state of the parents, are given by

$$\theta_{x|0} = p_{X|Z}(1|0) \qquad \theta_{x|1} = p_{X|Z}(1|1),$$

$$\theta_{y|0} = p_{Y|Z}(1|0) \qquad \theta_{y|1} = p_{Y|Z}(1|1),$$

$$\theta_z = p_Z(1).$$

The CPPs are fully specified by these five parameters in this example.

Expressed in these parameters, the probabilities of getting the four instantiations $u^{(1)}, u^{(2)}, u^{(3)}, u^{(4)}$ are:

$$p_U(u^{(1)}) = (1 - \theta_{x|0})\theta_{y|0}(1 - \theta_z)$$
$$p_U(u^{(2)}) = \theta_{x|0}\theta_{y|0}(1 - \theta_z)$$
$$p_U(u^{(3)}) = (1 - \theta_{x|1})(1 - \theta_{y|1})\theta_z$$
$$p_U(u^{(4)}) = (1 - \theta_{x|0})(1 - \theta_{y|0})(1 - \theta_z).$$

This gives

$$p_U(u^{(1)}) p_U(u^{(2)}) p_U(u^{(3)}) p_U(u^{(4)}) = \theta_{x|0}(1 - \theta_{x|0})^2(1 - \theta_{x|1})\theta_{y|0}^3(1 - \theta_{y|1})\theta_z(1 - \theta_z)^3.$$

This is to be maximized as a function of the five parameters in the expression. The maximization splits into the maximization of five separate likelihoods, which have already been considered in the thumb-tack example in Section (1.8). These are

$$L(\theta_{x|0}) = \theta_{x|0}(1 - \theta_{x|0})^2,$$

$$L(\theta_{x|1}) = 1 - \theta_{x|1},$$

$$L(\theta_{y|0}) = \theta_{y|0}^3,$$

$$L(\theta_{y|1}) = 1 - \theta_{y|1},$$

$$L(\theta_z) = (1 - \theta_z)^3\theta_z.$$

Maximizing these gives

$$\hat{\theta}_{x|0} = \frac{1}{3}, \ \hat{\theta}_{x|1} = 0, \ \hat{\theta}_{y|0} = 1, \ \hat{\theta}_{y|1} = 0, \ \hat{\theta}_z = \frac{1}{4}.$$

The following terminology is used: (X, Z) is called a *family* (child, parent). An *instantiation* (x, z) is a *family configuration*. z is called a *parent configuration*. Note, in the above example, that $z = 0$ appears three times and $(x, z) = (1, 0)$ appears once. Note that

$$\hat{\theta}_{x|z=0,MLE} = \frac{1}{3} = \frac{\text{frequency of family configuration } (x, z) = (1, 0)}{\text{frequency of parent configuration } z = 0}.$$

5.2   The maximum likelihood estimator for multinomial sampling

Consider a DAG over a set of variables $V = \{X_1, \ldots, X_d\}$, where Π_j denotes the set of parent variables of X_j. Suppose that for each $j = 1, \ldots, d$, the variables X_j takes values in the set $\mathcal{X}_j := \{x_j^{(1)}, \ldots, x_j^{(k_j)}\}$, and that the parent configurations Π_j take values in the set $\{\pi_j^{(q_1)}, \ldots, \pi_j^{(q_j)}\}$ for $j = 1, \ldots, d$, for j such that Π_j is non-empty. Let

$$\theta_{ijl} := p(\{X_j = x_j^{(i)}\}|\{\Pi_j = \pi_j^{(l)}\}).$$

In Section 5.3, it is shown that in any DAG with a set of complete instantiations, for all (i, j, l), the maximum likelihood estimate of θ_{ijl} will be of the form

$$\hat{\theta}_{MLE;ijl} = \frac{\text{frequency of the family configuration } (x_j^{(i)}, \pi_j^{(l)})}{\text{frequency of the parent configuration } \pi_j^{(l)}}.$$

To prepare the way, Section 5.2 considers multinomial sampling. Let $X_1, \ldots, X_n$ be independent, identically distributed random variables, each with the same distribution as X, where X takes values in a set $\mathcal{X} = \{x^{(1)}, \ldots, x^{(k)}\}$. (Consider, for example, an urn containing 25 balls of which six are red, six are blue and 13 are green. The random variable X denotes the resulting colour if one pulls out a ball at random, notes its colour and then puts it back. Let $x^{(1)}$ denote a red selection, $x^{(2)}$ a blue selection and $x^{(3)}$ a green selection, so here $k = 3$). Let

$$\theta_i = p_X(x^{(i)}),$$

so that $\theta_1 + \cdots + \theta_k = 1$. Set

$$\theta = (\theta_1, \ldots, \theta_k).$$

Let $\underline{X} = (X_1, \ldots, X_n)$, and let $\underline{x} = (x^{(i_1)}, \ldots, x^{(i_n)})$ denote the sequence drawn in n independent trials that are conditionally independent, given $\underline{\theta}$. Let

$$n_l = \text{number of times } x^{(l)} \text{ appears in } \underline{x}, \quad l = 1, \ldots, k$$

so that $n = n_1 + \cdots + n_k$, then

$$p_{\underline{X}}(\underline{x}|\underline{\theta}) = \theta_1^{n_1} \cdots, \theta_k^{n_k}.$$

Maximum likelihood   Let

$$\tilde{\Theta}_k = \left\{ (\theta_1, \theta_2, \ldots, \theta_k) \mid \theta_j \geq 0, \; j = 1, \ldots, k, \; \sum_{j=1}^{k} \theta_j = 1 \right\}$$

denote the parameter space.

Definition 5.2 (Likelihood function, Likelihood Estimate, Log Likelihood Function)
The likelihood function of the parameters $\underline{\theta}$ *is defined as*

$$L(\underline{\theta}|\underline{x}) = p_{\underline{X}}(\underline{x}|\underline{\theta}).$$

The maximum likelihood estimate $\widehat{\underline{\theta}}_{ME}$ *is defined as the value of $\underline{\theta}$ that maximizes $L(\underline{\theta}|\underline{x})$.
The* log likelihood function l *is defined as:*

$$l(\theta_1, \theta_2, \ldots, \theta_k) = \log p_{\underline{X}}(\underline{x} \mid \underline{\theta}),$$

where log *is used to denote the* natural *logarithm.*

The computation of $\widehat{\underline{\theta}}_{ME}$ is a maximization problem with constraints. Equivalently, the
maximum of $l(\theta_1, \theta_2, \ldots, \theta_k)$ may be computed, subject to the constraint $\theta_1 + \theta_2 + \cdots + \theta_k = 1$. This is best achieved by introducing the auxiliary function in the $k - 1$ free
variables

$$\tilde{l}(\theta_1, \theta_2, \ldots, \theta_{k-1}) = l(\theta_1, \theta_2, \ldots, 1 - (\theta_1 + \theta_2 + \cdots + \theta_{k-1})).$$

This gives

$$\tilde{l}(\theta_1, \theta_2, \ldots, \theta_{k-1}) = n_1 \cdot \log \theta_1 + n_2 \cdot \log \theta_2 + \cdots + n_k \cdot \log(1 - (\theta_1 + \theta_2 + \cdots + \theta_{k-1})).$$

The partial derivatives of $\tilde{l}(\theta_1, \theta_2, \ldots, \theta_{k-1})$ are taken with respect to each of θ_1,
$\theta_2, \ldots, \theta_{k-1}$ and the critical points occur at the values for which the partial derivatives
are all equal to zero. This gives the system of equations

$$\frac{\partial}{\partial \theta_1} \tilde{l}(\theta_1, \theta_2, \ldots, \theta_{k-1}) = \frac{n_1}{\theta_1} - \frac{n_k}{1 - (\theta_1 + \theta_2 + \ldots + \theta_{k-1})} = 0,$$

$$\vdots$$

$$\frac{\partial}{\partial \theta_{k-1}} \tilde{l}(\theta_1, \theta_2, \ldots, \theta_{k-1}) = \frac{n_{k-1}}{\theta_{k-1}} - \frac{n_k}{1 - (\theta_1 + \theta_2 + \ldots + \theta_{k-1})} = 0.$$

These equations are equivalent to the equalities

$$\frac{n_1}{\theta_1} = \frac{n_2}{\theta_2} = \cdots = \frac{n_k}{1 - (\theta_1 + \theta_2 + \cdots + \theta_{k-1})}.$$

To simplify notation, the common value of these ratios is denoted as λ. This gives

$$\theta_1 = \frac{n_1}{\lambda}, \quad \theta_2 = \frac{n_2}{\lambda}, \ldots, \theta_k = \frac{n_k}{\lambda}.$$

The constraint $\theta_1 + \theta_2 + \cdots + \theta_k = 1$ is now employed, giving

$$1 = \theta_1 + \theta_2 + \cdots + \theta_k = \frac{n_1}{\lambda} + \frac{n_2}{\lambda} + \cdots + \frac{n_k}{\lambda},$$

and

$$\lambda = n_1 + n_2 + \cdots + n_k = n.$$

This gives the solution

$$\widehat{\theta_i} = \frac{n_i}{n}, \qquad i = 1, \ldots, k.$$

It remains to show that this critical point yields a maximum. This could be achieved by checking the matrix of second order partial derivatives of $\widetilde{l}$. There is a simpler way to prove that the estimate found above maximizes the likelihood function, which requires the following two definitions and a lemma.

Definition 5.3 (Shannon Entropy) *The Shannon entropy, or entropy of a probability distribution* $\underline{\theta} = (\theta_1, \ldots, \theta_k)$, *where* $\theta_j \geq 0$, $j = 1, \ldots, k$ *and* $\theta_1 + \cdots + \theta_k = 1$ *is defined as*

$$H(\underline{\theta}) = -\sum_{j=1}^{k} \theta_j \log \theta_j.$$

Natural logarithms are used. In the definition of $H(\underline{\theta})$, *the definition* $0 \log 0 = 0$ *is used, obtained by continuous extension of the function* $x \log x$, $x > 0$.

Note that $H(\underline{\theta}) \geq 0$.

Definition 5.4 (Kullback-Leibler Divergence) *The Kullback-Leibler divergence between two discrete probability distributions* f *and* g *with the same state space* $\mathcal{X}$ *is defined as*

$$D(f|g) = \sum_{x \in \mathcal{X}} f(\{x\}) \log \frac{f(\{x\})}{g(\{x\})}.$$

If $g(\{x\}) = 0$ for $f(\{x\}) \neq 0$, then $f(\{x\}) \cdot \log 0 = +\infty$. If $D(f|g) = +\infty$, then there is at least one outcome x such that f and g may be distinguished without error.

Lemma 5.1 *For any two discrete probability distributions* f *and* g, *it holds that*

$$D(f|g) \geq 0.$$

Proof of Lemma 5.1 The proof uses Jensen's inequality,[1] namely, that for any convex function ϕ, $E[\phi(X)] \geq \phi(E[X])$. Note that $f(\{x\}) \geq 0$ for all $x \in \mathcal{X}$ and that $\sum_{x \in \mathcal{X}} f(\{x\}) = \sum_{x \in \mathcal{X}} g(\{x\}) = 1$. Using this, together with the fact that $-\log$ is convex, yields

$$D(f|g) = -\sum_{x \in \mathcal{X}} f(\{x\}) \log \left(\frac{g(\{x\})}{f(\{x\})} \right) \geq -\log \left(\sum_{x \in \mathcal{X}} f(\{x\}) \frac{g(\{x\})}{f(\{x\})} \right) = -\log 1 = 0.$$

$\square$

---

[1] J.L. Jensen (1859–1925) published this in *Acta Mathematica* 1906 **30**(1).

It is now proved that the candidate estimate given above maximizes the likelihood function.

Proposition 5.1 *The maximum likelihood estimate $\widehat{\underline{\theta}}_{ML}$ of $\underline{\theta}$ is*

$$\widehat{\underline{\theta}}_{ML} = \left(\frac{n_1}{n}, \frac{n_2}{n}, \ldots, \frac{n_k}{n} \right).$$

Proof of Proposition 5.1 The candidate solution $\widehat{\underline{\theta}}_{ML}$ belongs to $\tilde{\Theta}_k$ and is therefore feasible. Since $p_X\left(\mathbf{x}^{(n)} \mid \theta\right) = \prod_{i=1}^{k} \theta_i^{n_i}$, the following identity follows directly:

$$H\left(\widehat{\underline{\theta}}_{ML}\right) = -\frac{1}{n} \log p_X\left(\underline{x} \mid \widehat{\underline{\theta}}_{ML}\right), \tag{5.1}$$

where

$$H\left(\widehat{\underline{\theta}}_{ML}\right) = -\sum_{i=1}^{k} \widehat{\theta}_i \log \widehat{\theta}_i \tag{5.2}$$

is the *Shannon entropy* (for an empirical distribution) in natural logarithm, given in Definition 5.3.

For arbitrary $\underline{\theta} \in \tilde{\Theta}_k$, the following identity follows directly from Equation (5.2):

$$p_X\left(\underline{x} \mid \underline{\theta}\right) = \prod_{i=1}^{k} \theta_i^{n_i} = \exp\left\{ \sum_{i=1}^{k} n_i \log \theta_i \right\}$$

$$= \exp\left\{ n \sum_{i=1}^{k} \widehat{\theta}_i \log \theta_i \right\} \tag{5.3}$$

$$= \exp\left\{ n \sum_{i=1}^{k} \widehat{\theta}_i \log \widehat{\theta}_i - n \sum_{i=1}^{k} \widehat{\theta}_i \log \frac{\widehat{\theta}_i}{\theta_i} \right\}$$

$$= \exp\left\{ -n\left(H\left(\widehat{\underline{\theta}}_{ML}\right) + D\left(\widehat{\underline{\theta}}_{ML} | \underline{\theta}\right)\right) \right\}. \tag{5.4}$$

Thus, from Lemma 5.1, together with the fact that $D(f|f) = 0$ for any admissible f, it follows that

$$L(\underline{\theta}) := p_X\left(\underline{x} \mid \underline{\theta}\right) = \exp\left\{ -n\left(H\left(\widehat{\underline{\theta}}_{ML}\right) + D\left(\widehat{\underline{\theta}}_{ML} | \underline{\theta}\right)\right) \right\} \leq \exp\left\{ -nH\left(\widehat{\underline{\theta}}_{ML}\right) \right\}$$

$$= p_X\left(\underline{x} \mid \widehat{\underline{\theta}}_{ML}\right) = L(\widehat{\underline{\theta}}_{ML})$$

for every $\underline{\theta} \in \tilde{\Theta}_k$ and Proposition 5.1 is proved.   □

Mean posterior estimate The approach that has just been presented is the *maximum likelihood* method. An estimator derived from the Bayesian posterior distribution is now considered. First, a prior distribution is put over the parameter space. The Dirichlet distribution is useful here, because the integral can be computed explicitly to give a posterior density which is again a Dirichlet distribution. The definition of the Dirichlet distribution was given in Definition 1.13.

The *maximum posterior estimate*, the value of the parameter value which gives the maximum value for the posterior distribution, has already been discussed. The *mean posterior estimate* is the *expected value* of the posterior distribution. Here,

$$\hat{\theta}_{i,MEP} = \int \theta_i \pi(\theta_1, \ldots, \theta_k | \mathbf{x}, \alpha) d\theta_1 \ldots d\theta_k = \frac{n_i + \alpha_i}{\sum_{j=1}^{k} n_j + \sum_{j=1}^{k} \alpha_j}.$$

This computation is left as an exercise.

Remarks The following remarks are applicable to learning CPPs for any DAG $\mathcal{G} = (V, E)$ where $V = \{X_1, \ldots, X_d\}$ and the distribution of $p(X_i | \Pi_i)$ (i,e, of X_i conditioned on the parent set) is multinomial for each $i = 1, \ldots, d$.

1. The aim of statistics is to predict future outcomes based on past information. The parameters are only a tool to help this. In a fully Bayesian approach to statistics, the only estimate of the parameter, given data $\mathbf{x}$, is the entire posterior distribution $\pi_{\Theta | \mathbf{X}}(.|\mathbf{x})$, which is then used to compute the predictive distribution. But this approach, although intellectually satisfying, is not always practical, and with Bayesian networks it is often useful to have a point estimator. There are some situations where the MEP turns out to be more useful than the MLE. One example may be the situation of learning using information from a very large data warehouse. Even though the warehouse is large, it may happen that the size of the entire set of possible cases is very much larger (i.e. data is sparse). In other words, it is suspected that there may be positive cases, even if the data warehouse does not contain any examples. In such a situation, the sample yields $n_i = 0$. This would yield $\hat{\theta}_{i,MLE} = 0$, but if the a priori supposition that there may exist positive cases is modelled into the prior, then $\hat{\theta}_{i,MEP} > 0$. The strict positivity of the MEP sometimes turns out to be a valuable property in the mining of very large data sets, which are usually sparse. This makes it preferable to the MLE.

2. Comparing with $\hat{\theta}_{i,MLE} = \frac{n_i}{n}$, note that

$$\lim_{n \to +\infty} \frac{\hat{\theta}_{iMEP}}{\hat{\theta}_{iMLE}} = 1.$$

This is an asymptotic result and in the situation envisaged above, n may not be sufficiently large for the ratio to be close to 1.

5.3 MLE for the parameters in a DAG: the general setting

Notations Firstly, the necessary notation is developed. $V = \{X_1, \ldots, X_d\}$ denotes the collection of random variables under consideration and $\mathcal{X}$ denotes the set of all possible outcomes for the experiment. $\tilde{\Theta}$ denotes the parameter space and $\Omega = \tilde{\Theta} \times \mathcal{X}$ denotes the context of the experiment. $\mathcal{G} = (V, E)$ denotes the directed acyclic graph along which the probability function $p_{X_1, \ldots, X_d}$ is factorized.

In the situation considered here, the sample space for variable X_j is denoted

$$\mathcal{X}_j = \{x_j^{(1)}, \ldots, x_j^{(k_j)}\},$$

so that

$$\mathcal{X} = \mathcal{X}_1 \times \cdots \times \mathcal{X}_d.$$

That is,

$$\mathcal{X} = \{(x_1^{(j_1)}, \ldots, x_d^{(j_d)}) \mid j_1 \in (1, \ldots, k_1), \ldots, j_d \in (1, \ldots, k_d)\}.$$

A complete *instantiation* is an outcome $\underline{x} = (x_1^{(i_1)}, \ldots, x_d^{(i_d)})$. Note that here the vectors are taken as *rows*. This is because in a database of instantiation, it is usual that each column represents a variable and each row represents an instantiation. A sample of cases (also called records in the data mining literature) is given by a matrix

$$\mathbf{x} = \begin{pmatrix} \underline{x}_{(1)} \\ \vdots \\ \underline{x}_{(n)} \end{pmatrix},$$

where $\underline{x}_{(i)} = (x_{i,1}^{(j_1)}, \ldots, x_{i,d}^{(j_d)})$ denotes the ith instantiation. Usually, when data is stored, each column represents the outcomes for one particular variable. The CPPs of a Bayesian network are to be estimated on the basis of this data.

For each variable X_j, consider all possible instantiations of the parent set Π_j and label them $(\pi_j^{(l)})_{l=1}^{q_j}$. That is, $\pi_j^{(l)}$ denotes that the *parent* configuration of variable X_j is in state $\pi_j^{(l)}$ and there are q_j possible configurations of Π_j.

Example 5.1 For the DAG shown in Figure 5.1, where A, B, C are all binary variables, $\Pi(B) = (A, C)$ has four possible sets of instantiations:

$$\pi_B^{(1)} = (0, 0), \quad \pi_B^{(2)} = (0, 1), \quad \pi_B^{(3)} = (1, 0), \quad \pi_B^{(4)} = (1, 1).$$

Set

$$n_k(x_j^{(i)} | \pi_j^{(l)}) = \begin{cases} 1 & (x_j^{(i)}, \pi_j^{(l)}) \text{ is found in } \underline{x}_{(k)} \\ 0 & \text{otherwise,} \end{cases}$$

where $(x_j^{(i)}, \pi_j^{(l)})$ is a configuration of the *family* (X_j, Π_j). Let $\underline{\theta}$ denote the set of parameters defined as

$$\theta_{jil} = p(\{X_j = x_j^{(i)}\} | \{\Pi_j = \pi_j^{(l)}\}).$$

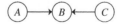

Figure 5.1 A collider.

for $j = 1, \ldots, d$, $i = 1, \ldots, k_j$, $l = 1, \ldots, q_j$, with given graph structure $\mathcal{G} = (V, E)$. Then the *joint probability* of a case $\underline{x}^{(k)}$ occurring may be written as

$$p_{\underline{X}|\Theta}(\underline{x}_{(k)}|\underline{\theta}, E) = \prod_{j=1}^{d} \prod_{l=1}^{q_j} \prod_{i=1}^{k_j} \theta_{jil}^{n_k(x_j^{(i)}|\pi_j^{(l)})}.$$

The following is a summary of the notation:

- d is the number of variables ($=$ the number of nodes).

- k_j is the number of possible states that the variable X_j can take.

- q_j is the number of possible parent configurations for variable X_j.

- $\theta_{jil} := p(\{X_j = x_j^{(i)}\}|\{\Pi_j = \pi_j^{(l)}\})$.   (5.5)

  That is, θ_{jil} is the conditional probability that variable X_j is in state i, given that the parent configuration is configuration l. The notation $\pi_j^{(l)}$ denotes that the parents of variable j are in state l.

Let $\mathbf{X} = \begin{pmatrix} \underline{X}_{(1)} \\ \vdots \\ \underline{X}_{(n)} \end{pmatrix}$. The cases $\mathbf{x} = \begin{pmatrix} \underline{x}_{(1)} \\ \vdots \\ \underline{x}_{(n)} \end{pmatrix}$ are considered to be independent obser-

vations, giving

$$p_{\mathbf{X}|\Theta}(\mathbf{x}|\underline{\theta}, \mathcal{G})$$

$$= \prod_{k=1}^{n} p_{\underline{X}|\Theta}(\mathbf{x}_{(k)}|\underline{\theta}, \mathcal{G}) = \prod_{j=1}^{d} \prod_{l=1}^{q_j} \prod_{i=1}^{k_j} \prod_{k=1}^{n} \theta_{jil}^{n_k(x_j^{(i)}|\pi_j^{(l)})} = \prod_{j=1}^{d} \prod_{l=1}^{q_j} \prod_{i=1}^{k_j} \theta_{jil}^{\sum_{k=1}^{n} n_k(x_j^{(i)}|\pi_j^{(l)})}.$$

Set

$$n(x_j^{(i)}|\pi_j^{(l)}) = \sum_{k=1}^{n} n_k(x_j^{(i)}|\pi_j^{(l)}).$$

This is simply the number of times the configuration $(x_j^{(i)}, \pi_j^{(l)})$ appears in $\mathbf{x} = \begin{pmatrix} \underline{x}_{(1)} \\ \vdots \\ \underline{x}_{(n)} \end{pmatrix}$.

The *likelihood* is therefore

$$L(\theta) = \prod_{k=1}^{n} p_{\underline{X}|\Theta}(\mathbf{x}^{(k)}|\theta, \mathcal{G}) = \prod_{j=1}^{d} \prod_{l=1}^{q_j} \prod_{i=1}^{k_j} \theta_{jil}^{n(x_j^{(i)}|\pi_j^{(l)})}.$$

The likelihood factorizes into local parent child factors and additionally to $d \times q_j$ separate maximum likelihood estimations all of the basic form treated in the preceding section. It follows that

$$\hat{\theta}_{MLE,jil} = \frac{n(x_j^{(i)}|\pi_j^{(l)})}{n(\pi_j^{(l)})},$$

where

$$n(\pi_j^{(l)}) = \sum_{i=1}^{k_j} n(x_j^{(i)}|\pi_j^{(l)})$$

is the frequency of configuration $\pi_j^{(l)}$ in $\mathbf{X}$. The maximum likelihood estimate of θ_{jil} is therefore as advertized:

$$\hat{\theta}_{jil} = \frac{\text{frequency of the family configuration}}{\text{frequency of the parent configuration}}.$$

Posterior distribution of the CPPs   The approach of computing the maximum likelihood estimator is now contrasted with the Bayesian approach. For convenience of writing,

$$\theta_{j.l} = \left(\theta_{j1l}, \ldots, \theta_{jk_jl}\right)$$

is used to denote the probability distribution over the states of X_j, given that $\pi_j^{(l)}$ is the parent configuration. The prior distribution over $\theta_{j.l}$ is taken to be

$$Dir(\alpha_{j1l}, \ldots, \alpha_{jk_jl}).$$

A standard computation using the Dirichlet integral yields that the posterior density is the Dirichlet density

$$\theta_{j.l}|(\underline{x}_{(1)}, \ldots, \underline{x}_{(n)}) \sim Dir(n(x_j^{(1)}|\pi_j^{(1)}) + \alpha_{j1l}, \ldots, n(x_j^{(k_j)}|\pi_j^{(1)}) + \alpha_{jk_jl}).$$

The *tables of counts* of family configurations at node j, e.g.

$$n(x_j^{(1)}|\pi_j^{(l)}), \ldots, n(x_j^{(k_j)}|\pi_j^{(l)})$$

is stored as a memory of past experience. The posterior distribution of $\theta_{j.l}$ depends only on counts of family configurations at node j and not on configurations at any other node. Therefore it follows from the discussion in Section 3.5.1 that

$$\left(n(x_j^{(1)}|\pi_j^{(l)}), \ldots, n(x_j^{(k_j)}|\pi_j^{(l)})\right)_{l=1}^{q_j}$$

is a predictive sufficient statistic for $(\theta_{j1l}, \ldots, \theta_{jk_jl})$.

Predictive distribution   The predictive distribution of a new case $\underline{x}_{(n+1)}$ may be computed using the posterior density. The basic idea was given in the thumb-tack modelling of Section 1.8. In the presence of a set of instantiations $\mathbf{x}$, θ_{jil}, defined in Equation (5.5), will be estimated by:

$$\tilde{\theta}_{jil} = p(\{X_{n+1,j} = x_j^{(i)}\}|\{\Pi_j = \pi_j^{(l)}\}, \{X = x\}).\tag{5.6}$$

This is the predictive conditional probability that variable X_j attains value $x_j^{(i)}$, given the parent configuration $\pi_j^{(l)}$ and the cases stored in $\mathbf{x}$. Then, by computations as before, with $\tilde{\theta}_{jil}$ defined by Equation (5.6),

$$\tilde{\theta}_{jil} = p(\{X_{n+1,j} = x_j^{(i)}\}|\{\Pi_j \pi_j^{(l)}\}, \{X = x\})$$

$$= \int_{S_{jl}} p(\{X_{n+1,j} = x_j^{(i)}\}|\{\underline{\Theta} = \underline{\theta}\})\pi_{\underline{\Theta}}(\underline{\theta}|\{\Pi_j = \pi_j^{(l)}\}, \{X = x\})d\underline{\theta}_{j.l}$$

$$= \int_{S_{jl}} \theta_{jil}\pi_{\Theta_{j.l}|X}(\theta_{j.l}|x; \alpha_{j.l})d\underline{\theta}_{j.l}$$

$$= \int_{S_{jl}} \theta_{jil} \frac{\Gamma(n(\pi_j^{(l)}) + \alpha_{j.l})}{\prod_{m=1}^{k_j} \Gamma(n(x_j^{(m)}|\pi_j^{(l)}) + \alpha_{jml})} \prod_{i=1}^{k_j} \theta_{jil}^{n(x_j^{(i)}|\pi_j^{(l)}) + \alpha_{jil}} d\underline{\theta}_{j.l}$$

$$= \frac{\Gamma(n(\pi_j^{(l)}) + \alpha_{j.l})}{\prod_{m=1}^{k_j} \Gamma(n(x_j^{(m)}|\pi_j^{(l)}) + \alpha_{jml})} \int_{S_{jl}} \prod_{m \neq l} \theta_{jml}^{n(x_j^{(m)}|\pi_j^{(l)})} \theta_{jil}^{n(x_j^{(i)}|\pi_j^{(l)}) + 1} d\underline{\theta}_{j.l}$$

$$= \frac{\Gamma(n(\pi_j^{(l)}) + \alpha_{j.l})}{\prod_{m=1}^{k_j} \Gamma(n(x_j^{(m)}|\pi_j^{(l)}) + \alpha_{jml})}$$

$$\times \frac{\Gamma(n(x_j^{(i)}|\pi_j^{(l)}) + \alpha_{jil} + 1) \prod_{m \neq i} \Gamma(n(x_j^{(i)}|\pi_j^{(l)}) + \alpha_{jil})}{\Gamma(n(\pi_j^{(l)}) + \alpha_{j.l} + 1)}$$

$$= \frac{n(x_j^{(i)}|\pi_j^{(l)}) + \alpha_{jil}}{n(\pi_j^{(l)}) + \sum_{i=1}^{k_j} \alpha_{jil}},$$

where S_{jl} is defined as

$$S_{jl} = \left\{ (\theta_{jil})_{i=1}^{k_j} | \theta_{jil} \geq 0, \ i = 1, \ldots, k_j, \ \sum_{i=1}^{k_j} \theta_{jil} = 1 \right\}.$$

5.4   Updating, missing data, fractional updating

Updating   Suppose the cases $\underline{x}_{(1)}, \ldots, \underline{x}_{(n)}$ are complete. Suppose next that $x_j^{(i)}$ and $\pi_j^{(l)}$ are observed in $\underline{x}_{(n+1)}$. Then, by Bayes' rule,

$$
\theta_{j.l} | (\underline{x}_{(1)}, \ldots, \underline{x}_{(n)}, (x_{n+1,j}^{(i)}, \pi_{n+1,j}^{(l)}))
$$
$$
\sim Dir(n^*(x_j^{(1)} | \pi_j^{(l)}) + \alpha_{j1l}, \ldots, n^*(x_j^{(k_j)} | \pi_j^{(l)}) + \alpha_{jk_j l}),
$$

where

$$
n^*(x_j^{(r)} | \pi_j^{(l)}) = \begin{cases} n(x_j^{(r)} | \pi_j^{(l)}) & r \neq i \\ n^*(x_j^{(i)} | \pi_j^{(l)}) = n(x_j^{(i)} | \pi_j^{(l)}) + 1 & r = i. \end{cases}
$$

The *virtual* sample size is updated as

$$
s^* = n(\pi_j^{(l)}) + 1 + \sum_{i=1}^{k_j} \alpha_{jil}.
$$

A missing instantiation   Suppose the instantiation at node j is missing in the new case; the parent configuration $\pi_j^{(l)}$ is present. Set

$$
e^* = (\underline{x}_{(1)}, \ldots, \underline{x}_{(n)}, \underline{x}_{(n+1)}).
$$

The distribution of the random variable $\theta_{j.l} | e^*$ is expressed as the *mixture of distributions*

$$
\sum_{i=1}^{k_j} w_i Dir(n(x_j^{(i)} | \pi_j^{(l)}) + \alpha_{j1l}, \ldots, n(x_j^{(i)} | \pi_j^{(l)}) + 1 + \alpha_{jil}, \ldots, n(x_j^{(k_j)} | \pi_j^{(l)}) + \alpha_{jk_j l}),
$$

where $w_i = p_X(\{(X_j, \Pi_j) = (x_j^{(i)}, \pi_j^{(l)})\} | e^*)$.

Updating: parent configuration and the state at node j are missing   Consider a new case $\underline{x}_{(n+1)}$ where both the state and the parent configuration of node j are missing. Then the distribution of $\theta_{j.l} | e^*$ is given as the mixture of distributions

$$
\sum_{i=1}^{k_j} v_i Dir(n(x_j^{(1)} | \pi_j^{(l)}) + \alpha_{j1l}, \ldots, n(x_j^{(i)} | \pi_j^{(l)}) + 1 + \alpha_{jil}, \ldots, n(x_j^{(k_j)} | \pi_j^{(l)}) + \alpha_{jk_j l})
$$
$$
+ Dir(n(x_j^{(1)} | \pi_j^{(l)}) + \alpha_{j1l}, \ldots, n(x_j^{(k_j)} | \pi_j^{(l)}) + \alpha_{jk_j l}) v^*,
$$

where

$$
v_i = p(\{(X_j, \Pi_j) = (x_j^{(i)}, \pi_j^{(l)})\} | e^*), \quad i = 1, \ldots, k_j
$$

and

$$
v^* = 1 - p(\{\Pi_j = \pi_j^{(l)}\} | e^*).
$$

Fractional updating   The preceding shows that adding new cases with missing values results in dealing with increasingly messy mixtures, with increasing numbers of components. One, perhaps naive, way of approximating this is to use only one Dirichlet density, with the parameters updated by

$$n^*(x_j^{(i)}|\pi_j^{(l)}) = n(x_j^{(i)}|\pi_j^{(l)}) + p(\{(X_j, \Pi_j) = (x_j^{(i)}, \pi_j^{(l)})\}|e^*), \quad i = 1, \ldots, k_j.$$

This is known as *fractional updating*.

Fading   If the parameters change with time, then information learnt a long time ago may not be so useful. A way to make the old cases less relevant is to have the sample size discounted by a fading factor q_F, a positive number less than one.

The *fading update* is

$$n \rightarrow q_F n(x_r^{(j)}|\pi_l^{(j)}) = n^*(x_j^{(r)}|\pi_j^{(l)}), \quad r \neq i$$

and

$$n \rightarrow q_F n(x_j^{(i)}|\pi_j^{(l)}) + 1 = n^*(x_j^{(i)}|\pi_j^{(l)}).$$

The virtual sample size is updated as

$$s_{n-1} \rightarrow q_F \left(n(\pi_j^{(l)}) + \sum_{i=1}^{k_j} \alpha_{jil} \right) + 1 = s_n.$$

The above may be written as

$$s_n = q_F s_{n-1} + 1, \quad s_0 = s.$$

Iteration gives

$$s_n = q_F^n s + \sum_{i=0}^n q_F^i = q_F^n s + \frac{1 - q_F^{n+1}}{1 - q_F}.$$

The limiting *effective maximal sample size* is therefore

$$s_* = \frac{1}{1 - q_F}.$$

Notes   Chapter 5 on learning of parameters (or estimation of parameters) is largely based on 'A Tutorial on Learning with Bayesian Networks' by D. Heckerman [25] and also [72]. Learning from incomplete data (Section 5.4) is discussed in [91]. Another treatment of learning is found in R.E. Neapolitan [34].

5.5   Exercises: Learning the conditional probability potentials

1. **Kullback-Leibler divergence** Let $\mathcal{X} = \{x_1, \cdots, x_L\}$ denote a space with L elements and let

$$\mathbf{f} := (f(x_1), \cdots, f(x_L))$$

and

$$\mathbf{g} := (g(x_1), \cdots, g(x_L))$$

be two probability distributions defined on $\mathcal{X}$. Recall that the Kullback-Leibler divergence between $\mathbf{f}$ and $\mathbf{g}$ is defined by

$$D(\mathbf{f} \mid \mathbf{g}) = \sum_{i=1}^{L} f(x_i) \log \frac{f(x_i)}{g(x_i)}, \tag{5.7}$$

with the conventions $0 \cdot \log \frac{0}{g(x_i)} = 0$ and $f(x_i) \log \frac{f(x_i)}{0} = \infty$. The logarithm is the natural logarithm unless otherwise stated.

Let $\mathcal{X} = \{0, 1\}$ and $0 \le p \le 1$ and $0 \le g \le 1$. Let $\mathbf{f} = (1 - p, p)$ and $\mathbf{g} = (1 - g, g)$ be the two Bernoulli distributions $\mathrm{Be}(p)$ and $\mathrm{Be}(g)$, respectively. Find the Kullback-Leibler divergence between them.

2. **Jensen's inequality** Let $\phi(x)$ be a convex function and X finite discrete real valued random variable, defined on a finite space $\mathcal{X}$. Prove, by induction, that

$$E[\phi(X)] \ge \phi(E[X]).$$

3. **Calibration of Kullback-Leibler divergence** Let $D(\mathbf{f} \mid \mathbf{g}) = k$ be the value of the Kullback distance between any two probability distributions defined on $\mathcal{X} = \{x_1, \ldots, x_n\}$. Let $h(x)$ solve

$$D(\mathrm{Be}(1/2) \mid \mathrm{Be}(h(x))) = x. \tag{5.8}$$

The function h is known as the *calibration*. The Kullback distance between $\mathbf{f}$ and $\mathbf{g}$ is the same as between a fair Bernoulli distribution $\mathrm{Be}(1/2)$ and a Bernoulli distribution $\mathrm{Be}(h(k))$. Prove that

$$h(k) = \frac{1}{2}\left(1 \pm \sqrt{1 - e^{-2k}}\right). \tag{5.9}$$

4. Let

$$\pi_\Theta(\theta) = \prod_{j,l} \pi_{\Theta^{j,l}}(\theta^{j,l})$$

denote the prior distribution over a Bayesian network, where

$$\pi_{\Theta^{j,l}} = \mathrm{Dir}(\alpha_{j1l}, \ldots, \alpha_{jk_jl}).$$

Let $\mathbf{X} = \begin{pmatrix} \underline{X}_{(1)} \\ \vdots \\ \underline{X}_{(n)} \end{pmatrix}$ denote the random matrix corresponding to n independent instan-

tiations and let $\underline{X}_{n+1}$ denote a random vector corresponding to an instantiation, independent of $\mathbf{X}$.

Prove that

$$p\left(\{X_{n+1,j} = x_j^i\}|\{\Pi_j = \pi_j^l\}, \{\mathbf{X} = \mathbf{x}\}\right) = \int_{S_{j,l}} \theta_{jil}\pi_{\Theta|\mathbf{X}}\left(\theta^{j,l}|\mathbf{x}; \alpha_{j,l}\right) d\theta^{j,l}$$

$$= \frac{n(x_j^i|\pi_j^l) + \alpha_{ijl}}{n(\pi_j^l) + \sum_{i=1}^{k_j} \alpha_{ijl}}.$$

5. This exercise is taken from [92]. The idea of using Kullback-Leibler in this way for a database is due to Jensen. Suppose one has a database C with n cases of configurations over a collection of variables V. Let $Sp(V)$ denote the set of possible configurations over V and let $\#(v)$ denote the number of cases of configuration v. Define $P^C(v) = \frac{\#(v)}{n}$. Let P^M denote a probability distribution over $Sp(V)$. Assume that $P^C(v) = 0$ if and only if $P^M(v) = 0$ and discount these configurations. Define $S^M(C) = -\sum_{c \in C} \log P^M(c)$.
Let d_K denote the Kullback-Leibler distance (namely, $d_K(\mathbf{f}|\mathbf{g}) = \sum_i f_i \log \frac{f_i}{g_i}$). Show that

$$S^M(C) - S^C(C) = nd_K(P^C|P^M).$$

6. Suppose that $p(A, B, C, D)$ factorizes along the DAG in Figure 5.2, where A, B, C and D are each binary variables, taking the values 1 or 0. Suppose there are 10 instantiations:

	$U^{(1)}$	$U^{(2)}$	$U^{(3)}$	$U^{(4)}$	$U^{(5)}$	$U^{(6)}$	$U^{(7)}$	$U^{(8)}$	$U^{(9)}$	$U^{(10)}$
A	1	1	0	1	0	0	0	0	1	1
B	1	1	0	0	1	0	1	1	1	1
C	0	0	1	0	0	1	0	0	1	0
D	0	1	1	0	0	1	1	0	1	1

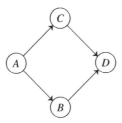

Figure 5.2 DAG for A, B, C, D.

(a) Find the maximum likelihood estimates for $\theta_B := p_{B|A}(1|0)$ and $\theta_D := p_{D|B,C}(1|1, 0)$.

(b) Suppose that θ_D has prior distribution $Be(\frac{1}{2}, 1)$ (Beta distribution with parameters $(\frac{1}{2}, 1)$). That is,

$$\pi(\theta_D) = \begin{cases} \frac{\Gamma(1.5)}{\Gamma(1)\Gamma(0.5)}\theta_D^{-0.5} & 0 \le \theta_D \le 1 \\ 0 & \theta_D \notin [0, 1] \end{cases}$$

Show that the posterior density $\pi(\theta_D|U)$ is given by

$$\pi(\theta_D|U) = \begin{cases} \frac{(6.5).(5.5).(4.5).(3.5)}{3!}\theta_D^{2.5}(1 - \theta_D)^3 & 0 \le \theta_D \le 1 \\ 0 & \theta_D \notin [0, 1]. \end{cases}$$

7. Consider the following model for the two variables A and B, where both A and B are binary variables, each taking values y/n (yes or no).
Let $\theta_a = p_A(y)$, $\theta_{b|y} = p_{B|A}(y|y)$ and $\theta_{b|n} = p_{B|A}(y|n)$. Suppose the parameters have prior distributions

$$\pi(\theta_a) = \frac{11!}{7!3!}\theta_a^7(1 - \theta_a)^3$$

$$\pi(\theta_{b|y}) = \frac{9!}{6!2!}\theta_{b|y}^6(1 - \theta_{b|y})^2$$

$$\pi(\theta_{b|n}) = \frac{6!}{2!3!}\theta_{b|n}^2(1 - \theta_{b|n})^3$$

Now suppose that a sequence of 20 instantiations is observations, with results

A	y	y	y	y	n	n	n	y	y	n	y	y	y	y	y	n	n	y	y	y
B	n	y	n	y	n	n	y	y	n	n	n	y	y	y	y	y	n	n	y	y

(a) Find the posterior distributions over the parameters where the updating is carried out without fading.

(b) Find the posterior distributions over the parameters for the same sequence, using a fading factor of 0.9.

8. Consider a Bayesian network over two binary variables A and B, where the directed acyclic graph is given in Figure 5.3 and A and B each take the values 0 or 1.

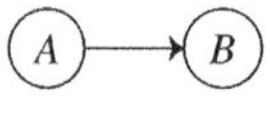

Figure 5.3   Directed acyclic graph on two variables.

Let $\theta_a = p_A(1)$, $\theta_{b|y} = p_{B|A}(1|1)$, $p_{B|A}(1|0)$. Let the prior distributions over the parameters be

$$\pi_a(\theta) = \begin{cases} 3\theta^2 & 0 \le \theta_a \le 1 \\ 0 & \theta \notin [0, 1], \end{cases}$$

$$\pi_{b|y}(\theta) = \begin{cases} 12\theta^2(1-\theta) & 0 \le \theta \le 1 \\ 0 & \theta \notin [0, 1], \end{cases}$$

$$\pi_{b|n}(\theta) = \begin{cases} 12\theta(1-\theta)^2 & 0 \le \theta \le 1 \\ 0 & \theta \notin [0, 1], \end{cases}$$

Suppose that there is a single instantiation, where $B = 1$ is observed, but A is unknown. Perform the approximate updating.

9. Consider a Bayesian network, with variables X_1, X_2, X_3, X_4, X_5, where all the variables are binary; variable X_j has state space $\mathcal{X}_j = \{0, 1\}$ for $j = 1, 2, 3, 4, 5$. Suppose the conditional probability tables are

$$p_{X_1}(1) = 0.4$$

$X_2\backslash X_1$	1	0
1	0.3	0.8
0	0.7	0.2

$p_{X_2|X_1} =$

$X_3\backslash X_1$	1	0
1	0.7	0.4
0	0.3	0.6

$p_{X_3|X_1} =$

$X_4\backslash X_2$	y	n
y	0.5	0.1
n	0.5	0.9

$p_{X_4|X_2} =$

$X_4\backslash X_3$	1	0
1	0.9	0.999
0	0.999	0.999

$p_{X_5|X_3,X_4}(1|.,.) =$

Suppose 100 trials are made on a network which is thought to have these probability tables. The numbers of times the corresponding configurations occurred is given below. Calculate the marginals from the observed probabilities below and compare with the 'theoretical' probabilities.

$X_1 X_2\backslash X_3 X_4 X_5$	111	110	101	100	011	010	001	000
11	4	0	5	0	1	0	2	0
10	2	0	16	0	1	0	8	0
01	9	1	10	0	14	0	16	0
00	0	0	4	0	0	0	7	0

10. Let the likelihood for $\underline{\theta} = (\theta_1, \ldots, \theta_L)$ with data $\mathbf{x}$ be given by

$$L\left(\underline{\theta}; \mathbf{x}\right) = \prod_{j=1}^{L} \theta_j^{n_j},$$

where n_j is the number of times the symbol x_j (in a finite alphabet with L symbols) is present in $\mathbf{x}$ and $\sum_{j=1}^{L} \theta_j = 1$. For the prior distribution over θ, a finite *Dirichlet mixture* is taken, given by

$$\pi_\Theta\left(\theta\right) = \sum_{i=1}^{k} \lambda_i Dir\left(\alpha^{(i)} q_1^{(i)}, \ldots, \alpha^{(i)} q_L^{(i)}\right),$$

where $\lambda_i \geq 0$, $\sum_{i=1}^{k} \lambda_i = 1$ (the mixture distribution), $\alpha^{(i)} > 0$, $q_j^{(i)} > 0$, $\sum_{i=1}^{L} q_j^{(i)} = 1$ for every i. Compute the mean posterior estimate $\widehat{\theta}_{j;MP}$ for $j = 1, \ldots, L$.

6

Learning the graph structure

As stated in the introduction to Chapter 5, there are two basic learning problems in Bayesian networks: learning the structure of the Bayesian network from the data; and, when the structure is established, learning the conditional probability potentials. Chapter 5 considered the second of these problems (under the assumption that the graph structure was given); Chapter 6 considers the first.

In real life applications, expert prior knowledge may often be incorporated when learning the structure. For example, in many domains of study there is a huge journal literature available in electronic form and an automated search of the literature may be carried out to derive text-based priors for Bayesian network structures, as described in [93].

This chapter restricts attention to methods of learning structures of networks known as 'tabula rasa', or 'blank slate' methods. This means that only data samples are available for estimating the structure. These methods fall generally into two categories: method that employ a score function and methods that are constraint based. After a cursory glance at possible prior distributions, the chapter considers a score function, the the data likelihood for the possible graph structures. The straightforward approach of maximizing the likelihood leads to a problem is, in general, not computationally feasible. There are two reasons for this: firstly, the number of possible DAGs grows super-exponentially in the number of nodes. Secondly, there are equivalence classes of network structures where, without additional information, each DAG within the equivalence class represents the data equally well. To tackle the first problem, the Chow-Liu tree may be used, where an effective solution for learning the best structure from a restricted class of DAGs is found by Kruskal's algorithm. The $K2$ algorithm presents another approach. The study then moves on to constraint based methods, which are based on the notions of faithful distributions, and testing for conditional independence.

Bayesian Networks: An Introduction   T. Koski, J. Noble
© 2009 John Wiley & Sons, Ltd

6.1   Assigning a probability distribution to the graph structure

For a Bayesian network with a directed acyclic graph $\mathcal{G} = (V, E)$, the edge set E is often referred to as the *structure* of the network. Let $\tilde{\mathcal{E}}$ denote the set of all possible edge sets that give a directed acyclic graph with node set V. In this chapter, it is assumed that E is unknown and has to be inferred from data, where the individual cases are assumed to be observations on random vectors that are independent, conditioned on the graph structure E and the parameters $\underline{\theta}$. The prior distribution over the parameter vectors $\underline{\theta}_{j,l}$ are taken from the family $Dir(\alpha_{j1l}, \ldots, \alpha_{jk_jl})$ for all nodes and parent configurations (j, l). As usual, $\mathbf{X} = \begin{pmatrix} \underline{X}_{(1)} \\ \vdots \\ \underline{X}_{(n)} \end{pmatrix}$ denotes the matrix where each row is an independent copy of $\underline{X}$ and E denotes the structure of the DAG. This has a prior distribution $p_{\mathcal{E}}$, which is the probability function for a random variable $\mathcal{E}$ which takes values in $\tilde{\mathcal{E}}$, the set of possible graph structures.

The prior distribution for the graph structure   There are several possible ways of constructing a prior distribution $p_{\mathcal{E}}$, as discussed in [94]. If it is known a priori that the graph structure lies within a subset $A \subseteq \tilde{\mathcal{E}}$, then an obvious choice is the uniform prior over A:

$$p_{\mathcal{E}}(E) = \begin{cases} \frac{1}{|A|} & \text{if } E \in A \\ 0 & \text{otherwise} \end{cases}$$

where $|A|$ is the number of elements in a subset $A \subseteq \tilde{\mathcal{E}}$.

Another simple choice is to assign to each pair of nodes X_i and X_j a probability distribution with three values such that

$$p\left(\{(X_i, X_j) \in E\}\right) + p\left(\{(X_j, X_i) \in E\}\right) + p\left(\{(X_i, X_j) \notin E\}, \{(X_j, X_i) \notin E\}\right) = 1.$$

Here again, uniform distribution over the three possibilities may be chosen. Then, for a given structure E, the prior probability is obtained by multiplying the appropriate probabilities over all the edges in E and normalizing to give $p_{\mathcal{E}}(E)$.

The likelihood for the graph structure   The *likelihood* for the graph structure, given data $\mathbf{x}$, is given by

$$p_{\mathbf{X}|\mathcal{E}}(\mathbf{x}|E) = \int p_{\mathbf{X}|\Theta,\mathcal{E}}(\mathbf{x}|\underline{\theta}, E)\pi_{\Theta|\mathcal{E}}(\underline{\theta}|E)d\underline{\theta},$$

where $\pi_{\Theta|\mathcal{E}}(.|E)$ denotes the prior distribution over the parameters $\underline{\theta}$, conditioned on the graph structure, and using the structure of the prior as the product of Dirichlet distributions,

$$p_{\mathbf{X}|\mathcal{E}}(\mathbf{x}|E) = \int \prod_{k=1}^{n} p_{\underline{X}|\Theta,\mathcal{E}}(\underline{x}_{(k)}|\underline{\theta}, E) \prod_{j=1}^{d} \prod_{l=1}^{q_j} \phi(\underline{\theta}_{j,l}, \underline{\alpha}_{j,l})d\underline{\theta}_{j,l}$$

where $\phi(\underline{\theta}_{j,l}, \underline{\alpha}_{j,l})$ is a compact way of referring to the Dirichlet density $\text{Dir}(\alpha_{j1l}, \ldots, \alpha_{jk_jl})$.

Because $p_{X|\Theta,\mathcal{E}}(\underline{x}|\underline{\theta}, E)$ has a convenient product form, computing the Dirichlet integral is straightforward and gives

$$p_{X|\mathcal{E}}(\mathbf{x}|E) = \prod_{j=1}^{d} \prod_{l=1}^{q_j} \frac{\Gamma(\sum_{i=1}^{k_j} \alpha_{jil})}{\Gamma(n(\pi_j^l) + \sum_{i=1}^{k_j} \alpha_{jil})} \prod_{i=1}^{k_j} \frac{\Gamma(n(x_j^i|\pi_j^l) + \alpha_{jil})}{\Gamma(\alpha_{jil})}. \tag{6.1}$$

This is the *Cooper-Herskovitz likelihood* for the graph structure. It was introduced and discussed by G.F. Cooper and E. Herskovitz in [95].

The Bayesian selection rule for a graph $\mathcal{G} = (V, E)$ uses the graph which maximizes the posterior probability

$$p_{\mathcal{E}|X}(E|\mathbf{x}) = \frac{p_{X|\mathcal{E}}(\mathbf{x}|E)p_{\mathcal{E}}(E)}{p_X(\mathbf{x})}, \tag{6.2}$$

where $p_{\mathcal{E}}$ is the prior probability over the space of edge sets. The *prior odds ratio* for two different edge sets E_1 and E_2 is defined as $\frac{p_{\mathcal{E}}(E_1)}{p_{\mathcal{E}}(E_2)}$ and the *posterior odds ratio* is defined as $\frac{p_{\mathcal{E}|X}(E_1|\mathbf{x})}{p_{\mathcal{E}|X}(E_2|\mathbf{x})}$. Equation (6.2) may then be expressed as

$$\text{Posterior odds} = \text{Likelihood ratio} \times \text{Prior odds.}$$

At a first glance, the learning procedure may appear fairly straightforward: there is only a finite number of different possible DAGs $\mathcal{G} = (V, E)$ with d nodes and hence only a finite number of values $p_{\mathcal{E}|X}(E|\mathbf{x})$ have to be computed. The edge set E that yields the maximum value is chosen. Let $N(d)$ denote the number of possible directed acyclic graphs on d nodes. Then, enumerating all possible graphs $(E_r)_{r=1}^{N(d)}$ gives

$$p_X(\mathbf{x}) = \sum_{r=1}^{N(d)} p_{X|\mathcal{E}}(\mathbf{x}|E_r)p_{\mathcal{E}}(E_r).$$

In [96], R.W. Robinson gave the following recursive function for computing the number $N(d)$ of acyclic directed graphs with d nodes:

$$N(d) = \sum_{i=1}^{d} (-1)^{i+1} \binom{d}{i} 2^{i(d-1)} N(d-i). \tag{6.3}$$

This number grows super-exponentially. For $d = 5$ it is $29\,000$ and for $d = 10$ it is approximately 4.2×10^{18}. Here $N(d)$ is a very large number, even for small values of d. Therefore, it is clearly not feasible to compute this sum, even for modest values of d. Computing the posterior distribution is an *NP-hard problem*; G.F. Cooper [82] proves that the inference problem is NP-hard. That means, worse than an NP-problem. This is discussed in [74]. Recently, M. Koivisto and K. Sood [97] constructed the first algorithm that had a complexity less than super-exponential for finding the posterior probability of a network.

Aside: P-, NP- and NP-hard problems A problem is assigned to the NP (non-deterministic polynomial time) class if it is verifiable in polynomial time by a non-deterministic Turing machine. (A non-deterministic Turing machine is a 'parallel' Turing machine which can take many computational paths simultaneously, with the restriction that the parallel Turing machines cannot communicate.) A P-problem (whose solution time is bounded by a polynomial) is always also NP. If a problem is known to be NP, and a solution to the problem is somehow known, then demonstrating the correctness of the solution can always be reduced to a single P (polynomial time) verification. A problem is NP-hard if an algorithm for solving it can be translated into one for solving any other NP-problem (non-deterministic polynomial time problem). NP-hard therefore means 'at least as hard as any NP-problem' although it might, in fact, be harder.

The prediction problem Given an $n \times d$ data matrix $\mathbf{x}$ representing n independent instantiations of a random vector $\underline{X} = (X_1, \ldots, X_d)$, the prediction problem is to compute the distribution $p_{\underline{X}_{(n+1)}|\mathbf{X}}(\underline{x}|\mathbf{x})$, the conditional probability distribution of the next observation. The article [98] introduces the *stochastic complexity distribution*.

Firstly, the maximum likelihood estimate $E(\mathbf{x})$ is defined as the value of E that maximizes

$$\hat{p}_{\mathbf{X}|\mathcal{E}}(\mathbf{x}|E) = \prod_{j=1}^{d}\prod_{l=1}^{q_j}\prod_{i=1}^{k_j}\hat{\theta}_{jil}^{n_k(x_j^{(i)}|\pi_j^{(l)})},$$

where $\hat{\theta}_{jil} = \frac{n(x_j^{(i)},\pi_j^l)}{n(\pi_j^l)}$; $n(x_j^{(i)}, \pi_j^l)$ denotes the number of times that the configuration $(x_j^{(i)}, \pi_j^l)$ appears in the data and $n(\pi_j^l)$ denotes the number of times that the 'parent' configuration π_j^l appears.

In principle, the likelihood has to be computed for *all* possible graph structures and the graph that maximizes it chosen. As discussed, locating the graph structure that obtains the maximum is an NP-hard problem.

The *stochastic complexity predictive distribution* $\mathcal{P}_{SC}$ based on n observations is then defined as

$$\mathcal{P}_{SC}^{(n)}(\mathbf{x}) = \frac{\hat{p}_{\mathbf{X}|\mathcal{E}}(\mathbf{x}|E(\mathbf{x}))}{\sum_{\mathbf{y}\in\mathcal{X}^{(n)}} \hat{p}_{\mathbf{X}|\mathcal{E}}(\mathbf{y}|E(\mathbf{y}))} = \frac{1}{F_n}\hat{p}_{\mathbf{X}|\mathcal{E}}(\mathbf{x}|E(\mathbf{x})).$$

This motivates the definition of the *stochastic complexity of* $\mathbf{x}$ *with respect to* $\tilde{\mathcal{E}}$, where $\tilde{\mathcal{E}}$ is the set of edge sets (or models) under consideration. It is defined as

$$S(\mathbf{x}|\tilde{\mathcal{E}}) = -\log \mathcal{P}_{SC}^{(n)}(\mathbf{x}) = -\log p_{\mathbf{X}|\mathcal{E}}(\mathbf{x}|E(\mathbf{x})) + k_n,$$

where $k_n = \log F_n$ is a constant. The aim is to predict the next observation $\underline{x}_{n+1}$ given the previous n observations. The stochastic complexity predictive distribution is defined as

$$\mathcal{P}_{SC}^{(n+1)}(\underline{x}|\mathbf{x}) = C \prod_{j=1}^{d}\prod_{l=1}^{q_j}\prod_{i=1}^{k_j}\tilde{\theta}_{jil}^{n_k(x_j^{(i)}|\pi_j^{(l)})},$$

where $n(x_j^{(i)}, \pi_j^{(l)})$, $n(\pi_j^{(l)})$ and $\tilde{\theta}_{jil}$ are computed from the extended data set, the $n + 1 \times d$ data matrix $\begin{pmatrix} \mathbf{x} \\ \hline \underline{x} \end{pmatrix}$.

6.2 Markov equivalence and consistency

Given a set of conditional independence relations, perhaps derived from data, this section describes how to determine whether or not there is a directed acyclic graph that is *faithful* (Definition 2.30) to the underlying probability distribution; that is, *consistent* with the set of independence relations. The method is constructive, in the sense that if there is a faithful directed acyclic graph, then the method constructs one particular faithful directed acyclic graph and indicates how to determine all the other DAGs that are faithful to the set of conditional independence statements. Different graphs that share exactly the same d-separation properties are said to be *Markov equivalent* (Definition 2.32).

Let $\mathcal{M}$ denote the complete set of conditional independence statements. That is, let $\mathcal{V}$ denote the set of all subsets of V (including ϕ and V) and let

$$\mathcal{M} = \{(X, Y, S) \in V \times V \times \mathcal{V} \mid X \perp Y | S, \quad X, Y \notin S\}.$$

Suppose that a set of conditional independence relations $\mathcal{M}$ has been obtained empirically from data. This section addresses two issues: firstly, whether or not there exists a directed acyclic graph $\mathcal{G}$ that is *consistent* with $\mathcal{M}$ (Definition 2.31), that is, whether or not there exists a directed acyclic graph $\mathcal{G}$ such that $X \perp Y \|_{\mathcal{G}} S \Leftrightarrow (X, Y, S) \in \mathcal{M}$; and secondly, how to find all the directed acyclic graphs that are consistent with a set of independence statements $\mathcal{M}$.

A set of conditional independence statements $\mathcal{M}$ for which there exists such a directed acyclic graph is said to be *DAG isomorphic*.

Definition 6.1 (DAG Isomorphic Conditional Probability Statements) *Let* $V = \{X_1, \ldots, X_n\}$ *be a collection of random variables and let* $\mathcal{M}$ *denote the entire collection of conditional independence statements: for all* $(X_i, X_j, S) \in V \times V \times \mathcal{V}: X_i, X_j \notin S$, $(X_i, X_j, S) \in \mathcal{M} \Leftrightarrow X_i \perp X_j | S$. *The collection* $\mathcal{M}$ *is said to be* DAG isomorphic *if there exists a DAG* $\mathcal{G} = (V, E)$ *such that* $\mathcal{G}$ *is consistent with* $\mathcal{M}$ *(Definition 2.31).*

The second of these issues is considered first; if a set of conditional independence statements $\mathcal{M}$ is DAG isomorphic, the following theorem gives a criterion for establishing whether or not $\mathcal{M}$ is consistent with a particular DAG $\mathcal{G}$.

Theorem 6.1 *Let* $V = \{X_1, \ldots, X_n\}$ *be a collection of random variables and let* $\mathcal{M}$ *denote the entire collection of conditional independence statements; for all* $(X_i, X_j, S) \in V \times V \times \mathcal{V}: X_i, X_j \notin S$, $(X_i, X_j, S) \in \mathcal{M}$ *if and only if* $X_i \perp X_j \|_{\mathcal{G}} S$. *Assume that* $\mathcal{M}$ *is DAG isomorphic (Definition 6.1). A DAG* $\mathcal{G} = (V, E)$ *is consistent with* $\mathcal{M}$ *(Definition 2.31) if and only if:*

1. $\forall S \in \mathcal{V}, \qquad X_i \sim X_j \Leftrightarrow (X_i, X_j, S) \notin \mathcal{M}.$

2. (X_i, X_j, X_k) forms an immorality in $\mathcal{G}$ if and only if $(X_i \not\sim X_k$, and $X_i \perp X_k|S \Longrightarrow X_j \notin S)$.

If there is a probability distribution p, for which every conditional independence statement in $\mathcal{M}$ holds for p and no others, then Theorem 6.1 gives necessary and sufficient conditions for a graph $\mathcal{G}$ to be *faithful* to p (Definition 2.30); the terms 'faithful' and 'consistent' are equivalent.

Proof of Theorem 6.1 The proof consists of three parts: Part 1 shows that the first condition is necessary, Part 2 shows that the second condition is necessary and Part 3 shows that both conditions taken together are sufficient.

Part 1: If $\mathcal{G}$ is consistent with $\mathcal{M}$, then $\forall S \in \mathcal{V}: X_i, X_j \notin S, X_i \sim X_j \Leftrightarrow (X_i, X_j, S) \notin \mathcal{M}$. Definition 2.31 states that $\mathcal{G}$ is consistent with $\mathcal{M}$ if and only if $(X_i, X_j, S) \in \mathcal{M} \Leftrightarrow X_i \perp X_j \|_{\mathcal{G}} S$. It is therefore sufficient to show that for any two distinct nodes X_i and $X_j \in \mathcal{G}$, $X_i \sim X_j$ if and only if they cannot be *d*-separated by any set of nodes $S \subset V$.

Clearly, if $X_i \sim X_j$, then following Definitions 2.17 and 2.18, there does not exist a set S that *d*-separates X_i from X_j. It remains to show that if $(X_i, X_j, S) \notin \mathcal{M}$ for all $S \subset V: X_i, X_j \notin S$, then $X_i \sim X_j$.

Let

$$S = \{Y \mid Y \text{ is an ancestor of } X_i \text{ or } X_j\}\backslash\{X_i, X_j\}.$$

Since X_i, X_j are not *d*-separated by any set, it follows that $(X_i, X_j, S) \notin \mathcal{M}$. Therefore, there is an S-active trail τ (Definition 2.16) connecting X_i and X_j in $\mathcal{G}$. Since the trail τ is active, all the collider connections along τ are either in S or have a descendant in S and all the fork or chain nodes are not in S. By the definition of S, every node that has a descendant in S is itself in S. Thus every collider node on the trail τ is in S. Every *other* node on τ (fork or chain) is an ancestor either of X_i or X_j or one of the collider nodes of the path. Hence every node on τ is in S with the exception of X_i and X_j and hence every node (other than X_i and X_j) is a collider node. It follows that the trail is either $X_i \rightarrow X_j$ or $X_j \rightarrow X_i$, or $X_i \rightarrow Y \leftarrow X_j$, where $Y \in S$. But, by construction of S, the third possibility implies that the graph has a cycle – since Y is a child of both X_i and X_j, there is a cycle if it is an ancestor of X_i and there is a cycle if it is an ancestor of X_j. It follows that $X_i \sim X_j$.

Part 2: Assume that $\mathcal{G}$ is consistent with $\mathcal{M}$, then (X_i, X_j, X_k) forms an immorality in $\mathcal{G}$ if and only if $(X_i \not\sim X_k$, and $\forall S \subset V: X_i, X_k \notin S, X_i \perp X_k|S \Longrightarrow X_j \notin S)$

Suppose (X_i, X_j, X_k) forms an immorality, then by definition $X_i \not\sim X_k$ and, by the basic separation properties of colliders, $X_i \perp X_k\|_{\mathcal{G}} S \Leftrightarrow X_j \notin S$.

For the other direction, suppose $X_i \not\sim X_k$ and $\forall S \subset V$, $X_i \perp X_k|S \Longrightarrow X_j \notin S$. Consider any trail τ between X_i and X_k containing X_j, where all the variables along the trail, except for X_i and X_k are instantiated. Let S denote the set containing all the variables along the trail except for X_i and X_k and no other variables, so that no other variables are instantiated. Then, since X_i is *not* independent of X_k given S, it follows from basic properties of instantiated connections that all connections along the trail are colliders and hence that S contains exactly one variable, X_j. It follows that (X_i, X_j, X_k) forms an immorality in $\mathcal{G}$.

Part 3: If the two conditions hold for a DAG $\mathcal{G} = (V, E)$, then it is consistent with $\mathcal{M}$. Since it is assumed that $\mathcal{M}$ is DAG isomorphic, there exists a DAG $\tilde{\mathcal{G}}$ that is consistent with $\mathcal{M}$. By parts 1) and 2), $\tilde{\mathcal{G}}$ satisfies both the conditions and hence has the same immoralities and skeleton as $\mathcal{G}$ By Lemma 4.3, it follows that for any S, a trail τ is S-active in $\mathcal{G}$ if and only if it is S-active in $\tilde{\mathcal{G}}$, so that $\mathcal{G}$ is consistent with $\mathcal{M}$.   □

A criterion for determining a collection of DAGs that are Markov equivalent has been established and this was used to determine conditions to characterize the set of directed acyclic graphs that are consistent with a given set of conditional probability statements, if such a directed acyclic graph exists.

6.2.1   Establishing the DAG isomorphic property

The last task of this section is to construct an algorithm that determines whether or not a given set of conditional probability statements is DAG isomorphic. The method is constructive; an algorithm is proposed that will return a DAG consistent with the conditional probability statements if the statements are DAG isomorphic and will declare that there is no such DAG otherwise. The algorithm is in three stages.

- Stage 1 examines the independence statements in $\mathcal{M}$ and tries to construct a consistent graph. If this stage fails to find a graph, then $\mathcal{M}$ is not DAG isomorphic.

- Stage 2 turns the graph into a directed acyclic graph, if this is possible.

- Stage 3 verifies whether or not the resulting directed acyclic graph is consistent with $\mathcal{M}$. If it is, then $\mathcal{M}$ is clearly DAG isomorphic, but if it is not, then there does not exist a consistent DAG.

Stage 1: Generate a graph from $\mathcal{M}$ if possible   This stage has three steps.

1. Consider all $(X_i, X_j, S) \in V \times V \times \mathcal{V}$, where $\mathcal{V}$ denotes the set of all subsets of V, such that $X_i \neq X_j$ and $S \subseteq V\backslash\{X_i, X_j\}$. Recall that (by definition) $\phi \in \mathcal{V}$, where ϕ denotes the empty set. Then set

$$X_i \sim X_j \Leftrightarrow \nexists S \in \mathcal{V} \mid (X_i, X_j, S) \in \mathcal{M}.$$

   Let $\mathcal{G} = (V, E)$ denote the undirected graph formed in this way. Let

$$S(X_i, X_j) = \{X_k \; : \; X_i \perp X_j | X_k\}.$$

2. For every pair of nodes X_i and X_j such that $X_i \nsim X_j$, test whether or not there is a node X_k such that $X_i \sim X_k$ and $X_j \sim X_k$ and $X_k \notin S(X_i, X_j)$.
   If there is such a node, then *direct* the edges $\langle X_i, X_k \rangle$ and $\langle X_j, X_k \rangle$ by removing them and replacing them with the directed edges (X_i, X_k) and (X_j, X_k) from E, so that X_k is a collider node, UNLESS the algorithm has already modified sufficient edges so that this involves changing the direction of a *directed* edge. If this is happens, then Stage 1 FAILS; the set of conditional independence statements is not DAG isomorphic. The proof of this follows the statement of the algorithm.

3. If the orientation of Step 2 is completed, then phase 1 SUCCEEDS and returns a graph $\tilde{\mathcal{G}}$, which is partially directed.

Stage 2: Turn $\mathcal{G}$ into a Directed Graph   The algorithm is as follows: Start with the partially directed graph obtained from Stage 1, call it $\tilde{\mathcal{G}}$. Let $\mathcal{G}$ denote the current state of the graph. While $\mathcal{G}$ contains undirected edges, repeat the following three steps.

1. Direct the graph $\mathcal{G}$ according to the following algorithm: run through all variables, labelled $\{1, \ldots, n\}$. For $j = 1$ to n, do:

    (a) **Rule 1:** If variable X_j is part of a structure shown in Figure 6.1 (that is if X_j takes the position of either A, B or C), where $A \not\sim C$; that is, if there is a structure $A \to B \leftrightarrow C$ and $A \not\sim C$, then replace $B \leftrightarrow C$ with the directed edge $B \to C$.

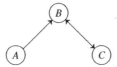

Figure 6.1   Structure for Rule 1.

    (b) **Rule 2:** If the variable X_j is part of a structure between the three variables (A, B, C) given in Figure 6.2, then replace $A \leftrightarrow C$ with the directed edge $A \to C$.

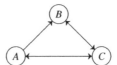

Figure 6.2   Structure for Rule 2.

    (c) **Rule 3:** If the variable X_j is part of an edge structure given in Figure 6.3, then replace $B \leftrightarrow D$ with $B \to D$.

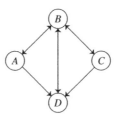

Figure 6.3   Structure for Rule 3.

    (d) **Rule 4:** If X_j is part of an edge structure given in Figure 6.4, then replace $A \leftrightarrow B$ with $A \to B$, and replace $C \leftrightarrow B$ with $C \to B$.

    IF any of pair of directed edges form a collider not present from the directed edges in $\tilde{\mathcal{G}}$, then the algorithm is terminated, and returns the value 'FAIL'; the set of conditional independence statements is not DAG isomorphic.

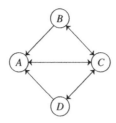

Figure 6.4 Structure for Rule 4.

Repeat step 1 of Stage 2 until either a directed acyclic graph is formed, or until no edges are directed during a full run, or until an additional collider connection or cycle is formed.
If a directed acyclic graph is returned, then the algorithm SUCCEEDS. Go to Stage 3.

2. If the resulting graph still has directed edges, then perform the following: Copy the current graph G and call the copy $\tilde{G}^*$. Select one of the undirected edges of G and choose a direction for it that does not introduce any directed cycles or new colliders. If this is possible, then TERMINATE with the resulting directed acyclic graph; the algorithm SUCCEEDS; go to Stage 3. If this is not possible, then discard $\tilde{G}^*$ and choose the opposite direction for the edge, then continue.

Stage 3: Check the Answer   Check that for every statement $(X_i, X_j, S) \in M$, $X_i \perp X_j \| _G S$ holds for the resulting directed acyclic graph G. This may be carried out using the algorithm to check for d-separation. If it does, then M is DAG isomorphic and G is a consistent DAG; otherwise M is not DAG isomorphic.

Proof that the Algorithm Determines the DAG Isomorphism Property The following additional lemma is needed.

Lemma 6.1 *Let M be a DAG isomorphic set of conditional independence statements. Suppose that X_i and X_k are two nodes such that $X_i \not\sim X_k$, but there is node X_j such that (X_i, X_j, X_k) is a trail. If there is an $S \subset V$ such that $X_j \notin S$ and $(X_i, X_k, S) \in M$, then for any $T \subset V$, $(X_i, X_k, T) \in M \Longrightarrow X_j \notin T$.*

Proof of Lemma 6.1 Suppose that X_j is a collider, chain or fork node between X_i and X_k and suppose that $X_i \perp X_k \| _G S$, with $X_j \notin S$. It follows that the connection is $X_i \rightarrow X_j \leftarrow X_k$. It follows that any set that contains X_j will activate the trail (X_i, X_j, X_k) between X_i and X_k. Hence, for any $T \subset V$, if $(X_i, X_k, T) \in M$, then $X_j \notin T$.   □

The next task is to show that the algorithm given above determines whether or not a set of conditional independence statements M is DAG isomorphic. That is, to show that if the algorithm fails to return a consistent DAG, then M is not DAG isomorphic.

Stage 1:   Theorem 6.1 shows that if M is DAG isomorphic, then any directed acyclic graph consistent with M will have the same skeleton and the same colliders as the graph produced by Stage 1. The first part of this stage produces the skeleton, the second part identifies the colliders that are present in all consistent graphs, if $\mathcal{M}$ is DAG isomorphic. It identifies a collider $X_i \rightarrow X_j \leftarrow X_k$ as soon as it finds a *single* $S \subset V$ such that $(X_i, X_k, S) \in \mathcal{M}$ with $X_j \notin S$. This decision is justified by Lemma 6.2.

The failure condition in Stage 1, Part 2 is necessary to prevent the algorithm from switching edges that are essential for a DAG consistent with $\mathcal{M}$; at each stage in the algorithm, the directions of all the edges that have already been directed are necessary.

Stage 2:   The second stage is to locate a directed acyclic graph that retains the same colliders and skeleton as the graph constructed in Stage 1. This is purely a task of graph theory and does not involve $\mathcal{M}$.

To show that the first part this part of the process is sound, it is sufficient to show that edges directed by the four rules are a logical consequence of the requirement that Stage 1 gives the skeleton and all the colliders.

- **Rule 1** The alternative directed edge $B \leftarrow C$ would create a new collider.

- **Rule 2** The alternative directed edge $A \leftarrow C$ would create a cycle.

- **Rule 3** Since $A \sim B$ and $B \sim C$, the edge $B \rightarrow D$ is permitted. If $D \rightarrow B$ were chosen, it would be necessary to also have both $A \rightarrow B$ and $C \rightarrow B$, to prevent a cycle. But then (A, B, C) would form a new immorality, so $D \rightarrow B$ is not permitted. Therefore the directed edge $B \rightarrow D$ is forced.

- **Rule 4** Firstly, $A \rightarrow B$ is forced to prevent the formation of a new immorality (D, A, B). If the edge $B \rightarrow C$ is used, this forces $C \rightarrow D$ to prevent formation of a new immorality (B, C, D), which results in a directed cycle (A, B, C, D, A). It follows that $B \rightarrow B \leftarrow C$ is forced.

Stage 3:   By the definition of 'consistency', any graph constructed from Stage 2 will be consistent with $\mathcal{M}$, if $\mathcal{M}$ is DAG isomorphic. Therefore, it is sufficient to check the statements in $\mathcal{M}$. If there is a statement $(X_i, X_j, S) \in \mathcal{M}$ such that X_i and X_j are not d-separated by S, then $\mathcal{M}$ is not DAG isomorphic.

Any graph $\mathcal{G}$ satisfying Part 1 of Stage 3 is therefore a perfect I-map for any probability distribution p satisfying the set of conditional independence statements $\mathcal{M}$. Part 2 of Stage 3 checks that the DAG does indeed satisfy the local directed Markov condition (Definition 2.28); this is necessary and sufficient to conclude that $\mathcal{G}$ is consistent with $\mathcal{M}$.   □

6.3   Reducing the size of the search

Finding the optimal structure from among all possible structures is (as stated earlier) an NP-hard problem. Furthermore, not only does one have to estimate the underlying probability distribution from a finite number of samples, one also has to store the resulting probability potentials in a *limited* amount of machine memory. In general, it is therefore not possible to consider *all* possible dependencies. Two methods of reducing of the problem, the Chow-Liu tree and the K2 algorithm, are now considered. These seem to have worked quite well in practice and applications of these approaches are discussed to illustrate this.

6.3.1   The Chow-Liu tree

The Chow-Liu tree assumes that each variable has at most one parent. Suppose there are d variables and possible dependencies between the variables. The Chow-Liu approach is to find $d - 1$ first order dependencies between the variables. That is, the distribution p over the random vector $\underline{X} = (X_1, \ldots X_d)$ is considered to be of the form

$$p_{X_1,\ldots,X_d} = \prod_{j=1}^{d} p_{X_{m_i} | X_{m_{i(j)}}},$$

where $(m_1, \ldots, m_d)$ is a permutation of $(1, \ldots, d)$ and $0 \leq i(j) \leq j$. Each variable may be conditioned on *at most one* of the variables; hence the term *first order tree dependence*.

When estimating the tree, the Kullback-Leibler divergence is used to determine how close two distributions are to each other. To compute the best fitting Chow-Liu tree, the potentials $\hat{p}_{X_i,X_j}$, $\hat{p}_{X_i}$ and $\hat{p}_{X_j}$ have to be computed, where $\hat{p}_{X_i,X_j}$, $\hat{p}_{X_i}$ and $\hat{p}_{X_j}$ denote the empirical estimates of the potentials p_{X_i,X_j}, p_{X_i} and p_{X_j} from data. The definition of *mutual information* is based on the Kullback-Leibler divergence.

Definition 6.2 (Mutual Information) *The mutual information between two variables X and Y is defined as*

$$I(X, Y) = \sum_{x,y} p_{X,Y}(x, y) \log \frac{p_{X,Y}(x, y)}{p_X(x) p_Y(y)}.$$

This is, of course, the Kullback-Leibler divergence between $p_{X,Y}$ and $p_X p_Y$:

$$I(X, Y) = D_{KL}(p_{X,Y} | p_X p_Y).$$

Let $\hat{I}$ denote the estimate of the mutual information from data; that is,

$$\hat{I}(X, Y) = \sum_{x,y} \hat{p}_{X,Y}(x, y) \log \frac{\hat{p}_{X,Y}(x, y)}{\hat{p}_X(x) \hat{p}_Y(y)}. \tag{6.4}$$

The idea is to find the *maximum weight dependence tree*; that is, a tree σ such that for any other tree σ',

$$\sum_{j=1}^{d} \hat{I}(X_j, X_{\sigma(j)}) \geq \sum_{j=1}^{d} \hat{I}(X_j, X_{\sigma'(j)}).$$

This uses Kruskal's algorithm, described below. Here $(j, \sigma(j))_{j=1}^{d}$ denotes the edge set of the maximal weight tree; $(j, \sigma'(j))_{j=1}^{d}$ denotes the edge set of any other admissible tree.

The optimization procedure   Kruskal's algorithm runs as follows:

1. The d variables yield $d(d - 1)/2$ edges. The edges are indexed in decreasing order, according to their weights $b_1, b_2, b_3, \ldots, b_{d(d-1)/2}$.

2. The edges b_1 and b_2 are selected. Then the edge b_3 is added, *if it does not form a cycle*.

3. This is repeated, through $b_4, \ldots b_{d(d-1)/2}$, in that order, adding edges if they do not form a cycle and discarding them if they form a cycle.

This procedure returns a unique tree if the weights are different. If two weights are equal, one may impose an arbitrary ordering. From the $d(d-1)/2$ edges, exactly $d-1$ will be chosen.

Lemma 6.2 *Kruskal's algorithm returns the tree with the maximum weight.*

Proof of Lemma 6.2 The result may be proved by induction. It is clearly true for two nodes. Assume that it is true for d nodes and consider a collection of $d+1$ nodes, labelled $(X_1, X_2, \ldots, X_{d+1})$, where they are ordered so that for each $j = 1, \ldots, d+1$, the maximal tree from $(X_1, \ldots, X_j)$ gives the maximal tree from any selection of j nodes from the full set of $d+1$ nodes. Let $b_{(i,j)}$ denote the weight of edge (i, j) for $1 \le i < j \le d+1$. Edges will be considered to be undirected. Let $T_j^{(d+1)}$ denote the maximal tree obtained by selecting j nodes from the $d+1$ and consider $T_{d+1}^{(d+1)}$.

Let Z denote the leaf node in $T_{d+1}^{(d+1)}$ such that among all leaf nodes in $T_{d+1}^{(d+1)}$ the edge (Z, Y) in $T_{d+1}^{(d+1)}$ has the smallest weight. Removing the node Z gives the maximal tree on d nodes from the set of $d+1$ nodes. This is seen as follows. Clearly, there is no tree with larger weight that can be formed with these d nodes, otherwise the tree on d nodes with larger weight, with the addition of the leaf (Z, Y) would be a tree on $d+1$ nodes with greater weight than T_{d+1}^{d+1}. It follows that $Z = X_{d+1}$ and hence that X_{d+1} is a leaf node of $T_{d+1}^{(d+1)}$.

By the inductive hypothesis, $T_d^{(d+1)}$ may be obtained by applying Kruskal's algorithm to the weights $(b_{(i,j)})_{1 \le i < j \le d}$. Now consider an application of Kruskal's algorithm to the weights $(b_{(i,j)})_{1 \le i < j \le d+1}$ and note that for any (i, j) with $i < j$ such that the undirected edge (X_i, X_j) forms part of the tree $T_d^{(d+1)}$, $b_{(i,d+1)} < b_{(i,j)}$ and $b_{(j,d+1)} < b_{(i,j)}$. Therefore, if the edges $(b_{(i,j)})_{1 \le i < j \le d+1}$ are listed according to their weight and the Kruskal algorithm applied, then all the edges used in $T_d^{(d+1)}$ will appear further up the list than any edge $(b_{(k,d+1)})_{k=1}^d$ and therefore all the edges of $T_d^{(d+1)}$ will be included by the algorithm before the edges $(b_{(k,d+1)})_{k=1}^d$ are considered. It follows that $T_{d+1}^{(d+1)}$ is the graph obtained by applying Kruskal's algorithm to the nodes $(X_1, \ldots, X_{d+1})$.   □

Application to pattern recognition   The example given in the article [99] considers the problem of machine recognition of handwritten numerals, $0, 1, 2, 3, 4, 5, 6, 7, 8, 9$. There are $c = 10$ pattern classes. Let a_i denote the numeral i. There is a prior distribution $p = (p_0, p_1, \ldots, p_9)$ over the numerals. The number is written on a 12×8 rectangle and 96 binary measurements are used to represent the numeral; 1 if the cell contains writing and 0 otherwise. In the example given in [99], 19 000 numerals produced by four inventory clerks were scanned. 7000 of these were employed as training examples, to find the best fitting trees and estimate the probabilities $p_0, \ldots, p_9$. The optimal trees for each of the 10 numerals were obtained. For the remaining numerals, the observation $\underline{x} = (x_1, \ldots, x_{96})$ was considered. Using Bayes' rule,

$$p(a_k | \underline{x}) = \frac{p(\underline{x}|a_k) p(a_k)}{p(\underline{x})} = \frac{p(\underline{x}|a_k) p_k}{p(\underline{x})},$$

the following classification rule was used: the numeral was declared to be of class a_k if $p_k p(\underline{x}|a_k) \geq p_i p(\underline{x}|a_i)$ for all $i \neq k$. Using the trees, the error rate was reduced from 0.09 to 0.04 compared with the model produced by assuming independence between the contents of the 96 cells.

6.3.2 The Chow-Liu tree: A predictive approach

The following predictive approach (and the computations) in the section are due to M. Gyllenberg and T.Koski [100], but the work turns out to be a special case of J. Suzuki [101]. Recall the definition of the *prior predictive distribution* (Definition 1.9). The predictive approach considers the prior distribution and a set of parameters so that it may be expressed in the form given in Equation (1.12), and uses this to construct the posterior distribution, conditioned on the observations.

The prediction problem for a *given* Chow-Liu tree is now considered. This is interpreted as the *likelihood* function for the tree structure and is used as part of a predictive technique for learning the optimal Chow-Liu tree.

Consider a Chow-Liu tree with d nodes, where X_1 is the root variable and $(X_{m(2)}, \ldots, X_{m(d)})$ are the parent variables for $(X_2, \ldots, X_d)$. That is, $\Pi_j = \{X_{m(j)}\}$, $j = 2, \ldots, d$. The distribution along a tree $\mathcal{T}$ factorizes as follows:

$$p_{\underline{X}}(.|\mathcal{T}) = p_{X_1} \prod_{j=2}^{d} p_{X_j|X_{m(j)}}.$$

Set

$$\theta_j = p_{X_j|X_{m(j)}}(1|1) \quad j = 2, \ldots, d$$

$$\phi_j = p_{X_j|X_{m(j)}}(1|0) \quad j = 2, \ldots, d$$

and

$$\theta_1 = p_{X_1}(1).$$

Now consider t complete, independent instantiations $\underline{x}_1, \ldots, \underline{x}_t$ of the variables in the tree network, where for each $j = 1, \ldots, t$, the row vector $\underline{x}_j = (x_{j1}, \ldots, x_{jd})$ denotes instantiation j. Let the matrix $\mathbf{x} = \begin{pmatrix} \underline{x}_1 \\ \vdots \\ \underline{x}_t \end{pmatrix}$ denote the complete set of independent instantiations and let $\mathbf{X}$ denote the random matrix where each row $\underline{X}_j$, $j = 1, \ldots, t$ is an independent copy of $\underline{X} = (X_1, \ldots, X_d)$. Then

$$p_{\mathbf{X}}(\mathbf{x}|\mathcal{T}) = \prod_{l=1}^{t} p_{\underline{X}}(\underline{x}_l|\mathcal{T})$$

$$= \theta_1^{n_1}(1 - \theta_1)^{t-n_1} \prod_{j=2}^{d} \theta_j^{n_j(1,1)}(1 - \theta_j)^{n_j(0,1)} \phi_j^{n_j(1,0)}(1 - \phi_j)^{n_j(0,0)},$$

where, for $j \geq 2$,

$$n_j(1, 1) = \sum_{l=1}^{t} x_{lj} x_{l,m(j)}, \quad n_j(1, 0) = \sum_{l=1}^{t} x_{lj}(1 - x_{l,m(j)})$$

$$n_j(0, 1) = \sum_{l=1}^{t} (1 - x_{lj}) x_{l,m(j)}, \quad n_j(0, 0) = \sum_{l=1}^{t} (1 - x_{lj})(1 - x_{l,m(j)}).$$

and

$$N_j(1) = \sum_{l=1}^{t} x_{lj}, \qquad N_j(0) = t - N_j(1), \qquad j = 1, \ldots, d.$$

Set $\mathbf{n} = \{(N_j(0), N_j(1))_{j=1}^{d}, (n_j(1, 1))_{j=2}^{d}, (n_j(1, 0))_{j=2}^{d}, (n_j(0, 1))_{j=2}^{d}, (n_j(0, 0))_{j=2}^{d}\}$.
Note that

$$N_{m(j)}(0) = n_j(0, 0) + n_j(1, 0) \qquad \text{and} \qquad N_{m(j)}(1) = n_j(1, 1) + n_j(0, 1).$$

The interpretation of these quantities is clear; for example, $n_j(1, 1)$ counts the number of rows of $\mathbf{x}$ in which the family configuration $(x_{kj}, x_{k,m(j)}) = (1, 1)$ appears. Regarded as a function of (θ, ϕ), where $\theta = (\theta_1, \ldots, \theta_d)$ and $\phi = (\phi_2, \ldots, \phi_d)$, this is the likelihood function:

$$L(\theta, \phi; \mathbf{x}) = \prod_{l=1}^{t} p_X(\underline{x}_l | T)$$

$$= \theta_1^{N_1}(1 - \theta_1)^{t - N_1} \prod_{j=2}^{d} \theta_j^{n_j(1,1)}(1 - \theta_j)^{n_j(0,1)} \phi_i^{n_i(1,0)}(1 - \phi_i)^{n_i(0,0)}. \qquad (6.5)$$

When the parameters (θ, ϕ) are included in the notation, the probability distribution may be written as

$$p_{X|\theta,\phi}(\mathbf{x}) = \prod_{j=1}^{d} p_{X_j | X_{m(j)}, \theta_j, \phi_j}(x_j | x_{m(j)}).$$

To compute the predictive distribution for a fixed tree, a prior distribution $g(\theta, \phi)$ is required over the parameter space. Then, using T to denote the tree structure, the prior predictive distribution for $\mathbf{X}$, defined in Equation (1.12), is given by

$$p_X(\mathbf{x}|T) = \int_{\Theta \times \Phi} \prod_{l=1}^{t} p_{X|\theta,\phi}(\underline{x}_l) g(\theta, \phi) d\theta d\phi,$$

where $\Theta \times \Phi$ denotes the parameter space for (θ, ϕ). It is convenient to choose

$$g(\theta, \phi) = \prod_{i=1}^{d} h(\theta_i) \prod_{i=2}^{d} k(\phi_i),$$

where the parameters are taken to be independent, the variables $(\theta_i)_{i=1}^{d}$ identically distributed and the variables $(\phi_i)_{i=2}^{d}$ are identically distributed. Such a model is known as *local meta independence* ([67] p. 191). For this model, it is clear that the probability distribution factorizes as

$$p_{\mathbf{X}}(\mathbf{x}|\mathcal{T}) = I_1 I_2 I_3,$$

where

$$I_1 = \int_0^1 \theta_1^{N_1}(1 - \theta_1)^{t-N_1} h(\theta_1)d\theta_1,$$

$$I_2 = \prod_{j=2}^{d} \int_0^1 \theta_j^{n(1,1)}(1 - \theta_j)^{n(0,1)} h(\theta_j)d\theta_j$$

$$I_3 = \prod_{j=2}^{d} \int_0^1 \phi_j^{n_j(1,0)}(1 - \phi_j)^{n_j(0,0)} k(\phi_j)d\phi_j.$$

If $h = k$, and h has a Beta density

$$h(\theta) = \begin{cases} \frac{\Gamma(\alpha_1+\alpha_2)}{\Gamma(\alpha_1)\Gamma(\alpha_2)} \theta^{\alpha_1-1}(1 - \theta)^{\alpha_2-1} & \theta \in [0, 1] \\ 0 & \theta \notin [0, 1], \end{cases}$$

the integrals may be computed in the usual way;

$$I_1 = \frac{\Gamma(\alpha_1 + \alpha_2)\Gamma(N_1 + \alpha_1)\Gamma(t - N_1 + \alpha_2)}{\Gamma(\alpha_1)\Gamma(\alpha_2)\Gamma(t + \alpha_1 + \alpha_2)},$$

$$I_2 = \prod_{j=2}^{d} \frac{\Gamma(\alpha_1 + \alpha_2)\Gamma(n_j(1, 1) + \alpha_1)\Gamma(n_j(0, 1) + \alpha_2)}{\Gamma(\alpha_1)\Gamma(\alpha_2)\Gamma(\alpha_1 + \alpha_2 + n_j(1, 1) + n_j(0, 1))},$$

$$I_3 = \prod_{j=2}^{d} \frac{\Gamma(\alpha_1 + \alpha_2)\Gamma(n_j(1, 0) + \alpha_1)\Gamma(n_j(0, 0) + \alpha_2)}{\Gamma(\alpha_1)\Gamma(\alpha_2)\Gamma(\alpha_1 + \alpha_2 + n_j(1, 0) + n_j(0, 0))}.$$

The learning of the tree structure $\mathcal{T}$ now involves finding the tree that maximizes $p_{\mathbf{X}}(\mathbf{x}|\mathcal{T})$ or, equivalently, minimizes $-\log p_{\mathbf{X}}(\mathbf{x}|\mathcal{T})$. This is equivalent to minimizing

$$F(\mathbf{n}) = -\log \Gamma(N_1(1) + \alpha_1) - \log \Gamma(t - N_1(1) + \alpha_2)$$

$$-\sum_{j=2}^{d} \log \Gamma(n_j(1, 1) + \alpha_1) - \sum_{j=2}^{d} \log \Gamma(n_j(0, 1) + \alpha_2)$$

$$+\sum_{j=2}^{d} \log \Gamma(\alpha_1 + \alpha_2 + N_{m(j)}(1))$$

$$-\sum_{j=2}^{d} \log \Gamma(n_j(1, 0) + \alpha_1) - \sum_{j=2}^{d} \log \Gamma(n_j(0, 0) + \alpha_2)$$

$$+\sum_{j=2}^{d} \log \Gamma(\alpha_1 + \alpha_2 + N_{m(j)}(0)).$$

Now consider, for example, the *Jeffreys' prior* [24], where $\alpha_1 = \alpha_2 = \frac{1}{2}$ and recall the standard formula:

$$\Gamma(n+1) = (2\pi)^{1/2} n^{n+\frac{1}{2}} e^{-n} e^{a(n)/12n},$$

where $0 \leq a_n \leq 1$. If n is a non-negative integer, this is simply Stirling's formula. It follows that, for any $n \geq 1$,

$$\log \Gamma\left(n + \frac{1}{2}\right) = \frac{1}{2}\log(2\pi) + n\log\left(n - \frac{1}{2}\right) - \left(n - \frac{1}{2}\right) + \frac{c_1(n)}{n}$$

and

$$\log \Gamma(n+1) = \left(n + \frac{1}{2}\right)\log n - n + \frac{c_2(n)}{n},$$

where $0 \leq c_i(n) \leq 1$. To good approximation, therefore, the optimal tree is found by minimizing

$$\tilde{F}(\mathbf{n}) = C(\mathbf{n}) - N_1(1)\log\left(N_1(1) - \frac{1}{2}\right) - N_1(0)\log\left(N_1(0) - \frac{1}{2}\right)$$

$$- \sum_{j=2}^{d} n_j(1,1)\log\left(n_j(1,1) - \frac{1}{2}\right) - \sum_{j=2}^{d} n_j(0,1)\log\left(n_j(0,1) - \frac{1}{2}\right)$$

$$+ \sum_{j=2}^{d}\left(N_{m(j)}(1) + \frac{1}{2}\right)\log N_{m(j)}(1)$$

$$- \sum_{j=2}^{d} n_j(1,0)\log\left(n_j(1,0) - \frac{1}{2}\right) - \sum_{j=2}^{d} n_j(0,0)\log\left(n_j(0,0) - \frac{1}{2}\right)$$

$$+ \sum_{j=2}^{d}\left(N_{m(j)}(0) + \frac{1}{2}\right)\log N_{m(j)}(0)$$

where C is bounded and the bound depends only on d. Now, for $j = 1, \ldots, d$ set $\hat{q}_j = \frac{N_j}{t}$ and, for $j \geq 2$, $\hat{p}_j(a,b) = \frac{n_j(a,b)}{N_{\Pi(j)}(b)}$ for $a = 0, 1$ and $b = 0, 1$ and $\hat{p}_j(a) = \frac{N_j(a)}{t}$. Recall the definition of the empirical mutual information, Equation (6.4). Let $\hat{I}_{j,\Pi(j)}$ denote the empirical mutual information between the variable X_j and its parent, so that

$$\hat{I}_{j,\Pi(j)} = \sum_{a=0}^{1}\sum_{b=0}^{1} \hat{p}_j(a,b)\log\frac{\hat{p}_j(a,b)}{\hat{q}_j(a)\hat{q}_{\Pi(j)}(b)}$$

and let H denote the function

$$H(x) = -x\log x - (1-x)\log(1-x).$$

Then, after a little computation,

$$-\log p(X^t|T) = \tilde{C} + t\sum_{j=1}^{d} H(\hat{q}_j) - t\sum_{j=2}^{d}\hat{I}_{j,\Pi(j)} + \frac{1}{2}\sum_{j=2}^{d}\left(\log N_{\Pi(j)}(1) + \log N_{\Pi(j)}(0)\right),$$

where $\tilde{C}$ is a bounded function. The problem therefore reduces to maximizing

$$t \sum_{j=2}^{d} \hat{I}_{j,\Pi(j)} - \frac{1}{2} \sum_{j=2}^{d} \left(\log N_{\Pi(j)}(1) + \log N_{\Pi(j)}(0) \right). \tag{6.6}$$

Recall that for the maximum likelihood approach, the problem was to maximize the mutual information, $\sum_{j=2}^{d} \hat{I}_{j,\Pi(j)}$, and this was carried out using Kruskal's algorithm. The problem of maximizing the Equation (6.6) may be tackled by a similar use of Kruskal's algorithm.   □

6.3.3   The K2 structural learning algorithm

This example is taken from the paper [102]. It shows an application to Bayesian network learning techniques for task execution in mobile robots. The task here is for the robot to locate an open door and travel through it.

The robot emits sonar pulses and is equipped with eight detectors, which detect the echoes. From this information, it has to decide where the door is located.

An action has to be taken: step to left, right, or straight ahead. This is the class variable and the class has to be determined by the signals received by the eight detectors. Since the signals are not independent of each other (the echoes may be created by the same object), the model is improved by incorporating a dependence structure.

In this experiment, the problem is to learn the *structure* of the Bayesian network and to estimate the probability potentials from the training database.

The K2 algorithm is employed to establish a suitable structure. The algorithm has to take into account that the number of parents of a node should not be high, since the size of the associated probability potentials increases exponentially according to the number of parents of a node. Therefore, the algorithm limits the number of parents a node can take. For the robot learning example, the maximum number is set to four, which is a value widely used. The size of the probability potentials cannot be too large, since the robot is expected to find the door and travel through it in real time.

The algorithm assumes that an order has been established for the d nodes $X_1, \ldots, X_d$ so that, for each i, the parent nodes Π_i for variable X_i are established among the nodes $X_1, \ldots, X_{i-1}$. For $j = 1, \ldots, i - 1$, the empirical Kullback-Leibler divergence between the two empirical probability distributions of $(X_1, \ldots, X_i)$, one determined by the graphs with and the other determined by the graph without the directed edge (i, j), is measured and the edge is retained if a) the divergence is sufficiently large and b) node i does not already have four parents.

The resulting algorithm is a *greedy algorithm* and therefore it cannot ensure that the net resulting from the learning process is the best fit.

The intensity of an echo may be modelled as a continuous random variable, but the variables are *discretised* for computational convenience. In general, it is not convenient to use a variable with more than 20 different values. For this reason, the multinomial distribution is by far the most useful in Bayesian networks.

When the K2 algorithm is used, the learnt structure depends completely on the random order of the variables generated before the learning process starts. This may have unfortunate consequences if an unfortunate order is used. In the Bayesian robotics

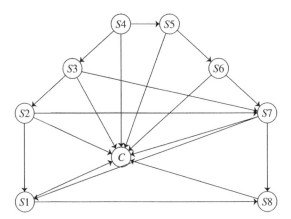

Figure 6.5 Network produced by the K2 algorithm. Here the nodes S_j represent the signals received by the sensors and C denotes the class variable, the action to be performed.

experiment therefore, in order to reduce the impact of the random order in the net structures learnt, the experiments were repeated 1000 times and nets with optimal values selected.

The Bayesian network learnt using the K2 algorithm and entropy evaluation metrics (i.e. the Kullback-Leibler divergence) provided the best classification accuracy. The resulting network for the eight variables is shown in Figure 6.5.

6.3.4 The MMHC algorithm

The maximum minimum hill climbing (MMHC) algorithm was introduced in [103]. It finds a directed acyclic graph structure for a probability distribution p, where there exists a graph $\mathcal{G}$, such that p and $\mathcal{G}$ are *faithful* to each other, following Definition 2.30. The problem is that it *requires* that there exists a graph faithful to the distribution, so there are situations where it will not work. The example of the trek (Section 2.10.1) gives an example of a relatively simple distribution over four binary variables for which there does not exist a faithful graph. The graph located will satisfy Definition 2.30, assuming that there are sufficient instantiations to determine that there is a significant association between two variables when an association exists. The key identity used is that, for a faithful Bayesian network $(\mathcal{G}, p)$, X and Y d-separated by a set of variables $\mathcal{Z}$ is equivalent to the statement that X and Y are conditionally independent given $\mathcal{Z}$. The algorithm then works by locating the conditional independence structure.

The maximum minimum parents and children algorithm   The MMPC algorithm, when run on a target variable T, identifies the existence of edges to and from T. It cannot determine the direction of the edges; it determines the *skeleton*. The algorithm works in three stages. Firstly, a forward stage starts with an empty graph, and adds in all possible edges. There are possibly too many edges after this stage. Secondly, a backward stage removes some of the edges. The resulting graph, after the second stage, will contain no false negatives, but may still contain some false positives. A third stage is implemented to remove the false positives. The algorithm runs as follows:

Stage 1:
- Order the variables $X_i \in V \backslash \{T\}$, $(X_i)_{i=1}^d$. Start with $\mathcal{Z}_1 = \phi$, the empty set.

- For $i = 1, \ldots d$, $X_i \in V$, $X_i \neq T$, check whether $X_i \perp T | \mathcal{Z}$. If it is not, let $\mathcal{Z}_{i+1} = \mathcal{Z}_i \cup \{X_i\}$. Otherwise, $\mathcal{Z}_{i+1} = \mathcal{Z}_i$.
  Set $\mathcal{Z} = \mathcal{Z}_d$.

Stage 2: Suppose that $\mathcal{Z}$ contains k variables. Label them $X_1, \ldots, X_k$. Let $\mathcal{Z}_k = \mathcal{Z}$. For $i = 0, \ldots, k-1$, check whether there exists a set $S \subseteq \mathcal{Z}_{k-i} \backslash \{X_{k-i}\}$ such that $T \perp X_{k-i} | S$. If there is, then $\mathcal{Z}_{k-i-1} = \mathcal{Z} \backslash \{X_{k-i}\}$, otherwise $\mathcal{Z}_{k-i-1} = \mathcal{Z}_{k-i}$.
Let $\mathcal{Z}_T = \mathcal{Z}_1$. This set contains all the variables which have an edge either to or from the variable T.

The algorithm may return false positives. Suppose a probability distribution may be represented by the DAG in Figure 6.6. Working from T, the node C may enter the output, and remain in the output.

This is because C is dependent on T, conditioned on all subsets of Ts parents and children; namely, ϕ (the empty set) and $\{A\}$. Note that the *collider* connection TAB, is opened when A is instantiated so that, when A is instantiated and B is uninstantiated, T is d-connected with C. For ϕ (the empty set), TAC is a chain connection, where A is uninstantiated, so that T is d-connected to C.

T and C are d-separated if and only if A and B are simultaneously instantiated; that is, $T \perp C | \{A, B\}$. But if B is independent from T given the empty set, so it will be removed from $\mathcal{Z}$. Therefore, the link TC will not be removed.

This is corrected by considering the parent/child sets of the other variables. When working from C, both A and B will be in the parent/child set, and $T \perp C | \{A, B\}$. This leads to the third stage of the algorithm.

Stage 3: Let $(\mathcal{Z}_X)_{X \in V}$ denote the parent/child sets for all the variables arrived at after Stage 2. Let $X_1, \ldots, X_j$ denote the set of variables in $\mathcal{Z}_T$, the parent child set for T arrived at after Stage 2.
Set $\mathcal{Y}_0 = \mathcal{Z}_T$. For $i = 1, \ldots, j$, set

$$\mathcal{Y}_i = \begin{cases} \mathcal{Y}_{i-1} \backslash \{X_i\} & T \notin \mathcal{Z}_{X_i} \\ \mathcal{Y}_{i-1} & T \in \mathcal{Z}_{X_i}. \end{cases}$$

Now set $\mathcal{Y}_T = \mathcal{Y}_j$. This returns the complete parent/child set for T.

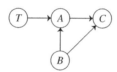

Figure 6.6 Min max parent child: A false positive.

Testing for conditional independence   Testing for conditional independence is carried out, quite simply, using the usual χ^2 test. To test whether or not $X \perp Y|Z$, let $n(x, y, z)$ denote the number of times $(X, Y, Z) = (x, y, z)$ appears in the data, $n(x, z)$, $n(y, z)$, $n(z)$ the number of instances of $(X, Z) = (x, z)$, $(Y, Z) = (y, z)$, $Z = z$ respectively. The G^2 statistic, which is standard, is defined as

$$G^2 = 2 \sum_{x,y,z} n(x, y, z) \log \frac{n(x, y, z)n(z)}{n(x, z)n(y, z)}.$$

Asymptotically, this is distributed as a χ^2 distribution on $(j_x - 1)(j_y - 1)j_z$ degrees of freedom, where j_x, j_y and j_z are the number of values that the random variables X, Y and Z respectively can take. Note that G tests independence based on characterization 4 of conditional independence, listed in Theorem 2.1.

Maximum minimum Hill Climbing algorithm   Having located the parent/child sets, the algorithm advocated by [103] works as follows: The nodes in V are labelled $(X_i)_{i=1}^d$ and, for $i = 1, \ldots, d$, $\mathcal{Z}_{X_i}$ denotes the edge set for node X_i obtained from the MMPC algorithm. Let $\mathcal{E} = \cup_{i=1}^d \mathcal{Z}_{X_i}$. The algorithm is as follows: denoting the current graph by $\mathcal{G}$,

- Start with the empty graph.

- At each stage, either *add* an edge in $\mathcal{E}$ to $\mathcal{G}$, or *delete* an edge from $\mathcal{G}$, or *reverse* an edge in $\mathcal{G}$, or leave the graph unaltered. From all the possibilities of 'add an edge', 'delete an edge', 'reverse an edge', 'leave the graph unaltered', choose the one that gives the greatest score; that is, the operation that produces the greatest reduction in the Kullback-Leibler divergence between the probability modelled along the graph and the empirical probability.

- Repeat until the score is not changed.

The algorithm may be modified as follows: instead of the best change, make the best change that results on a graph that has not already appeared. When 15 changes occur without an increase in the best score ever encountered during the search, the algorithm terminates. The DAG that produced the best score is then returned.

6.4   Monte Carlo methods for locating the graph structure

As usual, let $V = \{X_1, \ldots, X_d\}$ denote the set of variables. This section describes a Markov chain Monte Carlo method for locating the essential graph. If there is a large number of variables, the graph returned by the Markov chain Monte Carlo method may not necessarily be 'optimal', but it should represent the dependence structure to a good approximation.

A Monte Carlo algorithm generates a stochastic process that moves through the graph structures under consideration. Let $\mathbf{x}$ denote the data and let $\mathcal{E}^*$ denote the set of possible edge sets for *essential* graphs, then the idea is to generate a time homogeneous Markov chain $\{M(t) : t = 0, 1, 2, \ldots\}$, where $M(t) \in \mathcal{E}^*$ for each $t \geq 0$, with an equilibrium distribution $p_{\mathcal{E}}(.|\mathbf{x}) : \mathcal{E}^* \to [0, 1]$, defined by modifying Equation (6.2), to take into account the fact that $\mathcal{E}^*$ is the space of essential graphs. That is, a directed acyclic graph within the equivalence class is chosen, the probability given by Equation (6.2) is computed and then multiplied by the number of graphs in the equivalence class to give the probability of the essential graph.

The necessary results about Markov chains and Monte Carlo methods were discussed in Section 3.7. The Monte Carlo method finds constructs an irreducible aperiodic Markov chain that has $p_\mathcal{E}(.|\mathbf{x})$ (or a suitable approximation) as its invariant measure. By running the chain, an approximation $\hat{p}_\mathcal{E}(.|\mathbf{x})$ to this distribution may be computed and the edge set E that maximizes $\hat{p}_\mathcal{E}(.|\mathbf{x})$ is selected.

The Markov chain Monte Carlo model composition algorithm   The Markov chain Monte Carlo model composition algorithm, known as MC^3, and the augmented Markov chain Monte Carlo model composition (AMC^3) algorithm were introduced by Madigan and York, and Madigan, Andersson, Perlman and Volinsky in 1995 and 1996 respectively. They are described in [87]. The MC^3 algorithm constructs an aperiodic irreducible Markov chain $\{M(t), t = 1, 2, \ldots\}$ with state space $\mathcal{E}^*$ with an equilibrium distribution that approximates $p_\mathcal{E}(.|\mathbf{x})$, where $\mathbf{x}$ denotes the data matrix.

The difficulty with constructing Markov chains over the set of essential graphs is that if only a single edge is modified at a time, the chain is not irreducible. This is seen rather simply with the immorality $A \to B \leftarrow C$. This is an essential graph on three variables. Any alteration of a single edge (either by adding in one of (A, C), (C, A) or $\langle A, C \rangle$, or un-directing one of the directed edges, or changing the direction of an edge) gives a graph that is not an essential graph. It is therefore not possible to move in a single step from the immorality (A, B, C) (where B is the collider node) to a different essential graph on the variables (A, B, C). Filling in the details is left as an exercise (Exercise 2). Therefore, any Markov chain with state space the possible essential graphs for three variables will not be irreducible if at most one edge is altered at each transition.

The $(MC)^3$ algorithm therefore considers *triples* of nodes and works as follows. Let $M(0)$ be an edge set of an arbitrarily chosen essential graph. To move from $M(t)$ to $M(t + 1)$, do the following:

- Choose three nodes (X_i, X_j, X_k) at random, where $X_i \neq X_j$, $X_j \neq X_k$, $X_i \neq X_k$, taking any possible triple of nodes each with equal probability.

- Let E denote the current edge set. As usual, $E = D \cup U$ where D denotes the directed edges and U denotes the undirected edges. $\langle \alpha, \beta \rangle \in U$ if and only if both $(\alpha, \beta) \in E$ and $(\beta, \alpha) \in E$. $(\alpha, \beta) \in D$ if and only if both $(\alpha, \beta) \in E$ and $(\beta, \alpha) \notin E$. For F_{ij} and F_{jk} where F_{pq} is defined below, consider the 16 possible graphs generated by keeping all other edges the same and modifying any edges between the two pairs $[X_i, X_j]$ and $[X_j, X_k]$ (where $[\alpha, \beta]$ simply denotes the ordered pair of vertices) according to the four possibilities for each pair:

$$
F_{pq} = \begin{cases}
1 & (X_p, X_q) \notin E \quad (X_q, X_p) \notin E \quad (\text{i.e.} (X_p, X_q) \notin D, \\
& \qquad\qquad\qquad\qquad\qquad\qquad\qquad (X_q, X_p) \notin D, \ \langle X_p, X_q \rangle \notin U) \\
2 & (X_p, X_q) \notin E \quad (X_q, X_p) \in E \quad (\text{i.e.} (X_q, X_p) \in D) \\
3 & (X_p, X_q) \in E \quad (X_q, X_p) \notin E \quad (\text{i.e.} (X_p, X_q) \in D) \\
4 & (X_p, X_q) \in E \quad (X_q, X_p) \in E \quad (\text{i.e.} \langle X_p, X_q \rangle \in U).
\end{cases}
$$

Suppose the current state is E_0. Check each of the 16 graphs $E_0, E_1, \ldots, E_{15}$ (the current graph will be one of the possibilities) generated by all the possibilities of F_{ij} and F_{jk}. For each graph, check whether it is an essential graph, using the criteria of Theorem 4.5.

This part takes a fair amount of computation. It is necessary to keep track of both U and D, the undirected and directed nodes respectively. By keeping track of the chain components, it should be relatively easy to check

1. whether the graph is still a chain graph following the alteration of the two edges,

2. if it is, whether the chain components are still triangulated after alteration of the two edges,

3. that the alteration of the two edges has not introduced any induced sub-graph of three variables of the type shown in Figure 4.22 and

4. that every *directed* edge is strongly protected (the protection has not been removed from unaltered edges and that new directed edges are strongly protected).

For graph E_k, $k = 0, \ldots, 15$, assign a probability 0 if it is not a chain graph, otherwise compute a quantity proportional to $p_{\mathcal{E}|\mathbf{X}}(E_k|\mathbf{x})$, or a suitable approximation to this quantity.

One way to do this is to find a directed acyclic graph within the equivalence class of the essential graph with edge set E_k. If a uniform prior is considered over all possible directed acyclic graphs, then, from Equation (6.2), it is sufficient to use Equation (6.1) and multiply by the number of directed acyclic graphs within the equivalence class. In general, computing the exact number of directed acyclic graphs within an equivalence class is not straightforward; either all possible equivalent DAGs have to be constructed, or else a suitable approximation to this number has to be found. Let $x_k = p_{\mathbf{X}|\mathcal{E}}(D_k|\mathbf{x})$ using Equation (6.1), where D_k is a directed acyclic graph within the equivalence class. Set

$$P_{E_0, E_k} = \frac{x_k}{\sum_{m=0}^{15} x_m} \qquad k = 0, 1, \ldots, 15.$$

These are the one-step transition probability values for moving from $M(t) = E_0$ to $M(t + 1)$.

At first sight, this may look unpleasant. But the programming can be made more efficient by observing that graph $M(t)$ is already essential and that only two edges are being altered. This observation may be used both to help determine which of the 16 possibilities for $M(t + 1)$ are essential graphs and also to reduce the number of new quantities $n(x_j^i|\pi_j^i)$ and $n(\pi_j^i)$ that have to be computed at each stage. Writing the programme to run a Monte Carlo on directed acyclic graphs, changing a single edge at a time, is easier, but the problem with this is that the algorithm may run through several different DAGs that are all Markov equivalent. The selection procedure for the $(MC)^3$ algorithm ensures that the process moves between graphs that have essentially different Markov structures and hence improves 'irreducibility'.

To keep track of the number of DAGs within an equivalence class, the following modification of the MC^3 algorithm may be used:

1. Start with an empty graph.

2. It is required to store $M(t)$ for each $t = 1, \ldots, N$ (where N is the length of the Markov chain that is to be run) and at each stage it is required to store each DAG for which $M(t)$ is the essential graph.

3. For $l = 1, \ldots, \frac{1}{2}d(d - 1)(d - 2)$ (where d is the number of nodes) do:

   (a) Choose a triple of nodes (X_i, X_j, X_k) where $i < k$, $i \neq j$, $j \neq k$ at random, among the triples that have not been chosen before in the current cycle.

   (b) Consider all 16 possibilities of (F_{ij}, F_{jk}) (defined above) when applied to the essential graph and record those for which the new graph is an essential graph. For each of those that is an essential graph, apply each of the possible directed edge arrangements corresponding to (F_{ij}, F_{jk}) (exactly one if both F_{ij} and F_{jk} contain either no edge or a directed edge, two if one of them contains an undirected edge and the other either no edge or a directed edge and four if both of them contain undirected edges) and retain those that are directed acyclic graphs. This keeps track of the DAGs in the equivalence class.

   (c) For $E_0, \ldots, E_{16}$, set $x_k = 0$ if E_k is not a chain graph. Otherwise, choose a directed acyclic graph D_k within the equivalence class and compute $x_k = n(k) p_{X|\mathcal{E}}(\mathbf{x}|D_k)$, where $n(k)$ is the number of DAGs in the equivalence class of E_k and $p_{X|\mathcal{E}}(D_k|\mathbf{x})$ is computed using Equation (6.1).

   (d) Set
   $$P_{E_0, E_k} = \frac{x_k}{\sum_{m=0}^{15} x_k}, \qquad k = 1, \ldots, 15.$$

   These are the one-step transition probabilities for moving from $M(t)$ to $M(t + 1)$.

4. $M(t)$ is a Markov chain, but not time homogeneous, because the edges have not been chosen in the same way at each stage. But $N_n = M(\frac{n}{2}d(d - 1)(d - 2) + k)$ is a time homogeneous Markov chain for any choice of $k \in \{0, \ldots, \frac{1}{2}d(d - 1)(d - 2) - 1\}$. The whole sequence $M(t) : t = 1, 2, 3, \ldots, N$ may be taken to compute the estimate for the graph structure.

The $(MC)^3$ algorithm is computationally expensive, because it has to check whether each graph is an essential graph. The $(AMC)^3$ algorithm introduced by D. Madigan, S.A. Andersson and M.D. Perlman [87] makes more use of the graph structure in an attempt to reduce the computational complexity. Details are found in [87].

The number of DAGs per equivalence class was investigated by Pẽna in [104]. There does not exist a general method for exact computation of this; Pẽna develops Markov chain Monte Carlo techniques to estimate numbers of DAGs in an equivalence class in some situations.

6.5   Women in mathematics

The following example is taken from [87], concerning the attitudes of New Jersey high school students towards mathematics. The $(MC)^3$ algorithm applied to the 'Women in

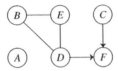

Figure 6.7 Essential graph for the 'Women in Mathematics' data, produced by AMC³ algorithm.

Mathematics' data given below returned the graph shown in Figure 6.7. It was well known that female students of high school age tended to take fewer mathematics courses than males. The Woman and Mathematics (WAM) Secondary School Lectureship Programme was designed both to encourage more interest in mathematics by females and to show 'positive role models'[1] by presenting lectures, all given by women in the mathematical sciences. A survey was carried out to evaluate one aspect of the WAM lectures. A total of 1190 students at eight high schools (four urban and four suburban) responded to a questionnaire inquiring about attitudes towards mathematics achievement and related topics. Although care was taken both in the selection of schools and in the assignment of students to either attendance or non attendance of lectures, this was not a formal experiment. The variables were: A – lecture attendance, B – gender, C – school type (suburban or urban), D – 'I'll need mathematics in my future work' (agree/disagree), E – subject preference (mathematical science/liberal arts), F – future plans (higher education/immediate job). The data obtained is given in Table 6.1, which may be found in [105]; the data is taken from [106].

This is an example with six variables, each binary, which may be implemented in MATLAB, using the algorithms outlined in the chapter; the computational requirements for a complete search are not too demanding.

The following results were obtained using a Markov chain Monte Carlo approaches. The *essential graph* selected by the Augmented Markov chain Monte Carlo model composition scheme, known as AMC³, is given in Figure 6.7. Note that this graph has two connected components; one component containing the variable A (whether or not the

Table 6.1 Data for 'Women in Mathematics'. Source: C.B. Lacampagne (1979) [106].

			suburban				urban			
	school	gender	female		male		female		male	
		lecture	y	n	y	n	y	n	y	n
future	preference	'need mathematics'								
college	mathematical	y	37	27	51	48	51	55	109	86
		n	16	11	10	19	24	28	21	25
	arts	y	16	15	7	6	32	34	30	31
		n	12	24	13	7	55	39	26	19
job	mathematical	y	10	8	12	15	2	1	9	5
		n	9	4	8	9	8	9	4	5
	arts	y	7	10	7	3	5	2	1	3
		n	8	4	6	4	10	9	3	6

---

[1] A person who serves as a model in a particular behavioural or social role for another person to emulate.

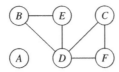

Figure 6.8   Moral graph for the 'Women in Mathematics' data, produced by an McMC algorithm.

students heard a lecture) and the other component containing the variables B, C, D, E and F.

This example illustrates that the optimal graph will not be connected if two sets of variables are independent of each other. The algorithm gives a graph where the variable A (lecture attendance) is independent of all the other variables in the model. The essential graph indicates that $C \perp D$, but that $C \not\perp D|F$.

The Markov chain Monte Carlo algorithm, introduced by J. Corander and T. Koski [107], returns the *moral graph*. It is computationally less expensive and therefore can be applied to larger sets of variables. The graph returned using this algorithm is given in Figure 6.8. It is the moral graph corresponding to the graph in Figure 6.7.

Notes   The Cooper-Herskovitz likelihood is taken from [95]. Statistical learning of graphical models from databases has been extensively discussed both in the computer science and statistics literature. Generally, the vast number of existing works agree on the main challenges related to such tasks, first of which is the super-exponential increase in the number of potential model structures as a function of the number of nodes, and the second obstacle being the equivalence of statistical models determined by different networks. The equivalence of Bayesian networks creates a set of equivalence classes in the space of DAGs. This constitutes a difficulty for efficiency of algorithms for structure learning, as an algorithm may be wasting resources searching within an equivalence class. For a study and references in this topic, the reader is referred to [107]. The Chow-Liu tree is taken from [99] and the robotics experiment is taken from [102]. The maximum minimum hill climbing algorithm is found in [103]. The treatise [89] contains an advanced treatment of the learning of graph structures and several new concepts for this.

6.6   Exercises: Learning the graph structure

1. Consider the DAGs in Figure 6.9. For each DAG in the figure, find all the equivalent DAGs. and find the essential graph for the equivalent graphs.

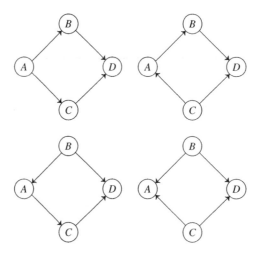

Figure 6.9   Four DAGs (see Exercise 1).

2. Consider a collider connection $A \to B \leftarrow C$. Is this an essential graph? List all the essential graphs on three variables. List all the graphs that may be obtained by altering one edge of the graph $A \to B \leftarrow C$ through either adding or removing a directed edge or an undirected edge, or from directing an undirected edge, or from 'un-directing' a directed edge. Which of these graphs are essential graphs?

3. (a) Let $\phi(\theta_{j.l}, \alpha_{j.l})$ denote the Dirichlet density $\mathrm{Dir}(\alpha_{j1l}, \dots, \alpha_{jk_jl})$. By performing the required integration, prove that the likelihood function for the graph structure, defined by

$$p_{X|\mathcal{E}}(\mathbf{x}|E) = \int \prod_{k=1}^{n} p_{\underline{X}_{(k)}|\Theta,\mathcal{E}}(\underline{x}_{(k)}|\theta, E) \prod_{j=1}^{d} \prod_{l=1}^{q_j} \phi(\theta_{j.l}, \alpha_{j.l}) d\theta_{j.l}$$

is given by

$$p_{X|\mathcal{E}}(\mathbf{x}|E) = \prod_{j=1}^{d} \prod_{l=1}^{q_j} \frac{\Gamma(\sum_{i=1}^{k_j} \alpha_{jil})}{\Gamma(n(\pi_j^{(l)}) + \sum_{i=1}^{k_j} \alpha_{jil})} \prod_{i=1}^{k_j} \frac{\Gamma(n(x_j^{(i)}|\pi_j^{(l)}) + \alpha_{jil})}{\Gamma(\alpha_{jil})}.$$

In your answer, state clearly the meaning of all the notation. You may use the identity: for any $a_1 > 0, \dots, a_n > 0$,

$$\int_0^1 \int_0^{1-\theta_1} \cdots \int_0^{1-(\theta_1+\ldots+\theta_{n-2})} \left(\prod_{j=1}^{n-1} \theta_j^{a_j-1} \right)$$

$$\times \left(1 - \sum_{j=1}^{n-1} \theta_j \right)^{a_n-1} d\theta_{n-1} \ldots d\theta_1 = \frac{\prod_{j=1}^n \Gamma(a_j)}{\Gamma(\sum_{j=1}^n a_j)}.$$

(b) What parameters $\alpha_{j,l}$ are used if a uniform prior is taken on every $\theta_{j,l}$? What does this give for the Cooper-Herskovitz likelihood? You may use $\Gamma(n) = (n-1)!$.

4. Chow-Liu Tree: MATLAB exercise. Generate three columns, $c1$, $c2$ and $c3$, each containing random samples of 50 $Be(1/2)$ observations. Here $Be(1/2)$ means Bernoulli trials, returning 0 with probability $1/2$ and 1 with probability $1/2$. Let $c4 = c1 + c2$ and let $c5 = c3 + c4$. Implement the Kruskal algorithm on the variables $c1, c2, c3, c4, c5$ and see which edges are chosen.

5. Chow-Liu Tree: MATLAB exercise. Download the data set from the URL address
http://archive.ics.uci.edu/ml/machine-learning-databases/zoo/zoo.data
A description of the data is found at the address http://archive.ics.uci.edu/ml/datasets/Zoo

The data set presents animal attributes: hair type, feather type, egg type, milk type; whether it is airborne, aquatic, a predator; has teeth, has a backbone, breathes, is venomous, has fins, legs, tail; is domestic, or cat-size. The last variable is a classification of the type of animal.

Firstly, compute the estimated probability distribution for the 17 variables, assuming that they are independent. What is the Kullback-Leibler distance between the empirical distribution and the estimate using the independence model? Secondly, use MATLAB to perform Kruskal's algorithm, to determine the optimal Chow-Liu tree. Calculate the estimated probability distribution, assuming that the distribution factorizes according to the Chow-Liu tree. Calculate the Kullback-Leibler distance between this estimate and the empirical probability distribution.

6. The K2 structural learning algorithm allows each variable to have at most four parents. It starts with an ordering of the nodes and, working through each node, chooses the best parents for that node, out of the nodes with a lower order, up to maximum of four parents. Write a MATLAB code to implement this on the 'zoo' data set.

There are $17! \simeq 3.56 \times 10^{14}$ ways to choose an ordering for the nodes. It is clearly not feasible to try all of them. Choose 100 at random. Compute the estimates of the probability distribution assuming that it may be factorized along the resulting trees.

Does the K2 algorithm give a significantly better approximation than the Chow-Liu tree?

7. Write a MATLAB code to implement an MMPC (Maximum Minimum Parents Children Algorithm) on the zoo data set. Next, try to implement the Maximum Minimum

Hill Climbing algorithm to find the graph structure. Is the resulting factorization significantly better than the factorizations obtained using the K2 algorithm or Chow-Liu algorithm? Use the Kullback-Leibler distance and an appropriate χ^2 test.

8. Download the Bayes Nets Toolbox for MATLAB from http://www.cs.ubc.ca/ ~murphyk/Software/BNT/bnt.html and see what results it gives for structure learning with the 'zoo' data set.

9. **Women in Mathematics** The 'Women in Mathematics' data provides an example on six variables to which the algorithms presented in the chapter may be applied.

    (a) Write a MATLAB code to establish the complete set of conditional independence statements of the form $X \perp Y$ or $X \perp Y|Z$ that seem to be supported by the data.

    The independence statements are obtained by computing for all variables (X, Y, Z) $\hat{p}_X$, $\hat{p}_{X,Y}$, $\hat{p}_{X,Y|Z}$. Since there are multiple hypothesis tests, the nominal significance level may be much lower than the true significance level, but we will ignore that problem in this example. Since all the variables are binary, the data supports the assertion $X \not\perp Y$ if

    $$\hat{I}(X, Y) := \sum_{x,y} \hat{p}_{X,Y} \log \frac{\hat{p}_{X,Y}(x, y)}{p_X(x)p_Y(y)} > \frac{1}{2}\chi_1(0.95)$$

    (where the nominal significance level for each test is 5%) and the data supports the assertion $X \not\perp Y|Z$ if

    $$\hat{I}(X, Y|Z = z) := \sum_{x,y} \hat{p}_{X,Y|Z} \log \frac{\hat{p}_{X,Y|Z}(x, y|z)}{p_{X|Z}(x|z)p_{Y|Z}(y|z)} > \frac{1}{2}\chi_1(0.95)$$

    for either of z.
    This is a 'toy' example, on only six binary variables. The conditional independence relations from conditioning only on no variables or a single variable, illustrates that the algorithm is unworkable in practice. There are already $6 \times 5 + 6 \times 5 \times 4 \times 2 = 270$ tests required to establish only those conditional independence relations that involve conditioning on zero or one variable.

    (b) Write a MATLAB code that implements the following steps:

    • For each pair (X, Y), add an undirected edge $\langle X, Y \rangle$ if either $X \not\perp Y$ or if there is a node $Z \neq X$ or Y such that $X \not\perp Y|Z$.

    • For each (X, Y), determine the sets

    $$S(X, Y) = \{Z \mid X \perp Y|Z\}.$$

    • For each pair X, Y such that $X \not\sim Y$, determine whether there is a node Z such that $X \sim Z$ and $Y \sim Z$ and $Z \notin S(X, Y)$. If there is such a node, then direct the edges $X \to Z$ and $Y \to Z$, so that Z is a collider node.

    (c) Does the algorithm SUCCEED or FAIL? If it succeeds, is the graph in Figure 6.7 obtained?

10. In MATLAB, implement the Minimum Maximum Parent Child and Minimum Maximum Hill Climbing algorithms on the 'Women in Mathematics' data.

11. In MATLAB, implement a Markov chain Monte Carlo algorithm on the 'Women in Mathematics' data set, where the search space is the space of directed acyclic graphs.

7

Parameters and sensitivity

Having specified the structure of a Bayesian network and the conditional probability potentials, the *parameters* of the network are the conditional probability values $(\theta_{jil})_{i=1}^{k_j}$ defined by Equation (5.5), or the update by Equation (5.6) if there is a data base of instantiations. These satisfy the condition $0 \leq \theta_{jil} \leq 1$ for all (j, i, l) and the constraints $\sum_{i=1}^{k_j} \theta_{jil} = 1$ for each (j, l), so that for each j, l variable/parent configuration, there are $k_j - 1$ free parameters.

In many practical applications, the full parameter set is too large for efficient updating. There are several ways to approach this problem. In Chapter 6, methods to find a *reduced* DAG (for example, the Chow-Liu tree) that describes the network adequately were discussed. In some situations there may be modelling constraints that make it inappropriate to remove any of the directed edges to assist computation. Instead, it may be appropriate to parametrize the conditional probabilities $(\theta_{jil})_{i=1}^{k_j}$ by a set of parameters $(t_1^{(jl)}, \ldots, t_{m_{jl}}^{(jl)})$ where $m_{jl} \leq k_j - 1$ which model the conditional probability distributions as functions of these parameters; $\theta_{jil}(t_1^{(jl)}, \ldots, t_{m_{jl}}^{(jl)})$. There are other situations where it may be appropriate to reduce the graph and parametrize the conditional probability potentials for the reduced graph.

After a suitable parametrization, a set of queries (defined in Section 3.3) may be processed and the parameters adjusted until the response of the Bayesian network to the test queries is in line with any constraints that are to be imposed.

Having parametrized the network in a suitable way, *sensitivity analysis* aims to describe changes in the network associated with small changes in parameter values.

The sensitivity of the network to the parameters strongly influences the accuracy and rates of convergence of numerical methods for estimating probability values associated with a Bayesian network.

Bayesian Networks: An Introduction   T. Koski, J. Noble
© 2009 John Wiley & Sons, Ltd

7.1   Changing parameters in a network

There are several types of parameter change that can be made. The simplest is where there is one free parameter under consideration. Alternatively, a single CPP may be varied. That is, all probabilities in one of the CPPs (conditional probability potentials), using the notation given by Equation (5.5) may be varied subject to the constraint that $\sum_{i=1}^{k_j} \theta_{jil} = 1$, while the other CPPs are kept fixed. A query constraint may be satisfied more efficiently if *several* of the CPPs are varied. The problem may be reduced using *proportional scaling* techniques, where the entries of a CPP depend linearly on a single parameter.

As usual, let Π_j denote the parent set for variable j and let $\pi_j^{(l)}$ denote configuration l of the parent set for variable j.

Definition 7.1 (Proportional Scaling Property) *A Bayesian network satisfies the proportional scaling property if for each conditional probability distribution $\theta_{j.l}$, where $\theta_{jil} = p(X_j = x_i^{(j)} | \Pi_j = \pi_l^{(j)})$, there is a parameter $t^{(jl)}$ such that*

$$p_{X_j|\Pi_j}(.|\pi_l^{(j)}) = (\alpha_{j1l} + \beta_{j1l} t^{(jl)}, \ldots, \alpha_{jk_jl} + \beta_{jk_jl} t^{(jl)}),$$

where $\sum_{m=1}^{k_j} \alpha_{jml} = 1$ and $\sum_{m=1}^{k_j} \beta_{jml} = 0$.

Theorem 7.1 *Let BN be a Bayesian network over a collection of variables U. Let t be a single parameter and let **e** be a collection of hard evidence potentials (Definition 3.1) entered into the BN; $p(\mathbf{e})$ the probability that this evidence is obtained. Assuming proportional scaling in that single parameter, then*

$$p(\mathbf{e})(t) = \alpha t + \beta$$

for two constants α and β.

Proof of Theorem 7.1 Let $\underline{X} = \{X_1, \ldots, X_d\}$. Let $p_{X_j|\Pi_j}(.|\pi_j^{(l)})$ denote the conditional probability distribution with parameter t. Let the evidence potentials be $\mathbf{e} = (e_1, \ldots, e_m)$. Recall that the $(e_k)_{k=1}^m$ are potentials containing 1s and 0s and recall the notation: for $j = 1, \ldots, d$, variable X_j takes values in state space $\mathcal{X}_j = (x_j^{(1)}, \ldots x_j^{(k_j)})$ and $\underline{X}$ takes values in state space $\mathcal{X} = \mathcal{X}_1 \times \cdots \times \mathcal{X}_d$. Recall the notation

$$p_{\underline{X};\mathbf{e}} = p_{\underline{X}} \prod_{k=1}^m e_k,$$

where $p_{\underline{X}}$ is the joint probability potential of $\underline{X}$ and multiplication is taken in the sense of multiplication of potentials (Definitions 2.23, 2.25 and 2.26). Then

$$p(\mathbf{e}) = \sum_{\underline{x} \in \mathcal{X}} p_{\underline{X};e} = \sum_{\underline{x} \in \mathcal{X}} p_{X_j|\Pi_j} \prod_{k \neq j} p_{X_k|\Pi_k} \prod_{l=1}^m e_l. \tag{7.1}$$

It is clear from the definition of proportional scaling, and from Equation (7.1), that t enters linearly. It therefore follows that

$$p(\mathbf{e})(t) = \alpha t + \beta.$$

□

Because the event '$\{A = a\}$' can be treated as *hard evidence*, it follows that $p(\{A = a\}, \mathbf{e})$ is also a linear function in t, say $p(\{A = a\}, \mathbf{e}) = \gamma t + \delta$. It follows that

$$p(\{A = a\}|\mathbf{e})(t) = \frac{p(\{A = a\}, \mathbf{e})}{p(\mathbf{e})} = \frac{\gamma t + \delta}{\alpha t + \beta}.$$

Example 7.1 Consider the Bayesian network called *Fire*.[1] The model is shown in Figure 7.1.

The network models the scenario of whether or not there is a fire in the building. Let F denote 'fire', T denote 'tampering', S 'smoke', A 'alarm', L 'leaving' and R 'report'. Now consider the following evidence $\mathbf{e} = \{report = true, smoke = false\}$. That is, the fire department receives a report that people are evacuating the building, but no smoke is observed. This evidence should make it more likely that the fire alarm has been tampered with than that there is a real fire. Let t denote 'true' and f denote 'false'. Suppose that the CPPs for this network are given by

$$p_F = \begin{array}{cc} t & f \\ \hline 0.01 & 0.99 \end{array}, \qquad p_T = \begin{array}{cc} t & f \\ \hline 0.02 & 0.98 \end{array},$$

$$p_{R|L} = \begin{array}{c|cc} L\backslash R & t & f \\ \hline t & 0.75 & 0.25 \\ f & 0.01 & 0.99 \end{array}$$

$$p_{S|F} = \begin{array}{c|cc} F\backslash S & t & f \\ \hline t & 0.9 & 0.1 \\ f & 0.01 & 0.99 \end{array}, \qquad p_{L|A} = \begin{array}{c|cc} A\backslash L & t & f \\ \hline t & 0.88 & 0.12 \\ f & 0.001 & 0.999 \end{array}$$

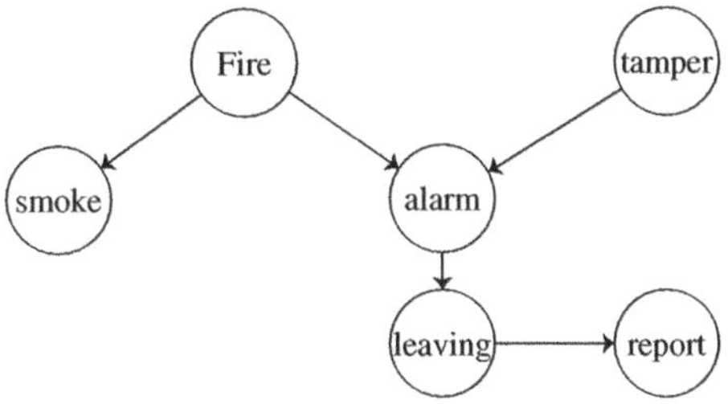

Figure 7.1   The DAG for the Bayesian Network 'Fire'.

[1] This Bayesian network is distributed with the evaluation version of the commercial HUGIN Graphical User Interface, by HUGIN Expert.

$$p_{A|F,T}(t|.,.) = \begin{array}{c|cc} F\backslash T & t & f \\ \hline t & 0.5 & 0.99 \\ f & 0.85 & 0.0001. \end{array}$$

Here,

$$p_T(t|e) = \frac{p_{T,R,S}(t,t,f)}{p_{R,S}(t,f)}.$$

Using the notation $\mathcal{X}_Z$ to denote the state space of a variable Z,

$$p_{T,R,S}(t,t,f) = p_T(t) \sum_{\mathcal{X}_L} p_{R|L}(t|.) \sum_{\mathcal{X}_A} p_{L|A} \sum_{\mathcal{X}_F} p_{A|T,F}(.|t,.) p_{S|F}(f|.) p_F$$

and

$$p_{R,S}(t,f) = \sum_{\mathcal{X}_T} p_T \sum_{\mathcal{X}_L} p_{R|L}(t|.) \sum_{\mathcal{X}_A} p_{L|A} \sum_{\mathcal{X}_F} p_{A|T,F} p_{S|F}(f|.) p_F.$$

Similarly,

$$p_F(t|e) = \frac{p_{F,R,S}(t,t,f)}{p_{R,S}(t,f)},$$

and

$$p_{F,R,S}(t,t,f) = p_F(t) p_{S|F}(f|t) \sum_{\mathcal{X}_T} p_T \sum_{\mathcal{X}_A} p_{A|T,F}(.|.,f) \sum_{\mathcal{X}_L} p_{L|A} p_{R|L}(t|.)$$

The computations are straightforward and give

$$p_T(t|e) = 0.501, \qquad p_F(t|e) = 0.0294.$$

Suppose that it is known from experience that the probability that the alarm has been tampered with should be no less than 0.65 given this evidence. The network should therefore be adjusted to accommodate. It is simplest to try changing only one network parameter. Suppose that the potential p_T is to be adjusted. Let $\theta = p_T(t)$. Let

$$\alpha = \sum_{\mathcal{X}_L} p_{R|L}(t|.) \sum_{\mathcal{X}_A} p_{L|A} \sum_{F} p_{A|T,F}(.|t,.) p_{S|F}(f|.) p_F$$

so that

$$p(\{T = t\}, e) = p_{T,R,S}(t,t,f) = \theta\alpha$$

and

$$\beta = \sum_{\mathcal{X}_L} p_{R|L}(t|.) \sum_{\mathcal{X}_A} p_{L|A} \sum_{\mathcal{X}_F} p_{A|T,F}(.|f,.) p_{S|F}(f|.) p(.),$$

so that

$$p(e) = p_{R,S}(t,f) = \theta\alpha + (1 - \theta)\beta.$$

Then α and β may be computed numerically and

$$p_T(t|e) = \frac{p(\{T = t\}, e)}{p(e)} = \frac{\alpha\theta}{(\alpha - \beta)\theta + \beta}.$$

The solution to the equation

$$\frac{\alpha\theta}{(\alpha - \beta)\theta} = 0.65$$

is $\theta = 0.0364$.

Similarly, let $\psi = p_{R|L}(t|t)$. Keeping all other potentials fixed, $p_T(t|e)$ may be computed as a function of ψ and the equation $p_T(t|e)(\psi) = 0.65$ has solution $\psi = 0.00471$.

For all other single parameter adjustments, the equation does not have a solution in the interval $[0, 1]$. Therefore, if only one parameter is to be adjusted, the constraint $p(\{tampering = t\}|e) = 0.65$ can be dealt with in either of the following two ways:

1. Increase $p_T(t)$ from 0.02 to greater than 0.0364, or

2. Decrease the probability of a *false* report, given that there is an evacuation, from 0.01 to less than 0.00471.

It turns out for this example that it is not possible to enforce the desired constraint by adjusting a *single* parameter in any of the CPPs of the variables *fire*, *smoke*, *alarm* and *leaving*.

7.2 Measures of divergence between probability distributions

In the 'fire' example, two suggestions were made for changing the parameters to accommodate the constraint. Firstly, it was proposed to raise θ from 0.02 to 0.0364, an increase of 0.0164, or a factor of 1.82. The other possibility for a single parameter adjustment was to reduce ψ from 0.01 to 0.00471. That is a difference of 0.00529, or a factor of $0.471 = \frac{1}{2.12}$. Clearly, the magnitude of the adjustment will depend on the way that the divergence between two probability distributions is computed.

A *distance* is a more specific measure of divergence, which satisfies the properties given in the following definition.

Definition 7.2 (Distance) *A measure of divergence d between probability distributions is a* distance *if it satisfies the following three properties: for any three probability distributions p_1, p_2 and p_3 over the same space $\mathcal{X} = (x_1, \ldots, x_k)$,*

- *Positivity: $d(p_1, p_2) \geq 0$. Furthermore, $d(p_1, p_2) = 0 \Leftrightarrow p_1 \equiv p_2$*

- *Symmetry: $d(p_1, p_2) = d(p_2, p_1)$*

- *Triangle Inequality: $d(p_1, p_3) \leq d(p_1, p_2) + d(p_2, p_3)$.*

Consider two common measures of divergence between probability distributions. Let p and q be two probability distributions over the same finite state space $\mathcal{X} = (x_1, \ldots, x_k)$ (Definition 1.2) and let $p_j = p(\{x_j\})$ and $q_j = q(\{x_j\})$ for $j = 1, \ldots, k$. The familiar *quadratic* or *Euclidean distance* is defined as

$$d_2(p, q) = \sqrt{\sum_{j=1}^{k}(p_j - q_j)^2}.$$

The Kullback-Leibler divergence between two probability distributions p and q over the same state space $\mathcal{X}$ is defined in Definition 5.4. In this case, it may be written as

$$d_{KL}(p|q) = \sum_{j=1}^{k} p_j \log \frac{p_j}{q_j}.$$

The Kullback-Leibler divergence is not a *distance* in the sense of Definition 7.2; it does not, in general, satisfy $d_{KL}(p|q) = d_{KL}(q|p)$. Let $p^{(1)} = (0.02, 0.98)$, $q^{(1)} = (0.0364, 0.9636)$, $p^{(2)} = (0.01, 0.99)$, $q^{(2)} = (0.00471, 0.99529)$. Then

$$d_2(p^{(1)}, q^{(1)}) = \sqrt{(0.02 - 0.0364)^2 + (0.98 - 0.9636)^2} = 0.0232$$

$$d_2(p^{(2)}, q^{(2)}) = \sqrt{(0.00471 - 0.01)^2 + (0.99529 - 0.99)^2} = 0.00748,$$

so the change represented by the second adjustment is less than one third of the change represented by the first if the change is measured using the quadratic distance measure.

$$d_{KL}(p^{(1)}|q^{(1)}) = 0.02 \log \frac{0.02}{0.0364} + 0.98 \log \frac{0.98}{0.9636} = 0.004562,$$

$$d_{KL}(p^{(2)}|q^{(2)}) = 0.01 \log \frac{0.01}{0.00471} + 0.99 \log \frac{0.99}{0.99529} = 0.00225,$$

so the change represented by the second adjustment is approximately one half of the change represented by the first. Clearly, different distance measures give different impressions of the relative importance of parameter changes. Furthermore, it is important to have a *reference point*; for example, if the Kullback-Leibler distance is being used, it is useful to know what a particular value of the Kullback-Leibler distance means in terms of a well known family of distributions.

7.3   The Chan-Darwiche distance measure

The problem with *both* the Kullback-Leibler and the quadratic distance measure is that they do not emphasize the *proportional* difference between two probability values when they are close to zero. In [66], the following distance measure is proposed.

Definition 7.3 (Chan-Darwiche Distance) *Let p and q be two probability functions over a finite state space $\mathcal{X}$. That is, $p : \mathcal{X} \to [0, 1]$ and $q : \mathcal{X} \to [0, 1]$, $\sum_{x \in \mathcal{X}} p(x) = 1$ and $\sum_{x \in \mathcal{X}} q(x) = 1$. The Chan-Darwiche distance is defined as*

$$D_{CD}(p, q) = \log \max_{x \in \mathcal{X}} \frac{q(x)}{p(x)} - \log \min_{x \in \mathcal{X}} \frac{q(x)}{p(x)},$$

where, by definition, $\frac{0}{0} = 1$ and $\frac{+\infty}{+\infty} = 1$. If p and q are two probability distributions (Definition 1.2) over a finite state space $\mathcal{X}$, then the Chan-Darwiche distance is defined as $D_{CD}(p, q)$, where p and q are taken as the probability functions in Definition 1.3.

Unlike the Kullback-Leibler divergence, the Chan-Darwiche distance is a *distance*; it satisfies the three requirements of Definition 7.2.

The *support* of a probability function defined on a finite state space; namely, those points where it is strictly positive (relating to outcomes that can happen) is important when comparing two different probability functions over the same state space.

Definition 7.4 (Support) *Let p be a probability function over a countable state space $\mathcal{X}$; that is, $p : \mathcal{X} \to [0, 1]$ and $\sum_{x \in \mathcal{X}} p(x) = 1$. The support of p is defined as the subset $\mathcal{S}_p \subseteq \mathcal{X}$ such that*

$$\mathcal{S}_p = \{x \in \mathcal{X} | p(x) > 0\}. \tag{7.2}$$

Theorem 7.2 *The Chan-Darwiche distance measure is a distance measure, in the sense that for any three probability functions p_1, p_2, p_3 over a state space $\mathcal{X}$, the following three properties hold:*

- *Positivity: $D_{CD}(p_1, p_2) \geq 0$ and $D_{CD}(p_1, p_2) = 0 \Leftrightarrow p_1 \equiv p_2$.*

- *Symmetry: $D_{CD}(p_1, p_2) = D_{CD}(p_2, p_1)$*

- *Triangle Inequality: $D_{CD}(p_1, p_2) + D_{CD}(p_2, p_3) \geq D_{CD}(p_1, p_3)$.*

Proof of Theorem 7.2 Positivity and symmetry are clear (Exercise 2). It only remains to prove the triangle inequality. Since the state space is discrete and *finite*, it follows that there exist $y, z \in \mathcal{X}$ such that

$$D_{CD}(p_1, p_3) = \log \max_{x \in \mathcal{X}} \frac{p_3(x)}{p_1(x)} - \log \min_{x \in \mathcal{X}} \frac{p_3(x)}{p_1(x)} = \log \frac{p_3(y)}{p_1(y)} - \log \frac{p_3(z)}{p_1(z)}$$

$$= \log \frac{p_3(y)}{p_2(y)} + \log \frac{p_2(y)}{p_1(y)} - \log \frac{p_3(z)}{p_2(z)} - \log \frac{p_2(z)}{p_1(z)}$$

$$= \left(\log \frac{p_3(y)}{p_2(y)} - \log \frac{p_3(z)}{p_2(z)} \right) + \left(\log \frac{p_2(y)}{p_1(y)} - \log \frac{p_2(z)}{p_1(z)} \right)$$

$$\leq \left(\log \max_{x \in \mathcal{X}} \frac{p_3(x)}{p_2(x)} - \log \min_{x \in \mathcal{X}} \frac{p_3(x)}{p_2(x)} \right) + \left(\log \max_{x \in \mathcal{X}} \frac{p_2(x)}{p_1(x)} - \log \min_{x \in \mathcal{X}} \frac{p_2(x)}{p_1(x)} \right)$$

$$= D_{CD}(p_1, p_2) + D_{CD}(p_2, p_3).$$

$\square$

This distance is relatively easy to compute. It has the advantage over the Kullback-Leibler divergence (which is not a true distance measure) that it may be used to obtain bounds on *odds ratios*.

Example 7.2: The Chan-Darwiche distance between two multivariate bernoulli distributions Consider d independent Bernoulli trials, $\underline{X} = (X_1, \ldots, X_d)$, where the 'success' probabilities for each trial may differ. The distribution of the random vector $\underline{X}$ is known as a *multivariate Bernoulli distribution*. This example considers the distance between two multivariate Bernoulli distributions where the 'success' probabilities for the two distributions are given by the vectors $\underline{p} = (p_1, \ldots, p_d)$ and $\underline{q} = (q_1, \ldots, q_d)$ respectively. From this, the Chan-Darwiche distance between two different models for tossing the thumb-tack (Section 5), and the Chan-Darwiche distance between two (univariate) Bernoulli distributions may be derived.

Let $\mathcal{X}$ be the binary hypercube; that is, $\mathcal{X} = \{0, 1\}^d$ and let $\underline{x} \in \mathcal{X}$ denote an element in $\mathcal{X}$. Then $\underline{x} = (x_i)_{i=1}^d$, where $x_i \in \{0, 1\}$. Let q and p be two *multivariate Bernoulli probability functions* over $\mathcal{X}$. That is, $q : \mathcal{X} \to [0, 1]$ and $p : \mathcal{X} \to [0, 1]$ are defined such that each $\underline{x} \in \mathcal{X}$,

$$q\left(\underline{x}\right) = \prod_{i=1}^d q_i^{x_i} \, (1 - q_i)^{1 - x_i}$$

and

$$p\left(\underline{x}\right) = \prod_{i=1}^d p_i^{x_i} \, (1 - p_i)^{1 - x_i}$$

where, for this example, it is assumed that $0 < q_i < 1$ and $0 < p_i < 1$ (i.e. the inequalities are strict) for all $i \in \{1, \ldots, d\}$.

Thus, the *likelihood ratio* between q and p is well defined and is given by

$$\mathrm{LR}\left(\underline{x}\right) = \frac{q\left(\underline{x}\right)}{p\left(\underline{x}\right)} = \prod_{i=1}^d \left(\frac{q_i}{p_i}\right)^{x_i} \left(\frac{1 - q_i}{1 - p_i}\right)^{1 - x_i}. \tag{7.3}$$

For each $i \in \{1, \ldots, d\}$, let m_i be defined as

$$m_i = \begin{cases} 1 & \text{if } \frac{q_i}{p_i} \ge \frac{1 - q_i}{1 - p_i} \\ 0 & \text{otherwise.} \end{cases} \tag{7.4}$$

Then $\underline{m} = (m_i)_{i=1}^d \in \mathcal{X}$ and, by construction, it follows from Equation (7.3) that for all $\underline{x} \in \mathcal{X}$,

$$\mathrm{LR}\left(\underline{x}\right) \le \mathrm{LR}\left(\underline{m}\right) = \prod_{i=1}^d \max\left(\frac{q_i}{p_i}, \frac{1 - q_i}{1 - p_i}\right). \tag{7.5}$$

Next let $\underline{\bar{m}}$ be the *binary complement* of $\underline{m}$ defined by Equation (7.4). That is, for each $i \in \{1, \ldots, d\}$, $\bar{m}_i = 1 - m_i$, giving $\bar{m}_i = 0$, if $m_i = 1$ and $\bar{m}_i = 1$ if $m_i = 0$. Then it holds that

$$\mathrm{LR}\left(\underline{x}\right) \ge \mathrm{LR}\left(\underline{\bar{m}}\right) = \prod_{i=1}^d \min\left(\frac{q_i}{p_i}, \frac{1 - q_i}{1 - p_i}\right). \tag{7.6}$$

It now follows from the definition of the Chan-Darwiche distance measure (Definition 7.3) and Equations (7.5) and (7.6) that

$$D_{DC}(p, q) = \log \mathrm{LR}\left(\underline{m}\right) - \log \mathrm{LR}\left(\underline{\bar{m}}\right)$$

$$= \sum_{i=1}^d \log \frac{\max\left(\frac{q_i}{p_i}, \frac{1 - q_i}{1 - p_i}\right)}{\min\left(\frac{q_i}{p_i}, \frac{1 - q_i}{1 - p_i}\right)}. \tag{7.7}$$

For i such that $m_i = 1$ it clearly holds that

$$\frac{\max\left(\frac{q_i}{p_i}, \frac{1-q_i}{1-p_i}\right)}{\min\left(\frac{q_i}{p_i}, \frac{1-q_i}{1-p_i}\right)} = \frac{\frac{q_i}{p_i}}{\frac{1-q_i}{1-p_i}} = \frac{O_{q,i}}{O_{p,i}},$$

where O denotes the odds;

$$O_{q,i} = \frac{q_i}{1-q_i}, \quad O_{p,i} = \frac{p_i}{1-p_i}.$$

Similarly for i such that $m_i = 0$ it holds that

$$\frac{\max\left(\frac{q_i}{p_i}, \frac{1-q_i}{1-p_i}\right)}{\min\left(\frac{q_i}{p_i}, \frac{1-q_i}{1-p_i}\right)} = \frac{O_{p,i}}{O_{q,i}},$$

from which it follows that

$$D_{DC}(p, q) = \sum_{i=1}^{d} \log \max\left(\frac{O_{q,i}}{O_{p,i}}, \frac{O_{p,i}}{O_{q,i}}\right). \tag{7.8}$$

The expression in Equation (7.8) gives two interesting special cases.

1. Let $q_i = q$ and $p_i = p$ for all i, and $0 < q < 1$ and $0 < p < 1$. This corresponds, for example, to d tosses of a thumb-tack when considering two different probabilities of head for tosses that are conditionally independent given this parameter. In this case,

$$q\left(\underline{x}\right) = q^k (1-q)^{d-k}, \qquad p\left(\underline{x}\right) = p^k (1-p)^{d-k},$$

where k is the number of digital ones in $\underline{x}$. By Equation (7.8)

$$D_{DC}(p, q) = d \log \max\left(\frac{O_q}{O_p}, \frac{O_p}{O_q}\right), \tag{7.9}$$

with obvious definitions of the odds. If, say, $\frac{O_q}{O_p} > \frac{O_p}{O_q}$, then

$$D_{DC}(p, q) = d\left(\log O_q - \log O_p\right),$$

which is a fairly neat formula.

2. Furthermore, taking $d = 1$ in the preceding gives the special case

$$D_{DC}(p, q) = \log \max\left(\frac{O_q}{O_p}, \frac{O_p}{O_q}\right), \tag{7.10}$$

This is the Chan-Darwiche distance between the Bernoulli distributions Be (q) and Be (p). □

Example 7.3: Chan-Darwiche distance for a fork with binary variables   Let (X, Y, Z) be three binary variables, each with state space $\{0, 1\}$, that satisfy $X \perp Y | Z$, so that their probability distribution may be factorized along a fork:

$$p_{X,Y,Z} = p_Z p_{X|Z} p_{Y|Z}. \tag{7.11}$$

Five parameters are required to specify the joint distribution:

$$\theta_{X|1} = p_{X|Z}(1|1), \ \ \theta_{X|0} = p_{X|Z}(1|0), \ \ \theta_{Y|1} = p_{Y|Z}(1|1), \ \ \theta_{Y|0} = p_{Y|Z}(1|0), \ \ \theta_Z = p_Z(1).$$

Assume that all of these probabilities lie *strictly* between zero and one. Then the factorization in Equation (7.11) may be written

$$p_{X,Y,Z} (x, y, z) = p_a \cdot p_b,$$

where

$$p_a = \left[\left(\theta_{X|1}^x (1 - \theta_{X|1})^{1-x} \right) \left(\theta_{Y|1}^y (1 - \theta_{Y|1})^{1-y} \right) \theta_Z \right]^z \qquad \forall (x, y, z) \in \{0, 1\}^3$$

and

$$p_b = \left[\left(\theta_{X|0}^x (1 - \theta_{X|0})^{1-x} \right) \left(\theta_{Y|0}^y (1 - \theta_{Y|0})^{1-y} \right) (1 - \theta_Z) \right]^{1-z} \qquad \forall (x, y, z) \in \{0, 1\}^3.$$

Let $q_{X,Y,Z}$ be another joint probability function where $X \perp Y | Z$, that may be factorized $q_{X,Y,Z} = q_Z q_{X|Z} q_{Y|Z}$. Then $q_{X,Y,Z}$ may be written as

$$q_{X,Y,Z} (x, y, z) = q_a \cdot q_b,$$

where, analogously

$$q_a = \left[\left(\psi_{X|1}^x (1 - \psi_{X|1})^{1-x} \right) \left(\psi_{Y|1}^y (1 - \psi_{Y|1})^{1-y} \right) \psi_Z \right]^z \qquad \forall (x, y, z) \in \{0, 1\}^3$$

and

$$q_b = \left[\left(\psi_{X|0}^x (1 - \psi_{X|0})^{1-x} \right) \left(\psi_{Y|0}^y (1 - \psi_{Y|0})^{1-y} \right) (1 - \psi_Z) \right]^{1-z}.$$

Assume that all these probabilities lie strictly between zero and one, so that p and q have the same support. Then

$$LR (x, y, z) = \frac{q_a \cdot q_b}{p_a \cdot p_b} \tag{7.12}$$

For $i = 0, 1$, set

$$O_q^{X|Z=i} = \frac{\psi_{X|i}}{1 - \psi_{X|i}}, \qquad O_p^{X|Z=i} = \frac{\theta_{X|i}}{1 - \theta_{X|i}}$$

and

$$O_q^{Y|Z=i} = \frac{\psi_{Y|i}}{1 - \psi_{Y|i}}, \qquad O_p^{Y|Z=i} = \frac{\theta_{Y|i}}{1 - \theta_{Y|i}}.$$

Then, following the same procedure as in the example above,

$$\text{LR}(x, y, z) \le A^z \cdot B^{1-z} \le \max(A, B). \tag{7.13}$$

where

$$A = \max\left(\frac{O_q^{X|Z=1}}{O_p^{X|Z=1}}, \frac{O_p^{X|Z=1}}{O_q^{X|Z=1}}\right) \cdot \max\left(\frac{O_q^{Y|Z=1}}{O_p^{Y|Z=1}}, \frac{O_p^{Y|Z=1}}{O_q^{Y|Z=1}}\right) \cdot \frac{\psi_Z}{\theta_Z}$$

$$B = \max\left(\frac{O_q^{X|Z=0}}{O_p^{X|Z=0}}, \frac{O_p^{X|Z=0}}{O_q^{X|Z=0}}\right) \cdot \max\left(\frac{O_q^{Y|Z=0}}{O_p^{Y|Z=0}}, \frac{O_p^{Y|Z=0}}{O_q^{Y|Z=0}}\right) \cdot \left(\frac{1 - \psi_Z}{1 - \theta_Z}\right).$$

Note that these inequalities determine one (or several) of the eight configurations (x, y, z) to maximize LR (x, y, z). Similarly, using the method of the example above,

$$\text{LR}(x, y, z) \ge C^z \cdot D^{1-z} \ge \min(C, D), \tag{7.14}$$

where

$$C = \min\left(\frac{O_q^{X|Z=1}}{O_p^{X|Z=1}}, \frac{O_p^{X|Z=1}}{O_q^{X|Z=1}}\right) \cdot \min\left(\frac{O_q^{Y|Z=1}}{O_p^{Y|Z=1}}, \frac{O_p^{Y|Z=1}}{O_q^{Y|Z=1}}\right) \cdot \frac{\psi_Z}{\theta_Z},$$

$$D = \min\left(\frac{O_q^{X|Z=0}}{O_p^{X|Z=0}}, \frac{O_p^{X|Z=0}}{O_q^{X|Z=0}}\right) \cdot \min\left(\frac{O_q^{Y|Z=0}}{O_p^{Y|Z=0}}, \frac{O_p^{Y|Z=0}}{O_q^{Y|Z=0}}\right) \cdot \left(\frac{1 - \psi_Z}{1 - \theta_Z}\right).$$

In other words, the Chan-Darwiche distance between the two distributions factorized along the fork is

$$D_{DC}(p, q) = \log \max(A, B) - \log \min(C, D). \qquad \square$$

Theorem 7.3 *Let p and q be two probability distributions (Definition 1.2) over the same finite state space X and let A and B be two subsets of X. Let $A^c = X \backslash A$ and $B^c = X \backslash B$. Let $O_p(A|B) = \frac{p(A|B)}{p(A^c|B)}$ and $O_q(A|B) = \frac{q(A|B)}{q(A^c|B)}$. Then*

$$e^{-D_{CD}(p,q)} \le \frac{O_q(A|B)}{O_p(A|B)} \le e^{D_{CD}(p,q)}.$$

The bound is sharp in the sense that for any pair of distributions (p, q) there are subsets A and B of X such that

$$\frac{O_q(A|B)}{O_p(A|B)} = \exp\{D_{CD}(p, q)\}, \qquad \frac{O_q(A^c|B)}{O_p(A^c|B)} = \exp\{-D_{CD}(p, q)\}.$$

Proof of Theorem 7.3 Recall Definitions 1.2 and 1.3. Without loss of generality, it may be assumed that p and q have the same support; that is, $p(x) > 0 \Leftrightarrow q(x) > 0$. Otherwise $D_{CD}(p, q) = +\infty$ and the statement is trivially true; for any $A, B \subseteq X, 0 \le$

$\frac{O_q(A|B)}{O_p(A|B)} \leq +\infty$. For p and q such that p and q have the same support, let $r(x) = \frac{q(x)}{p(x)}$.
For any two subsets $A, B \subseteq X$,

$$\frac{O_q(A|B)}{O_p(A|B)} = \frac{q(A|B)}{1 - q(A|B)} \frac{1 - p(A|B)}{p(A|B)} = \frac{q(AB)}{q(A^cB)} \frac{p(A^cB)}{p(AB)} = \frac{\sum_{x \in AB} q(x) \sum_{x \in A^cB} p(x)}{\sum_{x \in A^cB} q(x) \sum_{x \in AB} p(x)}$$

$$= \frac{\sum_{x \in AB} r(x)p(x) \sum_{x \in A^cB} p(x)}{\sum_{x \in A^cB} r(x)p(x) \sum_{x \in AB} p(x)} \leq \frac{\max_{z \in X} r(z) \sum_{x \in AB} p(x) \sum_{x \in A^cB} p(x)}{\min_{z \in X} r(z) \sum_{x \in A^cB} p(x) \sum_{x \in AB} p(x)}$$

$$= \frac{\max_{z \in X} r(z)}{\min_{z \in X} r(z)}.$$

Similarly,

$$\frac{O_q(A|B)}{O_p(A|B)} \geq \frac{\min_{z \in X} r(z)}{\max_{z \in X} r(z)}.$$

From the definition of $D_{CD}(p, q)$, it follows directly that

$$e^{D_{CD}(p,q)} = \frac{\max_{z \in X} r(z)}{\min_{z \in X} r(z)},$$

hence

$$e^{-D_{CD}(p,q)} \leq \frac{O_q(A|B)}{O_p(A|B)} \leq e^{D_{CD}(p,q)},$$

as required.   □

To prove that the bound is tight, consider x such that $r(x) = \max_{z \in X} r(z)$ and y such that $r(y) = \min_{z \in X} r(z)$. Set $A = \{x\}$ and $B = \{x, y\}$. Then

$$O_q(A|B) = \frac{r(x)p(x)}{r(y)p(y)}.$$

Since $O_p(A|B) = \frac{p(x)}{p(y)}$ and $e^{D_{CD}(p,q)} = \frac{\max_{z \in X} r(z)}{\min_{z \in X} r(z)}$, it follows that

$$\frac{O_q(A|B)}{O_p(A|B)} = e^{D_{CD}(p,q)}.$$

Similarly, let $C = \{y\}$, then

$$\frac{O_q(C|B)}{O_p(C|B)} = e^{-D_{CD}(p,q)}.$$ □

Theorem 7.3 may be used to obtain bounds on arbitrary queries $q(A|B)$ for the measure q in terms of $p(A|B)$.

Corollary 7.1 *Set* $d = D_{CD}(p, q)$, *then*

$$\frac{p(A|B)e^{-d}}{1 + (e^{-d} - 1)p(A|B)} \leq q(A|B) \leq \frac{p(A|B)e^d}{1 + (e^d - 1)p(A|B)}. \tag{7.15}$$

Proof of Corollary 7.1 This is straightforward and the details are left to the reader.   □

7.3.1   Comparison with the Kullback-Leibler divergence and euclidean distance

For binary variables, divergences are entirely characterized by the Kullback-Leibler measure. Consider two probability distributions p and q over $\{1, 2, 3\}$ defined by

$$p(1) = a, \qquad p(2) = b - a, \qquad p(3) = 1 - b$$
$$q(1) = ka, \qquad q(2) = b - ka, \qquad q(3) = 1 - b$$

Then

$$D_{KL}(p|q) = a \log \frac{1}{k} + (b - a) \log \frac{b - a}{b - ka} = -a \log k - (b - a) \log \frac{b - ka}{b - a}.$$

Consider the events $A = \{1\}$, $B = \{1, 2\}$, then $O_p(A|B) = \frac{a}{b-a}$ and $O_q(A|B) = \frac{ka}{b-ka}$ and the odds ratio is given by

$$\frac{O_q(A|B)}{O_p(A|B)} = \frac{k(b - a)}{b - ka}.$$

As $a \to 0$, $D_{KL}(p|q) \to 0$, while $\frac{O_q(A|B)}{O_p(A|B)} \to k$. It is therefore not possible to find a bound on the odds ratio in terms of the Kullback-Leibler divergence.

Similarly, in this example, the Euclidean distance is

$$d_2(p, q) = \sqrt{2}a(1 - k) \overset{a \to 0}{\longrightarrow} 0.$$

Neither the Kullback-Leibler divergence nor the Euclidean distance can be used to provide uniform bounds on the odds ratios; even if there is a large *relative* difference between pairs of probability values for p and q, they will be *ignored* if the absolute values of these probabilities are small.

The Chan-Darwiche distance in terms of the bayes factor   The Chan-Darwiche distance may be interpreted in terms of the *Bayes factor*, given in Definition 1.6. Recall that for two probability distributions p and q over a state space $\mathcal{X}$ and two events $A, B \subseteq \mathcal{X}$, the *Bayes factor* is defined as

$$F_{q,p}(A; B) := \frac{q(A)/q(B)}{p(A)/p(B)}. \tag{7.16}$$

Recall the notation given in Definition 1.3. Note that

$$D_{CD}(p, q) = \log \max_{x, y \in \mathcal{X}} F_{p,q}(\{x\}, \{y\}).$$

Theorem 7.4 *Let p and q be two probability distributions over the same state space $\mathcal{X}$. Let A and B be two events, then*

$$e^{-D_{CD}(p,q)} \le F_{q,p}(A, B) \le e^{D_{CD}(p,q)}.$$

Proof of Theorem 7.4 This is similar to the proof of Theorem 7.3. Note that

$$
F_{q,p}(A, B) = \frac{q(A)\, p(B)}{q(B)\, p(A)} = \frac{\sum_{x \in A} q(x) \sum_{x \in B} p(x)}{\sum_{x \in B} q(x) \sum_{x \in A} p(x)}
$$

$$
= \frac{\sum_{x \in A} r(x) p(x) \sum_{x \in B} p(x)}{\sum_{x \in B} r(x) p(x) \sum_{x \in A} p(x)} \leq \frac{\max_{z \in \mathcal{X}} r(z) \sum_{x \in A} p(x) \sum_{x \in B} p(x)}{\min_{z \in \mathcal{X}} r(z) \sum_{x \in B} p(x) \sum_{x \in A} p(x)}
$$

$$
= \frac{\max_{z \in \mathcal{X}} r(z)}{\min_{z \in \mathcal{X}} r(z)}.
$$

Similarly, $F(A, B) \geq \frac{\min_{z \in \mathcal{X}} r(z)}{\max_{z \in \mathcal{X}} r(z)}$ and the result follows. $\qquad \square$

7.3.2   Global bounds for queries

A key issue is to find bounds on the global effect of changing a parameter θ_{jil}; namely, bounds on an arbitrary query $p(A|B)$ where A and B are two events. The Chan-Darwiche distance satisfies the following important property.

Theorem 7.5 *Consider a Bayesian network, with probability distribution p. Suppose that the DAG remains the same and the only change to the CPPs is that $\theta_{j,l}$ is changed to $\tilde{\theta}_{j,l}$ for variable j, parent configuration $\pi_j^{(l)}$, resulting in a new probability distribution q. Then*

$$
d_{CD}(p, q) = d_{CD}(\theta_{j,l}, \tilde{\theta}_{j,l}).
$$

Proof of Theorem 7.5 This is straightforward from the construction. $\qquad \square$

Corollary 7.2 *Consider a DAG $\mathcal{G}$ and consider two probability distributions p and q which factorize according to $\mathcal{G}$. Suppose that q is obtained from p by changing $\theta_{j,l}$ to $\tilde{\theta}_{j,l}$ for one fixed (j, l), all other conditional probabilities remaining the same. Let $\pi_j^{(l)}$ denote parent configuration l for variable j and suppose that $p(\{\Pi_j = \pi_j^{(l)}\}) > 0$. Then for any two subsets A and B of $\mathcal{X}$,*

$$
e^{-d_{CD}(\theta_{j,l}, \tilde{\theta}_{j,l})} \leq \frac{O_q(A|B)}{O_p(A|B)} \leq e^{-d_{CD}(\theta_{j,l}, \tilde{\theta}_{j,l})}.
$$

Proof of Corollary 7.2 This follows directly from Theorem 7.5 and Theorem 7.3. $\qquad \square$

One feature of the Chan-Darwiche distance measure is that it can be used to bound odds ratios. Another important feature, following from Theorem 7.5 and Corollary 7.2 is that if the probability measure is changed locally, only local computations are required to obtain the Chan-Darwiche measure and to use this to bound changes in the values of queries. This is not the case with the Kullback-Leibler divergence. If the same change is applied to p to obtain q, then

$$
d_{KL}(p, q) = p(\{\Pi_j = \pi_j^{(l)}\}) d_{KL}(\theta_{j,l}, \tilde{\theta}_{j,l}).
$$

This follows almost directly from the definition and is left as an exercise (Exercise 1).

The optimality of proportional scaling   Consider one of the conditional probability distributions $(\theta_{j1l}, \ldots, \theta_{jk_jl})$ and suppose that θ_{j1l} is to be altered to a different value, denoted by $\tilde{\theta}_{j1l}$. Under *proportional scaling*, the probabilities of the other states are given by

$$\tilde{\theta}_{jil} = \frac{1 - \tilde{\theta}_{j1l}}{1 - \theta_{j1l}} \theta_{jil}.$$

Proportional scaling turns out to be *optimal* under the Chan-Darwiche distance measure.

Theorem 7.6 *Consider a probability distribution p factorized according to a DAG $\mathcal{G}$. Suppose the value θ_{j1l} is changed to $\tilde{\theta}_{j1l}$. Among the class of probability distributions $\mathcal{Q}$ factorized along $\mathcal{G}$ with $q(\{X_j = x_j^{(l)}\}|\{\Pi_j = \pi_j^{(l)}\}) = \tilde{\theta}_{j1l}$, $\min_{q \in \mathcal{Q}} D_{CD}(p, q)$ is obtained for q such that $\tilde{\theta}_{a.b} = \theta_{a.b}$ for all $(a, b) \neq (j, l)$ and*

$$\tilde{\theta}_{jil} = \frac{1 - \tilde{\theta}_{j1l}}{1 - \theta_{j1l}} \theta_{jil}.$$

Under proportional scaling, the Chan-Darwiche distance is then given by

$$D_{CD}(p, q) = |\log \tilde{\theta}_{j1l} - \log \theta_{j1l}| + |\log(1 - \tilde{\theta}_{j1l}) - \log(1 - \theta_{j1l})|.$$

Proof of Theorem 7.6 Let p be a distribution that factorizes along a DAG $\mathcal{G}$, with conditional probabilities $\theta_{aib} = p(X_a = x_i^{(a)}|\Pi_a = \pi_i^{(a)})$. Let q denote the distribution that factorizes along $\mathcal{G}$, with conditional probabilities

$$\tilde{\theta}_{aib} = q(\{X_a = x_a^{(i)}\}|\{\Pi_a = \pi_a^{(b)}\}),$$

where $\tilde{\theta}_{j1l}$ is given,

$$\tilde{\theta}_{aib} = \theta_{aib} \qquad (a, b) \neq (j, l)$$

and

$$\tilde{\theta}_{jil} = \frac{1 - \tilde{\theta}_{j1l}}{1 - \theta_{j1l}} \theta_{jil} \qquad i = 2, \ldots, k_j.$$

This is the distribution generated by the proportional scheme. Let r denote any other distribution that factorizes along $\mathcal{G}$ with $r(\{X_j = x_j^{(1)}\}|\{\Pi_j = \pi_j^{(b)}\}) = \tilde{\theta}_{j1b}$. The aim is to prove that $D_{CD}(p, r) \geq D_{CD}(p, q)$.

If $\theta_{j1l} = 1$ and $\tilde{\theta}_{j1l} < 1$, then there is a $\tilde{\theta}_{jkl} > 0$ with $\theta_{jkl} = 0$ and it follows that $D_{CD}(p, q) = D_{CD}(p, r) = +\infty$.

If $\theta_{j1l} = 0$ and $\tilde{\theta}_{j1l} > 0$, then it follows directly that $D_{CD}(p, q) = D_{CD}(p, r) = +\infty$. Consider $0 < \theta_{j1l} < 1$. Firstly, consider $\tilde{\theta}_{j1l} > \theta_{j1l}$. Then

$$\max_{x \in \mathcal{X}} \frac{q(x)}{p(x)} = \max \left(\frac{\tilde{\theta}_{j1l}}{\theta_{j1l}}, \frac{1 - \tilde{\theta}_{j1l}}{1 - \theta_{j1l}} \right) = \frac{\tilde{\theta}_{j1l}}{\theta_{j1l}}$$

and

$$\min_{x \in \mathcal{X}} \frac{q(x)}{p(x)} = \frac{1 - \tilde{\theta}_{j1l}}{1 - \theta_{j1l}}.$$

Similarly, if $\tilde{\theta}_{j1l} < \theta_{j1l}$, then $\max_{x \in \mathcal{X}} \frac{q(x)}{p(x)} = \frac{1-\tilde{\theta}_{j1l}}{1-\theta_{j1l}}$ and $\min_{x \in \mathcal{X}} \frac{q(x)}{p(x)} = \frac{\tilde{\theta}_{j1l}}{\theta_{j1l}}$, so

$$D_{CD}(p, q) = |\log \tilde{\theta}_{j1l} - \log \theta_{j1l}| + |\log(1 - \tilde{\theta}_{j1l}) - \log(1 - \theta_{j1l})|.$$

Now consider any other distribution r with $r(\{X_j = x_j^{(1)}\}|\{\Pi_j = \pi_j^{(b)}\}) = \tilde{\theta}_{j1b}$. It is clear, from the factorization along the DAG that if $\tilde{\theta}_{j1l} \geq \theta_{j1l}$ then $\max_{x \in \mathcal{X}} \frac{r(x)}{p(x)} \geq \frac{\tilde{\theta}_{j1l}}{\theta_{j1l}}$ and $\min_{x \in \mathcal{X}} \frac{r(x)}{p(x)} \leq \frac{1-\tilde{\theta}_{j1l}}{1-\theta_{j1l}}$; if $\tilde{\theta}_{j1l} \leq \theta_{j1l}$ then $\max_{x \in \mathcal{X}} \frac{r(x)}{p(x)} \geq \frac{1-\tilde{\theta}_{j1l}}{1-\theta_{j1l}}$ and $\min_{x \in \mathcal{X}} \frac{r(x)}{p(x)} \leq \frac{\tilde{\theta}_{j1l}}{\theta_{j1l}}$. In all cases

$$D_{CD}(p, q) \leq D_{CD}(p, r). \qquad \square$$

7.3.3   Applications to updating

Sections 1.4.2 and 1.4.3 considered the problem of incorporating soft evidence and proposed two methods, depending on the form in which the soft evidence was given: 'Jeffrey's rule' and 'Pearl's method of virtual evidence' respectively. The purpose of this section is to obtain bounds on the distance between the original and updated measures when these methods are applied.

Jeffrey's rule   Recall Jeffrey's rule, discussed in Section 3.2.1. Let p denote a probability distribution over a countable state space $\mathcal{X}$ and let q denote the distribution obtained by updating according to Jeffrey's rule. Then Theorem 7.3 may be applied to give the following bound.

Theorem 7.7 *Let p be a probability distribution over a countable state space $\mathcal{X}$ and let $G_1, \ldots, G_n$ be a collection of mutually exclusive and exhaustive events. Let $\lambda_j = p(G_j)$ for $j = 1, \ldots, n$. Let q denote the probability distribution such that $q(G_j) = \mu_j$ for $j = 1, \ldots, n$ and such that for any other event A,*

$$q(A) = \sum_{j=1}^{n} \mu_j p(A|G_j).$$

In other words, q is the Jeffrey's update of p, defined by $q(G_j) = \mu_j$, $j = 1, \ldots, n$. Then

$$d_{CD}(p, q) = \log \max_j \frac{\lambda_j}{\mu_j} - \log \min_j \frac{\mu_j}{\lambda_j}.$$
$$\square$$

Proof of Theorem 7.7   This is straightforward and the details are left to the reader (see Exercise 3, Chapter 7). $\qquad \square$

This immediately gives the following bound.

Corollary 7.3 *Let O_p and O_q denote the odds function before and after applying Jeffrey's rule. Let*

$$d = \log \max_j \frac{\lambda_j}{\mu_j} - \log \min_j \frac{\mu_j}{\lambda_j}.$$

Then for any two events A and B,

$$e^{-d} \leq \frac{O_q(A|B)}{O_p(A|B)} \leq e^d.$$

Under the Chan-Darwiche distance measure, Jeffrey's rule may be considered *optimal*, in the following sense.

Theorem 7.8 *Let p denote a probability distribution over $\mathcal{X}$ and let $G_1, \ldots, G_r$ denote a collection of mutually exclusive and exhaustive events. Let $\mu_j = p(G_j)$, let $\lambda_1, \ldots, \lambda_r$ be a collection of non-negative numbers such that $\sum_{j=1}^{r} \lambda_j = 1$ and let q be the probability distribution over $\mathcal{X}$ defined by*

$$q(x) = \sum_{j=1}^{r} \frac{\lambda_j}{\mu_j} p(x) \mathbf{1}_{G_j}(x), \qquad x \in \mathcal{X},$$

where

$$\mathbf{1}_{G_j}(x) = \begin{cases} 1 & x \in G_j \\ 0 & x \notin G_j. \end{cases}$$

Then $d_{CD}(p, q)$ minimizes $d_{CD}(p, r)$ subject to the constraint that r is a probability distribution over $\mathcal{X}$ such that $r(G_i) = \lambda_i$ for $i = 1, \ldots, r$.

Proof of Theorem 7.8 Let q denote the distribution generated by Jeffrey's rule and let r be any distribution that satisfies the constraint $r(G_j) = q(G_j) = \lambda_j$, $j = 1, \ldots, r$. If p and r do not have the same support (Definition 7.4), then $+\infty = D_{CD}(p, r) \geq D_{CD}(p, q)$. If they have the same support, let j denote the value such that $\frac{\lambda_j}{\mu_j} = \max_i \frac{\lambda_i}{\mu_i}$ and let k denote the value such that $\frac{\lambda_k}{\mu_k} = \min_i \frac{\lambda_i}{\mu_i}$. Let $\alpha = \max_{x \in \mathcal{X}} \frac{r(x)}{p(x)}$. Then

$$\alpha \mu_j = \alpha \sum_{x \in G_j} p(x) \geq \sum_{x \in G_j} \frac{r(x)}{p(x)} p(x) = r(G_j) = \lambda_j,$$

so that

$$\alpha \geq \frac{\lambda_j}{\mu_j}.$$

Similarly, set $\beta = \min_{x \in \mathcal{X}} \frac{r(x)}{p(x)}$, then a similar argument gives $\beta \leq \frac{\lambda_k}{\mu_k}$. It follows that the distance between p and r is

$$D_{CD}(p, r) = \log \max_{x \in \mathcal{X}} \frac{r(x)}{p(x)} - \log \min_{x \in \mathcal{X}} \frac{r(x)}{p(x)} = \log \alpha - \log \beta$$

$$\geq \log \frac{\lambda_j}{\mu_j} - \log \frac{\lambda_j}{\mu_j} = \log \max_i \frac{\lambda_i}{\mu_i} - \log \min_i \frac{\lambda_i}{\mu_i} = D_{CD}(p, q).$$

Therefore q gives the smallest distance. $\qquad\qquad\qquad\qquad\qquad\qquad \square$

Example 7.4 Consider Example 3.4, Section 3.2.1. It is taken from [65] and discussed in [66]. The probability may be updated using Jeffrey's rule to give, for example,

$$q_{S,C}(s, c_g) = \frac{\lambda_g}{\mu_g} p_{S,C}(s, c_g) = \frac{0.7}{0.3} \times 0.12 = 0.28.$$

Updating the whole distribution in this way gives

$$q_{S,C} = \begin{array}{c|ccc} S\backslash C & c_g & c_b & c_v \\ \hline s & 0.28 & 0.10 & 0.04 \\ s^c & 0.42 & 0.15 & 0.01 \end{array}.$$

Theorem 7.7 gives

$$d_{CD}(p, q) = \log\max_i \frac{\lambda_i}{\mu_i} - \log\min_i \frac{\lambda_i}{\mu_i} = \log\frac{0.7}{0.3} - \log\frac{0.05}{0.4} = 2.93,$$

while Corollary 7.3 gives

$$0.05 \leq \frac{O_q(c_g|s)}{O_p(c_g|s)} \leq 18.73.$$

This suggests that the distributions have changed dramatically. Note that $p_{C|S}(c_g|s) = \frac{0.12}{0.56} = 0.214$, while $q_{C|S}(c_g|s) = \frac{0.28}{0.42} = 0.667$. The change of probability has led to a dramatic change and

$$\frac{O_q(c_g|s)}{O_p(c_g|s)} = \frac{0.667/0.333}{0.214/0.786} = 7.34.$$

If the new distribution over colour were $q_C^* = (0.25, 0.25, 0.50)$ instead, then $d_{CD}(p, q^*) = 0.406$ and

$$0.666 \leq \frac{O_{q^*}(c_g|s)}{O_p(c_g|s)} \leq 1.5.$$

The evidence is weaker and the bounds are therefore tighter.   □

Now consider the following problem: the probability that the piece of cloth is green, given that it is sold tomorrow is, before updating, 0.214. What evidence would satisfy the constraint that the updated probability that the cloth is green, given that it is sold tomorrow, does not exceed 0.3?

By Corollary 7.1,

$$\frac{0.214e^{-d}}{1 + (e^{-d} - 1) \times 0.214} \leq q(c_g|s) \leq \frac{0.214e^d}{1 + (e^d - 1) \times 0.214}.$$

The constraint $q_{C|S}(c_g|s) \leq 0.3$ is satisfied if

$$\frac{0.214e^d}{1 + (e^d - 1) \times 0.214} \leq 0.3$$

giving $d \leq 0.454$. The current distribution over colour is $(\mu_g, \mu_b, \mu_v) = (0.3, 0.3, 0.4)$. The problem now reduces to finding $(\lambda_g, \lambda_b, \lambda_v)$ such that $q_{C|S}(c_g|s) = 0.3$ and

$$\log\max\left(\frac{\lambda_g}{0.3}, \frac{\lambda_b}{0.3}, \frac{\lambda_v}{0.4}\right) - \log\min\left(\frac{\lambda_g}{0.3}, \frac{\lambda_b}{0.3}, \frac{\lambda_v}{0.4}\right) = 0.454.$$

Since

$$q_{C,S}(c_j, s) = \frac{\lambda_j}{\mu_j} p_{C,S}(c_j, s), \qquad j = g, b, v$$

it follows that

$$q_{C|S}(c_g|s) = \frac{0.4\lambda_g}{0.4\lambda_g + 0.4\lambda_b + 0.4\lambda_v}.$$

With $p_{C|S}(c_g|s) = 0.3$,

$$0.28\lambda_g - 0.12\lambda_b - 0.24\lambda_v = 0,$$

with constraint $\lambda_g + \lambda_b + \lambda_v = 1$, so that $10\lambda_g - 3\lambda_v = 3$.
If the maximum and minimum are then given by λ_g and λ_v respectively, then

$$0.454 = \log \frac{\lambda_g}{0.3} - \log \frac{10\lambda_g - 3}{1.2}$$

giving

$$\lambda_g = \frac{3e^{0.454}}{10e^{0.454} - 4} = 0.402$$

$$\lambda_v = 0.34, \qquad \lambda_b = 0.258.$$

Finally, to check that the solution is valid, $\frac{0.4}{\lambda_v} = \frac{0.4}{0.34} = 1.176 > 1.163 = \frac{0.3}{0.258} = \frac{0.3}{\lambda_b}$. Similarly, $\frac{\lambda_g}{0.3}$ clearly gives the maximum in the first term. □

Pearl's method of virtual evidence Recall Pearl's method of virtual evidence, discussed in Section 1.4.3 and developed in Section 3.2.2.

Again, $d_{CD}(p, q)$ may be computed using only local information.

Theorem 7.9 *Let p be a probability distribution over a countable state space $\mathcal{X}$ and let $\lambda_1 = 1$ and $\lambda_2, \ldots, \lambda_r$ positive numbers. Let q denote a set function, defined over subsets of $\mathcal{X}$, such that $q(\phi) = 0$ (where ϕ denotes the empty set) and for any finite collection of disjoint subsets $A_1, \ldots, A_n \subset \mathcal{X}$, $q(\cup_{j=1}^n A_j) = \sum_{j=1}^n q(A_j)$, which satisfies*

$$q(x) = p(x) \sum_{j=1}^r \frac{\lambda_j}{\sum_{k=1}^r p(G_k)\lambda_k} 1_{G_j}(x), \qquad x \in \mathcal{X}.$$

Then q is a probability distribution over $\mathcal{X}$ and

$$d_{CD}(p, q) = \log \max_i \lambda_i - \log \min_i \lambda_i.$$

Proof of Theorem 7.9 Firstly, it is clear from the construction that $\sum_{x \in \mathcal{X}} q(x) = 1$ and that $q(x) \geq 0$ for all $x \in \mathcal{X}$ and (from the definition) that q satisfies the necessary additivity properties, so that q is a probability. From the definition,

$$\frac{q(x)}{p(x)} = \sum_j \frac{\lambda_j}{\sum_k p(G_k)\lambda_k} 1_{G_j}(x) \qquad x \in \mathcal{X}.$$

It follows that

$$D_{CD}(p, q) = \log \max_{x \in \mathcal{X}} \frac{q(x)}{p(x)} - \log \min_{x \in \mathcal{X}} \frac{q(x)}{p(x)}$$

$$= \log \max_{j} \frac{\lambda_j}{\sum_k p(G_k)\lambda_k} - \log \min_{j} \frac{\lambda_j}{\sum_k p(G_k)}$$

$$= \log \max_{j} \lambda_j - \log \min_{j} \lambda_j$$

as required.   □

This immediately gives the following bound.

Corollary 7.4 *Let O_q and O_p denote the odds functions associated with the probability measures defined in Theorem 7.9 and let*

$$d = D_{CD}(p, q) = \log \max_{i} \lambda_i - \log \min_{i} \lambda_i.$$

Then for any events $A, B \subseteq \mathcal{X}$,

$$e^{-d} \leq \frac{O_q(A|B)}{O_p(A|B)} \leq e^d.$$

Example 7.5 'Burglary' (Example 3.5, Section 3.2.2) may be developed to illustrate these results. The discussion follows [66]. Let A denote the event that the alarm goes off, B the event that a burglary takes place and let E denote the evidence of the telephone call from Jemima. According to Pearl's method, this evidence can be interpreted as

$$\lambda = \frac{p(E|A)}{p(E|A^c)} = 4.$$

Therefore, the distance between the original distribution p and the update $q(.) = p(.|E)$ derived according to Pearl's method is $D_{CD}(p, q) = \log 4 \simeq 1.386$. This distance may be used to bound $q(B)$, the probability of a burglary, after the update to incorporate the evidence. Using Equation (7.15),

$$\frac{p(B)e^{-d}}{1 + (e^{-d} - 1)p(B)} \leq q(B) \leq \frac{p(B)e^d}{1 + (e^d - 1)p(B)},$$

so that $2.50 \times 10^{-5} \leq q(B) \leq 4.00 \times 10^{-4}$. An application of Pearl's virtual evidence rule gives $q(B) = 3.85 \times 10^{-4}$.   □

7.4   Parameter changes to satisfy query constraints

The problem considered in this section is to decide whether an individual parameter is relevant to a given query constraint and, if it is, to compute the minimum amount of change needed to that parameter to enforce the constraint. The constraints considered are of the following form: Let $\underline{e}$ denote an instantiation of a collection of variables $\underline{E}$ (so that $\{\underline{E} = \underline{e}\}$ is a piece of hard evidence, and may be expressed as $\mathbf{e} = (e_1, \ldots, e_m)$ as in Definition 3.1) and let Y, Z denote two random variables such that $Y \notin \underline{E}$ and $Z \notin \underline{E}$.

The query constraints considered are of the following type:

$$p_{Y|\underline{E}}(y|\underline{e}) - p_{Z|\underline{E}}(z|\underline{e}) \geq \epsilon, \tag{7.17}$$

$$\frac{p_{Y|\underline{E}}(y|\underline{e})}{p_{Z|\underline{E}}(z|\underline{e})} \geq \epsilon. \tag{7.18}$$

The notation will be abbreviated by writing: $p(y|\underline{e})$ when the abbreviation is clear from the context.

Let $p_{\underline{X}}$ denote the probability function for a collection of variables $\underline{X} = (X_1, \ldots, X_d)$, which may be factorized along a graph $\mathcal{G} = (V, E)$ (where $V = \{X_1, \ldots, X_d\}$), with given conditional probability potentials, $\theta_{jil} = p_{X_j|\Pi_j}(x_j^{(i)}|\pi_j^{(l)})$. Then

$$p_{\underline{X}}(\underline{x}) = \prod_{j=1}^{d} \prod_{l=1}^{q_j} \prod_{i=1}^{k_j} \theta_{jil}^{n(x_i^{(j)}|\pi_l^{(j)})},$$

where $n(x_i^{(j)}|\pi_l^{(j)}) = 1$ if the child parent configuration appears in $\underline{x}$ and 0 otherwise. Suppose that the probabilities $(\theta_{j1l}, \ldots, \theta_{j,k_j,l})$ are parametrized by $(t_1^{(jl)}, \ldots, t_{m_j}^{(jl)})$, where $m_j \leq k_j - 1$. The following result holds.

Theorem 7.10 *Let $\underline{X} = (X_1, \ldots, X_d)$ denote a set of variables and let p be a probability distribution that factorizes along a DAG $\mathcal{G}$ with node set $V = \{X_1, \ldots, X_d\}$. with corresponding probability potentials $\theta_{jil} = p_{X_j|\Pi_j}(x_j^{(i)}|\pi_j^{(l)})$. Suppose that for each (j, l) the probabilities $(\theta_{j1l}, \ldots, \theta_{j,k_j,l})$ are parametrized by $(t_1^{(jl)}, \ldots, t_{m_{jl}}^{(jl)})$ where $m_{jl} \leq k_j - 1$. Let $\underline{E} = (X_{e_1}, \ldots X_{e_m})$ denote a subset of $\underline{X}$ and let $\underline{e} = (x_{e_1}^{(i_1)}, \ldots, x_{e_m}^{(i_m)})$ denote an instantiation of $\underline{E}$. Then for all $1 \leq k \leq m_{jl}$,*

$$\frac{\partial}{\partial t_k^{(jl)}} p_{\underline{E}}(\underline{e}) = \sum_{\alpha=1}^{k_j} \frac{p(\{\underline{E} = \underline{e}\}, \{X_j = x_j^{(\alpha)}\}, \{\Pi_j = \pi_j^{(l)}\})}{\theta_{j\alpha l}} \frac{\partial}{\partial t_k^{(jl)}} \theta_{j\alpha l}.$$

Proof of Theorem 7.10 Firstly,

$$p_{\underline{E}}(\underline{e}) = \sum_{il} p_{\underline{E}|X_j,\Pi_j}(\underline{e}|x_j^{(i)}, \pi_j^{(l)}) p_{X_j|\Pi_j}(x_j^{(i)}|\pi_j^{(l)}) p_{\Pi_j}(\pi_j^{(l)})$$

$$= \sum_{il} p_{\underline{E}|X_j,\Pi_j}(\underline{e}|x_j^{(i)}, \pi_j^{(l)}) \theta_{jil} p_{\Pi_j}(\pi_j^{(l)}).$$

It follows that

$$\frac{\partial}{\partial t_k^{(jl)}} p_{\underline{E}}(\underline{e}) = \sum_{i=1}^{k_j} p_{\underline{E}|X_j,\Pi_j}(\underline{e}|x_j^{(i)}, \pi_j^{(l)}) p_{\Pi_j}(\pi_j^{(l)}) \frac{\partial \theta_{jil}}{\partial t_k^{(jl)}}$$

$$= \sum_{i=1}^{k_j} \frac{p_{X_j,\Pi_j|\underline{E}}(x_j^{(i)}, \pi_j^{(l)}|\underline{e}) p_{\underline{E}}(\underline{e}) p_{\Pi_j}(\pi_j^{(l)})}{p_{X_j,\Pi_j}(x_j^{(i)}, \pi_j^{(l)})} \frac{\partial \theta_{jil}}{\partial t_k^{(jl)}}$$

$$= \sum_{i=1}^{k_j} \frac{p_{X_j, \Pi_j, \underline{E}}(x_i^{(j)}, \pi_l^{(j)}, \underline{e})}{p_{X_j|\Pi_j}(x_j^{(i)}|\pi_j^{(l)})} \frac{\partial \theta_{jil}}{\partial t_k^{(jl)}}$$

$$= \sum_{i=1}^{k_j} \frac{p_{X_j, \Pi_j, \underline{E}}(x_j^{(i)}, \pi_j^{(l)}, \underline{e})}{\theta_{jil}} \frac{\partial \theta_{jil}}{\partial t_k^{(jl)}}$$

as required.   □

Proportional scaling   Again, the complete set of variables is $\underline{X} = (X_1, \ldots, X_d)$, with a joint probability distribution p that may be factorized along a directed acyclic graph $\mathcal{G}$. Evidence is received on a subset of the variables $\underline{E} = (X_{e_1}, \ldots, X_{e_m})$. Consider a *proportional scaling* scheme, where each conditional probability distribution $(\theta_{j1l}, \ldots, \theta_{jk_jl})$ has exactly one parameter. Under proportional scaling, this may be represented as $\theta_{j1l} = t^{(jl)}$ and there are non-negative numbers $a_2^{(jl)}, \ldots, a_{k_j}^{(jl)}$ satisfying $\sum_{\alpha=2}^{k_j} a_\alpha^{(jl)} = 1$, such that

$$\theta_{j1l} = t^{(jl)}$$

$$\theta_{j\alpha l} = a_\alpha^{(jl)}(1 - t^{(jl)}), \qquad \alpha = 2, \ldots, k_j.$$

Then, an application of Theorem 7.10 in the simplified setting of proportional scaling immediately gives

$$\frac{\partial}{\partial t^{(jl)}} p_{\underline{E}}(\underline{e}) = \frac{p_{\underline{E}, X_j, \Pi_j}(\underline{e}, x_j^{(1)}, \pi_j^{(l)})}{\theta_{j1l}} - \sum_{\alpha=2}^{k_j} \frac{p_{\underline{E}, X_j, \Pi_j}(\underline{e}, x_j^{(\alpha)}, \pi_j^{(l)})}{\theta_{j\alpha l}} a_\alpha^{(jl)}. \qquad (7.19)$$

When a proportional scaling scheme is used, Theorem 7.1 gives

$$p_{\underline{E}}(\underline{e}) = \alpha + \beta t^{(jl)},$$

where α and β do not depend on $t^{(jl)}$. It follows that for any $t^{(jl)}$, $\frac{\partial}{\partial t^{(jl)}} p_{\underline{E}}(\underline{e}) = \beta$, where β is constant (i.e. it does not depend on $t^{(jl)}$). This observation makes it straightforward, under proportional scaling, to find the necessary change in a single parameter $t^{(jl)}$ (if such a parameter change is possible) to enforce a query constraint.

7.4.1   Binary variables

Assume that variable X_j is *binary*, with $p_{X_j|\Pi_j}(x_j^{(1)}|\pi_j^{(l)}) = t^{(jl)}$ and $p_{X_j|\Pi_j}(x_j^{(0)}|\pi_j^{(l)}) = 1 - t^{(jl)}$. Then Equation (7.19) reduces to:

$$\frac{\partial}{\partial t^{(jl)}} p_{\underline{E}}(\underline{e}) = \frac{p_{\underline{E}, X_j, \Pi_j}(\underline{e}, x_j^{(1)}, \pi_j^{(l)})}{t^{(jl)}} - \frac{p_{\underline{E}, X_j, \Pi_j}(\underline{e}, x_j^{(0)}, \pi_j^{(l)})}{1 - t^{(jl)}}. \qquad (7.20)$$

The statement $Y = y$, $\underline{E} = \underline{e}$ may be treated as hard evidence. By Theorem 7.1, it follows that there are real numbers λ, λ_y and λ_z such that

$$\lambda = \frac{\partial}{\partial t^{(jl)}} p_{\underline{E}}(\underline{e}) = \frac{p_{\underline{E},X_j,\Pi_j}(\underline{e}, x_j^{(1)}, \pi_j^{(l)})}{t^{(jl)}} - \frac{p_{\underline{E},X_j,\Pi_j}(\underline{e}, x_j^{(0)}, \pi_j^{(l)})}{1 - t^{(jl)}},$$

$$\lambda_y = \frac{\partial}{\partial t^{(jl)}} p_{Y,\underline{E}}(y, \underline{e}) = \frac{p_{Y,\underline{E},X_j,\Pi_j}(y, \underline{e}, x_j^{(1)}, \pi_j^{(l)})}{t^{(jl)}} - \frac{p_{Y,\underline{E},X_j}(y, \underline{e}, x_j^{(0)}, \pi_j^{(l)})}{1 - t^{(jl)}}$$

and

$$\lambda_z = \frac{\partial}{\partial t^{(jl)}} p_{Z,\underline{E}}(z, \underline{e}) = \frac{p_{Z,\underline{E},X_j,\Pi_j}(z, \underline{e}, x_j^{(1)}, \pi_j^{(l)})}{t^{(jl)}} - \frac{p_{Z,\underline{E},X_j,\Pi_j}(z, \underline{e}, x_j^{(0)}, \pi_j^{(l)})}{1 - t^{(jl)}}.$$

The following is a corollary of Theorem 7.10, which reduces to Equation (7.20) for the binary case.

Corollary 7.5 *To satisfy the constraint given by Equation (7.17), the parameter $t^{(jl)}$ has to be changed to $t^{(jl)} + \delta$, where δ satisfies*

$$p_{Y,\underline{E}}(y, \underline{e}) - p_{Z,\underline{E}}(z, \underline{e}) - \epsilon p_{\underline{E}}(\underline{e}) \geq \delta(-\lambda_y + \lambda_z + \epsilon\lambda). \tag{7.21}$$

To satisfy the constraint given by Equation (7.18), the parameter $t^{(jl)}$ has to be changed to $t^{(jl)} + \delta$, where

$$p_{Y,\underline{E}}(y, \underline{e}) - \epsilon p_{Z,\underline{E}}(z, \underline{e}) \geq \delta(-\lambda_y + \epsilon\lambda_z). \tag{7.22}$$

Proof of Corollary 7.5 Since $p_{Y|\underline{E}}(y|\underline{e}) = \frac{p_{Y,\underline{E}}(y,\underline{e})}{p_{\underline{E}}(\underline{e})}$, it follows that $p_{Y|\underline{E}}(y|\underline{e}) - p_{Z|\underline{E}}(z|\underline{e}) \geq \epsilon$ is equivalent to $p_{Y,\underline{E}}(y, \underline{e}) - p_{Z,\underline{E}}(z, \underline{e}) \geq \epsilon p_{\underline{E}}(\underline{e})$. A change in the constraint changes $p_{Y,\underline{E}}(y, \underline{e})$, $p_{Z,\underline{E}}(z, \underline{e})$ and $p_{\underline{E}}(\underline{e})$ to $p_{Y,\underline{E}}(y, \underline{e}) + \delta\lambda_y$, $p_{Z,\underline{E}}(z, \underline{e}) + \delta\lambda_z$ and $p_{\underline{E}}(\underline{e}) + \delta\lambda$ respectively. To enforce the difference constraint, it follows that δ satisfies

$$(p_{Y,\underline{E}}(y, \underline{e}) + \lambda_y\delta) - (p_{Z,\underline{E}}(z, \underline{e}) + \lambda_z\delta) \geq \epsilon(p_{\underline{E}}(\underline{e}) + \lambda\delta).$$

Equation (7.21) follows directly.

Similarly, to enforce the ratio constraint, the following inequality is required:

$$\frac{p_{Y,\underline{E}}(y, \underline{e}) + \lambda_y\delta}{p_{Z,\underline{E}}(z, \underline{e}) + \lambda_z\delta} \geq \epsilon.$$

Equation (7.22) now follows directly and the proof is complete. □

Proportional scaling If proportional scaling is being used, so that Equation (7.19) holds, then it is clear that Corollary 7.5 can be extended, with minor adjustment to the set of linear equations, to satisfy the constraints given by Equations (7.17) and (7.18).

7.5   The sensitivity of queries to parameter changes

In line with the Chan-Darwiche distance measure, the following definition will be used for 'sensitivity'.

Definition 7.5 (Sensitivity) *Let p denote a parametrized family of probability distributions, over a finite, discrete state space $\mathcal{X}$, parametrized by k parameters $(\theta_1, \ldots, \theta_k) \in \tilde{\Theta}$, where $\tilde{\Theta} \subseteq \mathbf{R}^k$ denotes the parameter space. Let $p^{(\theta_1, \ldots, \theta_k)}(.)$ denote the probability distribution when the parameters are fixed at $\theta_1, \ldots, \theta_k$. Then the sensitivity of p to parameter θ_j is defined as*

$$S_j(p)(\theta_1, \ldots, \theta_k) = \max_{\underline{x} \in \mathcal{X}} \frac{\partial}{\partial \theta_j} \log p^{(\theta_1, \ldots, \theta_k)}(\{\underline{x}\}) - \min_{\underline{x} \in \mathcal{X}} \frac{\partial}{\partial \theta_j} \log p^{(\theta_1, \ldots, \theta_k)}(\{\underline{x}\}).$$

Example 7.6   If p is a family of binary variables, with state space $\mathcal{X} = \{x_0, x_1\}$ and a single parameter θ, then

$$S(p)(\theta) = \left| \frac{\partial}{\partial \theta} \log \frac{p^{(\theta)}(\{x_1\})}{p^{(\theta)}(\{x_0\})} \right|.$$
□

This section restricts attention to a single parameter model. Consider a network with d variables, $\underline{X} = (X_1, \ldots, X_d)$ where one particular variable X_j is a binary variable. The other variables may be multivalued. Let

$$t^{(jl)} = p_{X_j | \Pi_j}(x_j^{(1)} | \pi_j^{(l)}).$$

Let $\underline{Y}$ denote a collection of variables, taken from $(X_1, \ldots, X_n)$ and let $\underline{Y} = \underline{y}$ denote an instantiation of these variables. Let $\underline{y}$ denote the event $\{\underline{Y} = \underline{y}\}$ and let $\underline{y}^c$ denote the event $\{\underline{Y} \neq \underline{y}\}$. Similarly, let $\underline{e}$ denote the event $\{\underline{E} = \underline{e}\}$, where $\underline{E}$ is a different sub-collection of variables from $\underline{X}$. From Definition 7.5, the sensitivity of a query $p(\underline{y}|\underline{e})$ to the parameter $t^{(jl)}$ is defined as

$$\left| \frac{\partial}{\partial t^{(jl)}} \log \frac{p(\underline{y}|\underline{e})}{p(\underline{y}^c|\underline{e})} \right|.$$

The following theorem provides a simple bound on the derivative in terms of $p(\underline{y}|\underline{e})$ and $t^{(jl)}$ only.

Theorem 7.11 *Suppose X_j is a binary variable taking values $x_j^{(1)}$ or $x_j^{(0)}$. Set*

$$t^{(jl)} = p_{X_j | \Pi_j}(x_j^{(1)} | \pi_j^{(l)}).$$

Then

$$\left| \frac{\partial}{\partial t^{(jl)}} p(\underline{y}|\underline{e}) \right| \leq \frac{p(\underline{y}|\underline{e})(1 - p(\underline{y}|\underline{e}))}{t^{(jl)}(1 - t^{(jl)})}. \tag{7.23}$$

The example given after the proof shows that this bound is *sharp*; there are situations where the derivative assumes the bound *exactly*.

Proof of Theorem 7.11 Firstly, $p(\underline{y}|\underline{e}) = \frac{p(\underline{y},\underline{e})}{p(\underline{e})}$, so that

$$\frac{\partial}{\partial t^{(jl)}} p(\underline{y}|\underline{e}) = \frac{1}{p(\underline{e})} \frac{\partial}{\partial t^{(jl)}} p(\underline{y},\underline{e}) - \frac{p(\underline{y}|\underline{e})}{p(\underline{e})} \frac{\partial}{\partial t^{(jl)}} p(\underline{e}).$$

Using this, Equation (7.20) gives

$$\frac{\partial}{\partial t^{(jl)}} p(\underline{y}|\underline{e})$$

$$= \frac{\left\{ (1 - t^{(jl)}) p(\underline{y}, x_j^{(1)}, \pi_j^{(l)}|\underline{e}) - t^{(jl)} p(\underline{y}, x_j^{(0)}, \pi_j^{(l)}|\underline{e}) \right\}}{t^{(jl)}(1 - t^{(jl)})} \tag{7.24}$$

$$- \frac{\left\{ (1 - t^{(jl)}) p(\underline{y}|\underline{e}) p(\{X_j = x_j^{(1)}\}, \{\Pi_j = \pi_j^{(l)}\}|\underline{e}) - t^{(jl)} p(\underline{y}|\underline{e}) p(\{X_j = x_j^{(0)}\}, \{\Pi_j = \pi_j^{(l)}\}|\underline{e}) \right\}}{t^{(jl)}(t - t^{(jl)})}$$

$$= \frac{(1 - t^{(jl)})(p(\underline{y}, \{X_j = x_j^{(1)}\}, \{\Pi_j = \pi_j^{(l)}\}|\underline{e}) - p(\underline{y}|\underline{e}) p(\{X_j = x_j^{(1)}\}, \{\Pi_j = \pi_j^{(l)}\}|\underline{e}))}{t^{(jl)}(1 - t^{(jl)})}$$

$$- \frac{t^{(jl)}(p(\underline{y}, \{X_j = x_j^{(0)}\}, \{\Pi_j = \pi_j^{(l)}\}|\underline{e}) - p(\underline{y}|\underline{e}) p(\{X_j = x_j^{(0)}\}, \{\Pi_j = \pi_j^{(l)}\}|\underline{e}))}{t^{(jl)}(t - t^{(jl)})}. \tag{7.25}$$

With the shorthand notation $\underline{y}^c$ to denote the event $\{Y \neq \underline{y}\}$,

$$p(\{X_j = x_j^{(1)}\}, \{\Pi_j = \pi_j^{(l)}\}, \underline{y}|\underline{e}) - p(\underline{y}|\underline{e}) p(\{X_j = x_j^{(1)}\}, \{\Pi_j = \pi_j^{(l)}\}|\underline{e})$$

$$\leq p(\{X_j = x_j^{(1)}\}, \{\Pi_j = \pi_j^{(l)}\}, \underline{y}|\underline{e}) - p(\underline{y}|\underline{e}) p(\{X_j = x_j^{(1)}\}, \{\Pi_j = \pi_j^{(l)}\}, \underline{y}|\underline{e})$$

$$= p(\{X_j = x_j^{(1)}\}, \{\Pi_j = \pi_j^{(l)}\}, \underline{y}|\underline{e})(1 - p(\underline{y}|\underline{e})$$

$$\leq p(\underline{y}|\underline{e})(1 - p(\underline{y}|\underline{e}))$$

and

$$p(\underline{y}|\underline{e}) p(\{X_j = x_j^{(1)}\}, \{\Pi_j = \pi_j^{(l)}\}|\underline{e}) - p(\{X_j = x_j^{(1)}\}, \{\Pi_j = \pi_j^{(l)}\}, \underline{y}|\underline{e})$$

$$= (1 - p(\underline{y}^c|\underline{e})) p(\{X_j = x_j^{(1)}\}, \{\Pi_j = \pi_j^{(l)}\}|\underline{e})$$

$$- p(\{X_j = x_j^{(1)}\}, \{\Pi_j = \pi_j^{(l)}\}|\underline{e}) + p(\{X_j = x_j^{(1)}\}, \{\Pi_j = \pi_j^{(l)}\}, \underline{y}^c|\underline{e})$$

$$= p(\{X_j = x_j^{(1)}\}, \{\Pi_j = \pi_j^{(l)}, \underline{y}^c|\underline{e}) - p(\underline{y}^c|\underline{e}) p(\{X_j = x_j^{(1)}\}, \{\Pi_j = \pi_j^{(l)}|\underline{e})$$

$$= p(\{X_j = x_j^{(1)}\}, \{\Pi_j = \pi_j^{(l)}\}, \underline{y}^c|\underline{e})(1 - p(\underline{y}^c|\underline{e}))$$

$$\leq p(\underline{y}^c|\underline{e})(1 - p(\underline{y}^c|\underline{e}))$$

$$= (1 - p(\underline{y}|\underline{e})) p(\underline{y}|\underline{e})$$

From this, it follows directly from Equation (7.25) that

$$\left| \frac{\partial}{\partial t^{(jl)}} p(\underline{y}|\underline{e}) \right| \leq \frac{p(\underline{y}|\underline{e})(1 - p(\underline{y}|\underline{e}))}{t^{(jl)}(1 - t^{(jl)})}.$$

The proof of Theorem 7.11 is complete.   □

Corollary 7.6 *The sensitivity of $p(y|e)$ to the parameter $t^{(jl)}$ is bounded by*

$$\left| \frac{\partial}{\partial t^{(jl)}} \log \frac{p(y|e)}{p(y^c|e)} \right| \leq \frac{1}{t^{(jl)}(1 - t^{(jl)})}. \tag{7.26}$$

Proof of Corollary 7.6 Immediate.   □

It is clear that the *worst* situation from a robustness point of view arises when the parameter value $t^{(jl)}$ is close to either 0 or 1, while the query takes values that are close to neither 0 nor 1.

Example 7.7 This example shows that the bounds given by inequalities (7.23) and (7.26) are tight, in the sense that there are examples where it is attained. Consider the network given in Figure 7.2, where X and Y are binary variables taking values from (x_0, x_1) and (y_0, y_1) respectively. $p_X(x_0) = \theta_x$ and $p_Y(y_0) = \theta_y$. Suppose that E is a *deterministic* binary variable; that is, $p(\{E = e\}|\{X = Y\}) = 1$ and $p(\{E = e\}|\{X \neq Y\}) = 0$.
    The probability potentials are

$$p_X = \begin{array}{cc} x_0 & x_1 \\ \hline \theta_x & 1 - \theta_x \end{array} \qquad p_Y = \begin{array}{cc} y_0 & y_1 \\ \hline \theta_y & 1 - \theta_y \end{array}$$

$$p_{E|X,Y}(e|.,.) = \begin{array}{c|cc} X\backslash Y & y_0 & y_1 \\ \hline x_0 & 1 & 0 \\ x_1 & 0 & 1 \end{array}$$

It follows that

$$p_{Y|E}(y_0|e) = \frac{p_{Y|E}(y_0, e)}{p_E(e)} = \frac{p_Y(y_0) \sum_x p_X(x) p_{E|X,Y}(e|x, y_0)}{\sum_{x,y} p_X(x) p_Y(y) p_{E|X,Y}(e|x, y)}$$

$$= \frac{\theta_y \theta_x}{\theta_y \theta_x + (1 - \theta_y)(1 - \theta_x)}.$$

It follows that

$$\frac{\partial}{\partial \theta_x} p_{Y|E}(y_0|e) = \frac{\theta_y(1 - \theta_y)}{(\theta_x \theta_y + (1 - \theta_x)(1 - \theta_y))^2}$$

while

$$\frac{p_{Y|E}(y_0|e)(1 - p_{Y|E}(y_0|e))}{\theta_x(1 - \theta_x)} = \frac{\theta_y \theta_x (1 - \theta_y)(1 - \theta_x)}{(\theta_x \theta_y + (1 - \theta_x)(1 - \theta_y))^2 \theta_x (1 - \theta_x)}$$

$$= \frac{\theta_y(1 - \theta_y)}{(\theta_x \theta_y + (1 - \theta_x)(1 - \theta_y))^2},$$

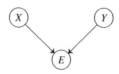

Figure 7.2   The network for Example 7.7.

so that

$$\frac{\partial}{\partial \theta_x} p_{Y|E}(y_0|e) = \frac{\theta_y(1 - \theta_y)}{(\theta_x \theta_y + (1 - \theta_x)(1 - \theta_y))^2}$$

showing that the bound (7.23) is achieved.   □
For the bound (7.26), note from the above that

$$\frac{\partial}{\partial \theta_x} p_{Y|E}(y_0|e) = \frac{p_{Y|E}(y_0|e) p_{Y|E}(y_1|e)}{\theta_x(1 - \theta_x)}$$

so that

$$\frac{\partial}{\partial \theta_x} \log p_{Y|E}(y_0|e) = \frac{p_{Y|E}(y_1|e)}{\theta_x(1 - \theta_x)}$$

and, because $p_{Y|E}(y_0|e) + p_{Y|E}(y_1|e) = 1$,

$$\frac{\partial}{\partial \theta_x} p_{Y|E}(y_1|e) = -\frac{\partial}{\partial \theta_x} p_{Y|E}(y_0|e) = -\frac{p_{Y|E}(y_0|e) p_{Y|E}(y_1|e)}{\theta_x(1 - \theta_x)}$$

so that

$$\frac{\partial}{\partial \theta_x} \log \frac{p_{Y|E}(y_0|e)}{p_{Y|E}(y_1|e)} = \frac{1}{\theta_x(1 - \theta_x)},$$

so that equality is achieved in bound (7.26).   □
Although a small change in a parameter $t^{(jl)}$ can lead to a large change in a query $p(y|e)$, the change is not so large if instead the change in *odds* are considered.

Theorem 7.12 *Let p be a parametrized family of probability distributions, factorized along the same DAG, with a single parameter θ. Let X_j be a binary variable and let $\theta = p_{X_j|\Pi_j}^{(\theta)}(x_j^{(0)}|\pi_j^{(l)})$; all the other CPPs remain fixed and let $O_\theta = \frac{\theta}{1-\theta}$. Consider a parameter change from $\theta = t$ to $\theta = s$. Note that $O_t = \frac{t}{1-t}$ and $O_s = \frac{s}{1-s}$. Let $p^{(\theta)}(y|e)$ denote the probability value of a query when θ is the parameter value. Let $\tilde{O}_\theta(y|e) = \frac{p^{(\theta)}(y|e)}{1-p^{(\theta)}(y|e)}$. Then*

$$\frac{O_t}{O_s} \leq \frac{\tilde{O}_s(y|e)}{\tilde{O}_t(y|e)} \leq \frac{O_s}{O_t} \qquad s \geq t$$

$$\frac{O_s}{O_t} \leq \frac{\tilde{O}_s(y|e)}{\tilde{O}_t(y|e)} \leq \frac{O_t}{O_s} \qquad s \leq t.$$

This gives the bound

$$\left| \log \tilde{O}_s(y|e) - \log \tilde{O}_t(y|e) \right| \leq |\log O_s - \log O_t|.$$

Proof of Theorem 7.12 Let x denote the probability of the query $p(y|e)$ when the value of the parameter $t^{(jl)}$ is z.
Note that, for $0 < a \leq b < 1$,

$$\int_a^b \frac{dx}{x(1 - x)} = \int_a^b \frac{dx}{x} + \int_a^b \frac{dx}{1 - x} = \log \frac{b}{a} \frac{1 - a}{1 - b}.$$

Then, for $t^{(jl)} \leq s^{(jl)}$, Equation (7.23) gives

$$-\int_t^s \frac{dz}{z(1-z)} \leq \int_{p_t(\underline{y}|e)}^{p_s(\underline{y}|e)} \frac{dx}{x(1-x)} \leq \int_t^s \frac{dz}{z(1-z)},$$

so that

$$-\log \frac{s}{t}\frac{1-t}{1-s} \leq \log \frac{p_s(\underline{y}|e)}{p_t(\underline{y}|e)}\frac{1-p_s(\underline{y}|e)}{1-p_t(\underline{y}|e)} \leq \log \frac{s}{t}\frac{1-t}{1-s}$$

giving immediately that

$$\frac{O_t}{O_s} \leq \frac{\tilde{O}_s(\underline{y}|e)}{\tilde{O}_t(\underline{y}|e)} \leq \frac{O_s}{O_t}.$$

For $s \leq t$ the argument is similar and gives

$$\frac{O_s}{O_t} \leq \frac{\tilde{O}_s(\underline{y}|e)}{\tilde{O}_t(\underline{y}|e)} \leq \frac{O_t}{O_s}.$$

In both cases

$$\left|\log \tilde{O}_s(\underline{y}|e) - \log \tilde{O}_t(\underline{y}|e)\right| \leq |\log O_s - \log O_t|$$

and the result follows.   □

Notes   The observation that the probability of evidence is a linear function of any single parameter in the model and hence that the conditional probability is the ratio of two linear functions is due to E. Castillo, J.M. Gutiérrez and A.S. Hadi [108] and [109]. See also V.M. Coupé and L.C. van der Gaag [110]. The most significant developments in sensitivity analysis, which comprise practically the whole chapter, were introduced by H. Chan and A. Darwiche in the article [111] and developed in the article [66], where Chan-Darwiche distance measure was introduced, and the article [81], which discusses the application to Jeffrey's update rule and Pearl's method of virtual evidence.

7.6   Exercises: Parameters and sensitivity

1. Let $\mathcal{G} = (V, E)$ be a directed acyclic graph, where $V = (X_1, \ldots, X_d)$, and let p and q be two probability distribution factorized along $\mathcal{G}$. Suppose that the conditional probability tables for p and q are the same except for one single (j, l) variable/parent configuration, where p has table $\theta_{j,l}$ and q has table $\tilde{\theta}_{j,l}$. Let d_{KL} denote the Kullback-Leibler distance. Show that

$$d_{KL}(p, q) = p(\{\Pi_j = \pi_j^{(l)}\}) d_{KL}(\theta_{j,l}, \tilde{\theta}_{j,l}).$$

2. Let $D_{CD}(p, q)$ denote the Chan-Darwiche distance between two probability distributions. Prove that for any two probability distributions p and q, defined on the same finite state space $\mathcal{X}$,

   - $D_{CD}(p, q) \geq 0$ with equality if and only if $p(x) = q(x)$ for all $x \in \mathcal{X}$.

   - $D_{CD}(p, q) = D_{CD}(q, p)$.

3. Let p be a probability distribution over a countable state space $\mathcal{X}$ and let $G_1, \ldots, G_n$ be a collection of mutually exclusive and exhaustive events. Let $\lambda_j = p(G_j)$ for $j = 1, \ldots, n$. Let q denote the probability distribution such that $q(G_j) = \mu_j$ for $j = 1, \ldots, n$ and such that for any other event A,

$$q(A) = \sum_{j=1}^{n} \mu_j p(A|G_j).$$

   In other words, q is the Jeffrey's update of p, defined by $q(G_j) = \mu_j$, $j = 1, \ldots, n$. Prove that

$$d_{CD}(p, q) = \log \max_j \frac{\lambda_j}{\mu_j} - \log \min_{x \in \mathcal{X}} \frac{\mu_j}{\lambda_j},$$

   where d_{CD} denotes the Chan-Darwiche distance.

4. Consider a Bernoulli trial, with probability function $p_X(.|t)$ defined by

$$p_X(x|t) = t^x (1 - t)^{1-x}, \qquad x = 0, 1, \qquad t \in [0, 1].$$

   Recall the definition of sensitivity, Definition 7.5. Compute the sensitivity with respect to the parameter t.

5. Find the calibration $cd(k)$ of the Chan-Darwiche distance by suitably reformulating the concept of calibration as found in Exercise 3, Chapter 5. You should obtain

$$cd(k) = \frac{e^{\pm k}}{1 + e^{\pm k}}.$$

   Why is this a reasonable calibration? Compare by plotting $h(k)$, the calibration of the Kullback-Leibler divergence in Example 3, Chapter 5, together with the calibration $cd(k)$ and comment on any differences you observe.

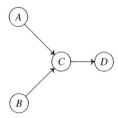

Figure 7.3 DAG on four variables.

6. Consider the DAG given in Figure 7.3, where (A, B, C, D) are all binary variables with probability tables

$$p_A = \frac{1 \quad 0}{s \quad 1-s}, \quad p_B = \frac{1 \quad 0}{t \quad 1-t}$$

$$p_{C|A,B}(1|.,.) = \begin{array}{c|cc} A\backslash B & 1 & 0 \\ \hline 1 & 1 & 0.8 \\ 0 & 0.8 & 0 \end{array}$$

Compute the sensitivity, with respect to s and with respect to t, for $p_{A|C}(1|1)$.

7. A test to detect whether a person has the Green Monkey Disease virus is being developed. Laboratory tests show that if a person has the condition, then the test gives a positive result with probability 0.99. If a person does not have the condition, then the test gives a positive result with probability 0.02. The proportion of the population that has the condition is t, where t is unknown, but suspected to be around 1%. Compute the sensitivity of the probability that a randomly chosen person has the disease given a positive test result with respect to this parameter. Compute the probability that the person has the disease, given a positive test result, for $t = 0.01$ and $t = 0.02$.

8. Following the example (Section 7.3) of the Chan-Darwiche distance for two distributions factorized along a fork, find an expression for the Chan-Darwiche distance between two distributions factorized along the same Chow-Liu tree.

9. Suppose that the probability distribution $p_{A,B,C,D}$ can be factorized along the the DAG given in Figure 7.3. Suppose that $p_A(1) = p_B(1) = 0.5$,

$$p_{D|C} = \begin{array}{c|cc} C\backslash D & 0 & 1 \\ \hline 0 & 0.7 & 0.3 \\ 1 & 0.3 & 0.7 \end{array}$$

and

$$p_{C|A,B}(1|.,.) = \begin{array}{c|cc} A\backslash B & 0 & 1 \\ \hline 0 & 0 & t \\ 1 & 1 & 1 \end{array}$$

Is it possible to find a value of t such that $p_{A|D} = 0.8$?

10. Suppose the probability distribution of (A, B, C, D) may be factorized along the DAG given in Figure 7.3.

All variables are binary, taking values 0 or 1. Suppose, after training examples, $p_A(1) = p_B(1) = 0.5$,

$$p_{D|C} = \begin{array}{c|cc} C\backslash D & 0 & 1 \\ \hline 0 & 0.6 & 0.4 \\ 1 & 0.2 & 0.8 \end{array}$$

Suppose

$$p_{C|A,B}(1|.,.) = \begin{array}{c|cc} A\backslash B & 0 & 1 \\ \hline 0 & 0 & 1-t \\ 1 & 1-s & 1-st \end{array}$$

(a) Compute the range of values that $p_{A|D}(1|1)$ can take.

(b) The *gradient descent method* is a standard numerical procedure. Suppose the parameters are t and a constraint (in the example below, $p_{A|D}(1|1) = 0.6$) has to be satisfied. If one wants to find parameter values t as close as possible to prescribed initial values t_0 according to a certain distance measure, one starts by choosing a *step length* δ. One then finds values of the parameters for which the constraints are satisfied. (In the example below, this can be done by computing s_1 and t_1 such that (s_0, t_1) and (s_1, t_0) satisfy the constraint and letting $(s^*, t^*) = (s_0, t_1)$ or (s_1, t_0), whichever gives the smallest value of $d((s, t), (s_0, t_0))$.

Next, if there are constraints, such as $f(t) = c$ (where t is the set of parameters), the gradient of the constraint $\nabla_t f(t)|_{t^*}$ is computed. From the *unit* vectors v such that $(v, \nabla_t f(t)|_{t^*}) = 0$ (that is $|v| = 1$), the one that gives the lowest value (i.e. negative number with largest absolute value) of $(v, d_t(t, t_0)|_{t=t^*})$ is chosen. The parameters are then updated; $t^* \to t^* + \delta v$. Then, compute the appropriate vector v and update again. Continue until the change in the distance $d(t^*, t_0)$ is less than a prescribed value.

Write a MATLAB code that employs a Gradient Descent Method, with step length 0.02, to compute the values of (s, t) as close as possible to prescribed initial values (s_0, t_0), which satisfies the constraint $p_{A|D}(1|1) = 0.6$, where the distance used is the Chan-Darwiche distance.

8

Graphical models and exponential families

This chapter introduces exponential families of distributions, focusing on links with convex analysis and specifically with the theory of conjugate duality. This is then applied to updating the probability distribution in a graphical model in the light of new information. One of the key features of exponential families, with mean parameters, is the relative ease with which the entropy and Kullback-Leibler distance between two members of the family may be computed.

8.1 Introduction to exponential families

The notations are as before, except that continuous random variables will also be considered for modelling the outcome of an experiment. Let $V = \{X_1, \ldots, X_d\}$ denote the random variables. For $j = 1, \ldots, d$, $\mathcal{X}_j$ will denote the state space for variable X_j. If X_j is continuous, then $\mathcal{X}_j \subseteq \mathbf{R}$ (the real numbers). If X_j is discrete, then $\mathcal{X}_j = \{x_j^{(1)}, \ldots, x_j^{(k_j)}\}$, where k_j is possibly $+\infty$. The notation $\underline{X} = (X_1, \ldots, X_d)$ denotes the row vector; when data is presented in a matrix, it is usual that each column presents a different attribute and each row represents an independent instantiation. An instantiation of $\underline{X}$ will be denoted $\underline{x} \in \mathcal{X}_1 \times \ldots \mathcal{X}_d \equiv \mathcal{X}$ (when no subscript is employed, $\mathcal{X}$ denotes the product space, which is the state space of the row vector $\underline{X}$).

An *exponential family* is a family of probability distributions satisfying certain properties, listed in Definition 8.1 below. For the purposes of Bayesian networks, the emphasis is on discrete variables.

Definition 8.1 (Exponential Family) *An* exponential family *is a family of probability distributions defined by a probability function* $p_{\underline{X}}(.|\underline{\theta})$ *if* $\underline{X}$ *are discrete variables, or a probability density function* $\pi_{\underline{X}}(.|\underline{\theta})$ *for continuous variables, indexed by a parameter set* $\tilde{\Theta} \subseteq \mathbf{R}^p$ *(where p is possibly infinite), where there is a function* $\Phi : \mathcal{X} \to \mathbf{R}^p$, *a function* $A : \tilde{\Theta} \to \mathbf{R}$ *and a function* $h : \mathcal{X} \to \mathbf{R}$ *such that*

$$p_{\underline{X}}(\underline{x}|\underline{\theta}) = \exp\{\langle \underline{\theta}, \Phi(\underline{x}) \rangle - A(\underline{\theta})\}h(\underline{x})$$

if $\underline{X}$ *is a discrete random vector and*

$$\pi_{\underline{X}}(\underline{x}|\underline{\theta}) = \exp\{\langle \underline{\theta}, \Phi(\underline{x}) \rangle - A(\underline{\theta})\}h(\underline{x})$$

if $\underline{X}$ *is a continuous random vector.*

It is convenient to use the notation $\mathcal{I}$ to denote the indexing set for the parameters; $\underline{\theta} = (\theta_\alpha)_{\alpha \in \mathcal{I}}$. Then Φ denotes a collection of functions $\Phi = (\phi_\alpha)_{\alpha \in \mathcal{I}}$, where $\phi_\alpha : \mathcal{X} \to \mathbf{R}$. The inner product notation is defined as

$$\langle \underline{\theta}, \Phi(\underline{x}) \rangle = \sum_{\alpha \in \mathcal{I}} \theta_\alpha \phi_\alpha(\underline{x}).$$

The parameters in the vector $\underline{\theta}$ *are known as the* canonical parameters *or* exponential parameters.

Attention will be restricted to distributions where $|\mathcal{I}| = p < +\infty$; namely, $\mathcal{I}$ has a finite number, p, of elements.

Since $\sum_{\mathcal{X}} p_{\underline{X}}(\underline{x}|\underline{\theta}) = 1$ for discrete variables and $\int_{\mathcal{X}} \pi_{\underline{X}}(\underline{x}|\underline{\theta})d\underline{x} = 1$ for continuous variables, it follows that the quantity A, known as the *log partition function*, is given by the expression

$$A(\underline{\theta}) = \log \int_{\mathcal{X}} \exp\{\langle \underline{\theta}, \Phi(\underline{x}) \rangle\}h(\underline{x})d\underline{x}$$

for continuous variables and

$$A(\underline{\theta}) = \log \sum_{\mathcal{X}} \exp\{\langle \underline{\theta}, \Phi(\underline{x}) \rangle\} h(\underline{x})$$

for discrete variables. It is assumed that h, $\underline{\theta}$ and Φ satisfy appropriate conditions so that A is finite.

Set

$$P(\underline{x}; \underline{\theta}) = \frac{p_{\underline{X}}(\underline{x}|\underline{\theta})}{h(\underline{x})}. \tag{8.1}$$

With the set of functions Φ fixed, each parameter vector $\underline{\theta}$ indexes a particular probability function $p_{\underline{X}}(.|\underline{\theta})$ belonging to the family. The exponential parameters of interest belong to the parameter space, which is the set

$$\tilde{\Theta} = \{\underline{\theta} \in \mathbf{R}^p | A(\underline{\theta}) < +\infty\}. \tag{8.2}$$

It will be seen shortly that A is a convex function of $\underline{\theta}$.

Definition 8.2 (Regular Families) *An exponential family for which the domain $\tilde{\Theta}$ of Equation (8.2) is an open set is known as a* regular *family.*

Attention will be restricted to regular families.

Definition 8.3 (Minimal Representation) *An exponential family, defined using a collection of functions Φ for which there is no linear combination $\langle \underline{a}, \Phi(\underline{x}) \rangle = \sum_{\alpha \in \mathcal{I}} a_\alpha \phi_\alpha(\underline{x})$ equal to a constant is known as a* minimal *representation.*

For a minimal representation, there is a unique parameter vector $\underline{\theta}$ associated with each distribution.

Definition 8.4 (Over-complete) *An* over-complete representation *is a representation that is not minimal; there is a linear combination of the elements of Φ which yields a constant.*

When the representation is over-complete, there exists an affine subset of parameter vectors $\underline{\theta}$, each associated with the same distribution.

Recall the definition of sufficiency, given in Definition 3.3. The following lemma is crucial. Its proof is left as an exercise.

Lemma 8.1 *Let $\underline{X} = (X_1, \ldots, X_d)$ be a random vector with joint probability function*

$$p_{\underline{X}}(\underline{x}|\underline{\theta}) = \exp\{\langle \underline{\theta}, \Phi(\underline{x}) \rangle - A(\underline{\theta})\}h(\underline{x}), \qquad \underline{X} \in \mathcal{X}$$

then $\Phi(\underline{X})$, which will be denoted Φ, is a Bayesian sufficient statistic for $\underline{\theta}$. If the representation is minimal, then $\Phi(\underline{X})$ is a minimal sufficient statistic for $\underline{\theta}$.

Proof of Lemma 8.1 See Exercise 2, Chapter 8. $\qquad\qquad\qquad\qquad\qquad\qquad$ □

8.2   Standard examples of exponential families

The purpose of this section is to take some basic distributions, which are well known, and illustrate that they satisfy the definition of an exponential family.

Bernoulli   Consider the random variable X, taking values 0 or 1, with probability function $p_X(1) = p$, $p_X(0) = 1 - p$. This may be written as

$$p_X(x) = \begin{cases} p^x(1-p)^{1-x} & x \in \{0, 1\} \\ 0 & \text{other } x. \end{cases}$$

Then

$$p_X(x) = \exp\left\{ x \log\left(\frac{p}{1-p}\right) + \log(1-p) \right\}$$
$$= \exp\{x\theta + \log(1-p)\}$$
$$= \exp\{x\theta - \log(1 + e^\theta)\},$$

where $\theta = \log\left(\frac{p}{1-p}\right)$.

Note: Change of notation   In Chapter 1.9, the quantity now denoted by p was denoted by θ. The quantity θ no longer denotes a probability; it now denotes the *canonical parameter*.

In the language of exponential families, $\mathcal{X} = \{0, 1\}$, $\Phi = \{\phi\}$ where $\phi(x) = x$, $h(0) = h(1) = 1$,

$$p_X(0|\theta) = e^{-A(\theta)}, \quad p_X(1|\theta) = e^{\theta - A(\theta)}$$

In other words

$$\log p_X(x|\theta) = \theta x - A(\theta),$$

which gives

$$1 = p_X(0|\theta) + p_X(1|\theta) = e^{-A(\theta)}(1 + e^{\theta})$$

so that

$$A(\theta) = \log(1 + \exp\{\theta\}).$$

Gaussian   Recall that the one-dimensional Gaussian density is of the form

$$\pi(x|\mu, \sigma) = \frac{1}{\sqrt{2\pi}\sigma} \exp\left\{-\frac{(x - \mu)^2}{2\sigma^2}\right\}.$$

This may be expressed in terms of an exponential family as follows: $\mathcal{X} = \mathbf{R}$, $h(x) = 1$, $\Phi = \{\phi_1, \phi_2\}$ where $\phi_1(x) = x$ and $\phi_2(x) = -x^2$.

$$\log \pi(x|\underline{\theta}) = \theta_1 x - \theta_2 x^2 - A(\underline{\theta})$$

where

$$1 = e^{-A(\underline{\theta})} \int_{-\infty}^{\infty} e^{\theta_1 x - \theta_2 x^2} dx.$$

The partition function is therefore

$$A(\underline{\theta}) = \frac{1}{2}\log \pi - \frac{1}{2}\log \theta_2 + \frac{\theta_1^2}{4\theta_2^2}$$

and the parameter space is

$$\tilde{\Theta} = \{(\theta_1, \theta_2) \in \mathbf{R}^2 | \theta_2 > 0\}.$$

Note that in the 'usual' notation

$$\theta_1 = \frac{\mu}{\sigma^2}, \qquad \theta_2 = \frac{1}{\sigma^2}.$$

Exponential   Recall that an exponential density is of the form

$$\pi(x|\lambda) = \begin{cases} \lambda e^{-\lambda x} & x \geq 0 \\ 0 & x < 0. \end{cases}$$

This is an exponential family, taking $\mathcal{X} = (0, +\infty)$, $h(x) = 1$, $\Phi = \phi$, where $\phi(x) = -x$, $\theta = \lambda$, so that $e^{-A(\theta)} = \theta$, yielding $A(\theta) = -\log \theta$, $\Theta = (0, +\infty)$.

Poisson   Recall that the probability function p for a Poisson distribution with parameter μ is given by

$$p(x|\mu) = \frac{\mu^x}{x!} e^{-\mu}, \qquad x = 0, 1, 2, \ldots$$

This is an exponential family with $h(x) = \frac{1}{x!}$, $\theta = \log \mu$ so that $p(x|\mu) = P(x; \theta)h(x)$, where

$$P(x; \theta) = e^{x\theta - e^\theta}.$$

This gives $A(\theta) = \exp\{\theta\}$. Since $\mu \geq 0$ and $\theta = \log \mu$, it follows that $\tilde{\Theta} = \mathbf{R}$.

Beta   Recall that the probability density function for a Beta distribution is given by

$$\pi(x|\alpha, \beta) = \begin{cases} \frac{\Gamma(\alpha+\beta)}{\Gamma(\alpha)\Gamma(\beta)} x^{\alpha-1}(1-x)^{\beta-1} & x \in [0, 1] \\ 0 & \text{other } x. \end{cases}$$

This is an exponential family, with $\mathcal{X} = (0, 1)$, $h \equiv 1$, $\alpha - 1 = \theta_1$, $\beta - 1 = \theta_2$, $\Phi = \{\phi_1, \phi_2\}$ where $\phi_1(x) = \log x$, $\phi_2(x) = \log(1-x)$. Then

$$\log \pi(x|\underline{\theta}) = \theta_1 \log x + \theta_2 \log(1-x) - A(\underline{\theta}),$$

where the partition function A is given by

$$A(\underline{\theta}) = \log \Gamma(\theta_1 + 1) + \log \Gamma(\theta_2 + 1) - \log \Gamma(\theta_1 + \theta_2 + 2)$$

and the parameter space is $\tilde{\Theta} = (-1, \infty)^2$.

8.3 Graphical models and exponential families

The scalar examples described in Section 8.2 serve as building blocks for the construction of exponential families, which have an underlying graphical structure.

Example 8.1: Sigmoid belief network model   The *sigmoid belief network model*, described below, was introduced by R. Neal in [112]. It is an exponential family, with an underlying graphical structure.

Consider a directed acyclic graph $\mathcal{G} = (V, E)$, where $V = \{X_1, \ldots, X_d\}$ is the set of variables, along which the probability distribution of $\underline{X} = (X_1, \ldots, X_d)$ may be factorized. Suppose that for each $X_j \in V$, $j = 1, \ldots, d$, the random variable X_j takes values 0 or 1, each with probability $1/2$. For any two components X_s and X_t of the random vector $\underline{X}$, component X_s has a direct causal effect on X_t only if $(X_s, X_t) \in E$.

The following notation will be used:

$$\tilde{V} = \{1, \ldots, d\}, \qquad \tilde{E} = \{(s, t)|(X_s, X_t) \in E\}.$$

The probability distribution over the possible configurations is modelled by an exponential family with probability function $p_{\underline{X}}(.|\underline{\theta})$ of the form

$$p_{\underline{X}}(x|\underline{\theta}) = \exp\left\{ \sum_{s=1}^{d} \theta_s x_s + \sum_{(s,t)\in\tilde{E}} \theta_{(s,t)} x_s x_t - A(\underline{\theta}) \right\}.$$

The notation Π_i denotes the parent set of node X_i and $\pi_i(\underline{x})$ denotes the instantiation of Π_i corresponding to the instantiation $\{\underline{X} = \underline{x}\}$, this may be rewritten as

$$p_{\underline{X}}(\underline{x}|\underline{\theta}) = \prod_{i=1}^{d} p_{X_i|\Pi_i}(x_i|\pi_i(\underline{x}), \underline{\theta}),$$

where (clearly)

$$p_{X_i|\Pi_i}(x_i|\pi_i(\underline{x}), \underline{\theta}) = \frac{\exp\left\{x_i\left(\theta_i + \sum_{x_j \in \pi_i(\underline{x})} \theta_{(ij)}x_j\right)\right\}}{1 + \exp\left\{\theta_i + \sum_{x_j \in \pi_i(\underline{x})} \theta_{(ij)}x_j\right\}},$$

where the notation $x_j \in \pi_i(\underline{x})$ is clear. The index set is $\mathcal{I} = \tilde{V} \cup \tilde{E}$. The domain $\Theta = \mathbf{R}^n$, where $n = |\mathcal{I}|$. Since the sum that defines $A(\underline{\theta})$ is finite for all $\underline{\theta} \in \mathbf{R}^n$, it follows that the family is regular. It is *minimal*, since there is no linear combination of the functions equal to a constant.

This model may be generalized. For example, one may consider higher order interactions. To include coupling of triples (X_s, X_t, X_u), one would add a monomial $x_s x_t x_u$ with corresponding exponential parameter $\theta_{(s,t,u)}$. More generally, the set $\mathcal{C}$ of indices of interacting variables may be considered, giving

$$p_{\underline{X}}(\underline{x}|\underline{\theta}) = \exp\left\{\sum_{C \in \mathcal{C}} \theta_{(C)} \prod_{s \in C} X_s - A(\underline{\theta})\right\}.$$

8.4   Noisy 'or' as an exponential family

This section outlines some ideas of propositional logic in the presence of noise. The noisy 'or' gate is a device that reduces the sizes of the conditional probability tables and can hence lead to sharper probabilistic inference. The section ends with an example illustrating that the basic structures of the noisy 'or' gate can be expressed as exponential families.

Disjunction in propositional logic   In logic, the 'or' disjunction of two propositions p and q is denoted by $p \vee q$ and is defined by the truth table

p	q	$p \vee q$
1	1	1
1	0	1
0	1	1
0	0	0

Here $1 =$ the proposition is true, $0 =$ the proposition is false. For example, if p and q are the causes of some effect (e.g. a sore throat) and the presence of either or both of them will make the effect occur.

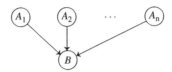

Figure 8.1 A logical 'or' gate.

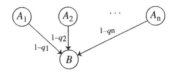

Figure 8.2 Noisy 'or' junction.

Noisy 'or' as a causal network Consider the DAG given in Figure 8.1 where $B = A_1 \vee A_2 \vee \ldots \vee A_n$. This is the logical 'or' and there is no noise.

The noise then enters, as in the DAG given in Figure 8.2, by considering that if any of the variables A_i, $i = 1, \ldots, n$ is present, then B is present *unless something has inhibited it*.

Noisy 'or': inhibitors Consider the DAG in Figure 8.2, where q_i denotes the probability that the impact of A_i is inhibited.

All variables are binary, and take value 1 if the cause, or effect, is present and 0 otherwise. In other words, $p_{B|A_i}(0|1) = q_i$. The assumption from the DAG is that all the inhibitors are independent. This implies that

$$p_{B|A_1,\ldots,A_n}(0|a_1, \ldots, a_n) = \prod_{j \in Y} q_j,$$

where $Y = \{j \in \{1, \ldots, n\}|a_j = 1\}$. This may be described by a *noisy 'or' gate*.

Noisy 'or' gate The noisy 'or' can be modelled directly, introducing the variables B_i $i = 1, \ldots, n$, where B_i takes the value 1 if the cause A_i is on and it is not inhibited and 0 otherwise. The corresponding DAG is given in Figure 8.3.

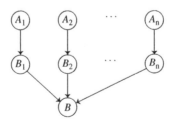

Figure 8.3 Noisy 'Or' Gate.

where

$$p_{B|B_1,\ldots,B_n}(1|b_1, \ldots, b_n) = b_1 \vee \ldots \vee b_n.$$

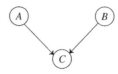

Figure 8.4   A collider connection.

The $B_1, \ldots, B_n$ are introduced as mutually independent inhibitors, and

$$p_{B_i | A_i}(0 | 1) = q_i,$$

yielding the result given above.

Example 8.2   Consider the collider connection given in Figure 8.4.

Suppose that the variables A, B and C are binary, taking values 0 or 1 and that the conditional probabilities of C given $\{A = 1\}$ and $\{B = 1\}$ are $p_{C|A}(1|1) = 0.7$ and $p_{C|B}(1|1) = 0.8$. Suppose, furthermore, that they satisfy the assumptions of the 'noisy or' gate; namely, that $\{A = 1\}$ causes $\{C = 1\}$ unless an inhibitor prevents it; this inhibitor functions with probability 0.3, that $\{B = 1\}$ causes $\{C = 1\}$ unless an inhibitor prevents it; this inhibitor functions with probability 0.2. Then this network may be considered as a 'noisy or' gate, with additional variables X and Y,

$$p_{C|X,Y}(1|1, 1) = p_{C|X,Y}(1|1, 0) = p_{C|X,Y}(1|0, 1) = 1, \quad p_{C|X,Y}(1|0, 0) = 0$$

and

$$p_{X|A}(1|1) = 0.7, \quad p_{X|A}(1|0) = 0, \quad p_{Y|B}(1|1) = 0.8, \quad p_{Y|B}(1|0) = 0.$$

Example 8.3: Noisy 'or' as an exponential family   The QMR–DT (Quick Medical Reference–Decision Theoretic) database is a large scale probabilistic data base that is intended to be used as a diagnostic aid in the domain of internal medicine. It is a *bipartite* graphical model; that is, a graphical model where the nodes may be of one of two types. The upper layer of nodes (the parents) represent diseases and the lower layer of nodes represent symptoms. There are approximately 600 disease nodes and 4000 symptom nodes in the database.

An *evidence*, or *finding*, will be a set of observed symptoms, denoted by a vector of length 4000, each entry being a 1 or 0 depending upon whether or not the symptom is present or absent. This will be denoted f, which is an instantiation of the random vector $\underline{F}$. The vector $\underline{d}$ will be used to represents the diseases; this is considered as an instantiation of the random vector $\underline{D}$. Let d_j denote component j of vector $\underline{d}$ and let f_j denote component j of vector $\underline{f}$. Then, if the occurrence of various diseases are taken to be independent of each other, the following factorization holds:

$$p_{\underline{F},\underline{D}}(\underline{f}, \underline{d}) = p_{\underline{F}|\underline{D}}(\underline{f}|\underline{d}) p_{\underline{D}}(\underline{d}) = \prod_i p_{F_i|\underline{D}}(f_i|\underline{d}) \prod_j p_{D_j}(d_j).$$

This may be represented by a *noisy 'or'* model. Let q_{i0} denote the probability that symptom i is present in the absence of any disease and q_{ij} the probability that disease j induces symptom i, then the probability that symptom i is absent, given a vector of diseases $\underline{d}$ is

$$p_{F_i|\underline{D}}(0|\underline{d}) = (1 - q_{i0}) \prod_j (1 - q_{ij})^{d_j}.$$

The noisy 'or' may then be rewritten in an exponential form:

$$p_{F_i|\underline{D}}(0|\underline{d}) = \exp\left\{ -\sum_j \theta_{ij} d_j - \theta_{i0} \right\},$$

where $\theta_{ij} \equiv \log(1 - q_{ij})$ are the transformed parameters.

8.5   Properties of the log partition function

Firstly, some basic properties of the log partition function $A(\underline{\theta})$ are discussed, which are then developed using convex analysis, discussed in [113]. Let $E_{\underline{\theta}}[.]$ denote expectation with respect to $p(.|\underline{\theta})$ for discrete variables, or $\pi(.|\underline{\theta})$ for continuous variables. Of particular importance is the idea that the vector μ, where $\mu_i := E_{\underline{\theta}}[\phi_i(\underline{X})]$ provides an alternative parametrisation of the exponential family. Here expectation is defined as

$$E_{\underline{\theta}}[f(\underline{X})] = \int_{\mathcal{X}} \pi_{\underline{X}}(\underline{x}|\underline{\theta}) f(\underline{x}) d\underline{x}$$

if $\underline{X}$ is a continuous random vector and

$$E_{\underline{\theta}}[f(\underline{X})] = \sum_{\underline{x} \in \mathcal{X}} p_{\underline{X}}(\underline{x}|\underline{\theta}) f(\underline{x})$$

if $\underline{X}$ is a discrete random vector. Recall that, for discrete variables,

$$A(\underline{\theta}) = \log \sum_{\underline{x} \in \mathcal{X}} e^{\langle \underline{\theta}, \Phi(\underline{x}) \rangle} h(\underline{x}). \tag{8.3}$$

Provided expectations and variances exist, it follows that

$$\frac{\partial}{\partial \theta_\alpha} A(\underline{\theta}) = \sum_{\underline{x} \in \mathcal{X}} e^{\langle \underline{\theta}, \Phi(\underline{x}) \rangle - A(\underline{\theta})} \phi_\alpha(\underline{x}) h(\underline{x}) = E_{\underline{\theta}}[\phi_\alpha(\underline{X})]. \tag{8.4}$$

Taking second derivatives yields

$$\frac{\partial}{\partial \theta_\alpha \partial \theta_\beta} A(\underline{\theta}) = E_{\underline{\theta}}[\phi_\alpha(\underline{X})\phi_\beta(\underline{X})] - E_{\underline{\theta}}[\phi_\alpha(\underline{X})]E_{\underline{\theta}}[\phi_\alpha(\underline{X})] = \text{Cov}_{\underline{\theta}}(\phi_\alpha(\underline{X}), \phi_\beta(\underline{X})).$$

It is and easy to show, and a standard fact, that any covariance matrix is non-negative definite. It now follows that, on $\tilde{\Theta}$, A is a convex function.

Mapping to mean parameters   Given a vector of functions Φ, set $F(\theta) = E_\theta[\Phi(\underline{X})]$ and let $\mathcal{M} = F(\tilde{\Theta})$. For an arbitrary exponential family defined by

$$p_{\underline{X}}(\underline{x}|\underline{\theta}) = \exp\left\{\langle\underline{\theta}, \Phi(\underline{x})\rangle - A(\underline{\theta})\right\} h(\underline{x}),$$

a mapping $\Lambda : \tilde{\Theta} \rightarrow \mathcal{M}$ may be defined as follows:

$$\Lambda(\underline{\theta}) := E_{\underline{\theta}}[\Phi(\underline{X})].$$

To each $\underline{\theta} \in \tilde{\Theta}$, the mapping Λ associates a vector of *mean parameters* $\underline{\mu} = \Lambda(\underline{\theta})$ belonging to the set $\mathcal{M}$. Note that, by Equation (8.4),

$$\Lambda(\underline{\theta}) = \nabla A(\underline{\theta}).$$

The mapping Λ is one to one, and hence invertible on its image, when the representation is minimal. The image of $\tilde{\Theta}$ is the interior of $\mathcal{M}$.

Example 8.4   Consider a Bernoulli random variable X with state space $\{0, 1\}$. That is, $p_X(0) = 1 - p$ and $p_X(1) = p$. Now consider an over-complete exponential representation

$$p_X(x|\underline{\theta}) = \exp\left\{\theta_0(1 - x) + \theta_1 x - A(\theta_0, \theta_1)\right\}$$

so that

$$A(\theta_0, \theta_1) = \log\left(e^{\theta_0} + e^{\theta_1}\right).$$

Here $\tilde{\Theta} = \mathbf{R}^2$. $\phi_0(x) = 1 - x$ and $\phi_1(x) = x$.

$$\frac{\partial}{\partial\theta_0} A(\underline{\theta}) = e^{\theta_0 - A(\theta_0, \theta_1)} = 1 - p = \mu_0$$

$$\frac{\partial}{\partial\theta_1} A(\underline{\theta}) = e^{\theta_1 - A(\theta_0, \theta_1)} = p = \mu_1.$$

The set $\mathcal{M}$ of *mean parameters* is the simplex $\{(\mu_0, \mu_1) \in \mathbf{R}_+ \times \mathbf{R}_+ | \mu_0 + \mu_1 = 1\}$. For any fixed $\underline{\mu} = (\mu_0, \mu_1)$ where $\mu_0 \geq 0$, $\mu_1 \geq 0$, $\mu_0 + \mu_1 = 1$, the inverse image is,

$$\Lambda^{-1}(\underline{\mu}) = \left\{(\theta_0, \theta_1) \in \mathbf{R}^2 \,\middle|\, \frac{e^{\theta_0}}{e^{\theta_0} + e^{\theta_1}} = \mu_0\right\}$$

which may be rewritten as

$$\Lambda^{-1}(\underline{\mu}) = \left\{(\theta_0, \theta_1) \in \mathbf{R}^2 \,\middle|\, \theta_1 - \theta_0 = \log\frac{\mu_1}{\mu_0}\right\}.$$

In an over-parametrized, or over-complete representation, there is no longer a bijection between $\tilde{\Theta}$ and $\Lambda(\tilde{\Theta})$. Instead, there is a bijection between elements of $\Lambda(\tilde{\Theta})$ and affine subsets of $\tilde{\Theta}$. A pair $(\underline{\theta}, \underline{\mu})$ is said to be *dually coupled* if $\underline{\mu} = \Lambda(\underline{\theta})$, and hence $\underline{\theta} \in \Lambda^{-1}(\underline{\mu})$.

8.6   Fenchel Legendre conjugate

The *Fenchel Legendre conjugate* of the log partition function A is defined as follows:

$$A^*(\underline{\mu}) := \sup_{\underline{\theta} \in \tilde{\Theta}} \left\{ \langle \underline{\mu}, \underline{\theta} \rangle - A(\underline{\theta}) \right\}. \tag{8.5}$$

The choice of notation is deliberately suggestive; the variables in the Fenchel Legendre dual turn out to have interpretation as the mean parameters. Recall the definition of P given by Equation (8.1); namely, if $p_{\underline{X}}(\underline{x}|\underline{\theta})$ is the probability function (or density function), then

$$P(\underline{x}; \underline{\theta}) = \frac{p_{\underline{X}}(\underline{x}|\underline{\theta})}{h(\underline{x})}.$$

Definition 8.5 (Boltzmann-Shannon Entropy) *The Boltzmann-Shannon entropy of $p_{\underline{X}}(\underline{x}|\underline{\theta})$ with respect to h is defined as*

$$H(p_{\underline{X}}(\underline{x}|\underline{\theta})) = -E_{\underline{\theta}}[\log P(\underline{x}; \underline{\theta})].$$

The following is the main result of the chapter.

Theorem 8.1 *For any $\underline{\mu} \in \mathcal{M}$, let $\underline{\theta}(\underline{\mu}) \in \Lambda^{-1}(\underline{\mu})$. Then*

$$A^*(\underline{\mu}) = -H(p_{\underline{X}}(\underline{x}|\underline{\theta}(\underline{\mu})).$$

In terms of this dual, for $\underline{\theta} \in \tilde{\Theta}$, the log partition satisfies be expressed:

$$A(\underline{\theta}) = \sup_{\underline{\mu} \in \mathcal{M}} \{ \langle \underline{\theta}, \underline{\mu} \rangle - A^*(\underline{\mu}) \}. \tag{8.6}$$

Proof of Theorem 8.7 From the definition $\underline{\mu} = E_{\underline{\theta}}[\Phi(\underline{X})]$, it follows that

$$-H(p_{\underline{X}}(\underline{x}|\underline{\theta})) = E_{\underline{\theta}}[\log P(\underline{X}; \underline{\theta})] = E_{\underline{\theta}}[\langle \underline{\theta}, \Phi(\underline{X}) \rangle] - A(\underline{\theta}) = \langle \underline{\theta}, \underline{\mu} \rangle - A(\underline{\theta}). \tag{8.7}$$

Consider the function

$$F(\underline{\mu}, \underline{\theta}) = \langle \underline{\mu}, \underline{\theta} \rangle - A(\underline{\theta}).$$

Let $\underline{\theta}(\underline{\mu})$ denote a value of $\underline{\theta}$ that maximizes $F(\underline{\mu}, \underline{\theta})$ if such a value exists in $\tilde{\Theta}$. The result follows directly by using the definition given by Equation (8.5) together with Equation (8.7). Otherwise, let $\underline{\theta}^{(n)}(\underline{\mu})$ denote a sequence such that $\lim_{n \to +\infty} F(\underline{\mu}, \underline{\theta}^{(n)}(\underline{\mu})) = A^*(\underline{\mu})$. The first statement of the theorem follows directly from this.

For the second part, choose $\underline{\theta} \in \tilde{\Theta}$ and choose $\underline{\mu}(\underline{\theta}) = \nabla_{\underline{\theta}} A(\underline{\theta})$. By the definition of $\mathcal{M}$, note that $\underline{\mu}(\underline{\theta}) \in \mathcal{M}$. Since A is convex, it follows that $\underline{\mu}(\underline{\theta})$ maximizes $\langle \underline{\theta}, \underline{\mu} \rangle - A(\underline{\theta})$, so that

$$A(\underline{\theta}) = \langle \underline{\mu}(\underline{\theta}), \underline{\theta} \rangle - A^*(\underline{\mu}(\underline{\theta})).$$

But, from the definition of $A^*(\underline{\mu})$, it follows that for all $\underline{\mu} \in \mathcal{M}$,

$$A(\underline{\theta}) \geq \langle \underline{\mu}, \underline{\theta} \rangle - A^*(\underline{\mu}).$$

From this,

$$A(\underline{\theta}) = \sup_{\mu \in \mathcal{M}} \{\langle \underline{\mu}, \underline{\theta} \rangle - A^*(\underline{\mu})\}$$

and Theorem 8.1 is established. □

The conjugate dual pair (A, A^*) for the families of exponential variables given before are now given.

Bernoulli   Recall that $A(\theta) = \log(1 + \exp\{\theta\})$ for $\theta \in \mathbf{R}$. It follows that

$$A^*(\mu) = \sup_{\theta \in \mathbf{R}}\{\theta\mu - \log(1 + e^{\theta})\}$$

The supremum is attained for $\theta(\mu)$ satisfying

$$\mu = \frac{e^{\theta(\mu)}}{1 + e^{\theta(\mu)}}.$$

It follows that

$$e^{\theta(\mu)} = \frac{\mu}{1 - \mu}$$

and

$$\theta(\mu) = \log \mu - \log(1 - \mu)$$

so that

$$A^*(\mu) = \mu \log \mu - \mu \log(1 - \mu) - \log(1 + \frac{\mu}{1 - \mu}),$$

which gives

$$A^*(\mu) = \mu \log \mu + (1 - \mu) \log(1 - \mu).$$

Gaussian   Recall that $\tilde{\Theta} = \{(\theta_1, \theta_2) | \theta_2 > 0\}$ and

$$A(\underline{\theta}) = \frac{1}{2} \log \pi - \frac{1}{2} \log \theta_2 + \frac{\theta_1^2}{4\theta_2}.$$

$$A^*(\underline{\mu}) = \sup_{\underline{\theta} \in \tilde{\Theta}}\{\theta_1 \mu_1 + \theta_2 \mu_2 - \frac{1}{2} \log \pi + \frac{1}{2} \ln \theta_2 - \frac{\theta_1^2}{4\theta_2}\}.$$

This is maximized when

$$\begin{cases} \mu_1 - \frac{\theta_1(\mu)}{2\theta_2(\mu)} = 0 \\ \mu_2 + \frac{1}{2\theta_2(\mu)} + \frac{\theta_1^2(\mu)}{4\theta_2^2(\mu)} = 0, \end{cases}$$

which gives

$$\begin{cases} \theta_2(\mu_1, \mu_2) = -\frac{1}{2(\mu_1^2 + \mu_2)} \\ \theta_1(\mu) = -\frac{\mu_1}{\mu_1^2 + \mu_2} \end{cases}$$

and
$$A^*(\mu_1, \mu_2) = -\frac{1}{2} - \frac{1}{2} \log \pi - \frac{1}{2} \log(-2(\mu_1^2 + \mu_2)).$$

Note that
$$\mathcal{M} = \{(\mu_1, \mu_2) | \mu_1^2 + \mu_2 < 0\}.$$

Exponential distribution Recall that $\tilde{\Theta} = (0, +\infty)$ and that $A(\theta) = -\log(\theta)$. By a straightforward computation,

$$A^*(\mu) = -1 - \log(-\mu)$$

and
$$\mathcal{M} = (-\infty, 0).$$

Poisson distribution Recall that $\tilde{\Theta} = \mathbf{R}$ and that $A(\theta) = \exp\{\theta\}$. It is a straightforward computation to see that
$$A^*(\mu) = \mu \log \mu - \mu$$

and that
$$\mathcal{M} = (0, +\infty).$$

8.7 Kullback-Leibler divergence

Recall Definition 5.4, the Kullback-Leibler distance between two probability distributions p and q over a finite state space $\mathcal{X}$;

$$D_{KL}(q|p) = \sum_{x \in \mathcal{X}} q(\{x\}) \log \frac{q(\{x\})}{p(\{x\})}.$$

When $\mathcal{X} = \mathbf{Z}^d$ and p and q are two probability functions, this may be written as

$$D_{KL}(q|p) = E_q \left[\log \frac{q(\underline{X})}{p(\underline{X})} \right], \tag{8.8}$$

where $\underline{X}$ is a random vector taking values in $\mathbf{Z}^d$ and E_q denotes expectation with respect the probability distribution by q (i.e. q is the probability function of $\underline{X}$). The definition of Kullback-Leibler may be extended to continuous distributions using Equation (8.8), where q and p denote the respective density functions. In this case, Equation (8.8) is taken as

$$D_{KL}(q|p) = \int_{\mathbf{R}^d} q(\underline{x}) \log \frac{q(\underline{x})}{p(\underline{x})} d\underline{x}.$$

When q and p are members of the same exponential family, the Kullback-Leibler divergence may be computed in terms of the parameters. The key result, for expressing the divergence in terms of the partition function, is *Fenchel's inequality* given in Equation (8.9), which can be seen directly from the definition of $A^*(\underline{\mu})$.

$$A(\underline{\theta}) + A^*(\underline{\mu}) \geq \langle \underline{\mu}, \underline{\theta} \rangle, \tag{8.9}$$

with equality if and only if $\underline{\mu} = \Lambda(\underline{\theta})$ and $\underline{\theta} \in \Lambda^{-1}(\underline{\mu})$. That is, for $\underline{\mu} = \Lambda(\underline{\theta})$ and $\underline{\theta} \in \Lambda^{-1}(\underline{\mu})$,

$$A(\underline{\theta}) + A^*(\underline{\mu}) = \langle \underline{\mu}, \underline{\theta} \rangle. \tag{8.10}$$

Consider an exponential family of distributions, and consider two exponential parameter vectors, $\underline{\theta}_1 \in \tilde{\Theta}$ and $\underline{\theta}_2 \in \tilde{\Theta}$. When distributions are from the same exponential family, the notation $D(\underline{\theta}_1|\underline{\theta}_2)$ is used to denote $D_{KL}(p(.|\theta_1)|p(.|\theta_2))$. Set $\underline{\mu}_i = \Lambda(\underline{\theta}_i)$. Using the parameter to denote the distribution with respect to which the expectation is taken, note that

$$D(\underline{\theta}_1|\underline{\theta}_2) = E_{\theta_1}\left[\log \frac{p(\underline{X}|\theta_1)}{p(\underline{X}|\theta_2)}\right] = A(\underline{\theta}_2) - A(\underline{\theta}_1) - \langle \underline{\mu}_1, \underline{\theta}_2 - \underline{\theta}_1 \rangle. \tag{8.11}$$

The representation of the Kullback-Leibler divergence given in Equation (8.11) is known as the *primal form* of the Kullback-Leibler divergence.

Taking $\underline{\mu}_1 = \Lambda(\underline{\theta}_1)$ and applying Equation (8.10), the Kullback-Leibler divergence may also be written

$$D(\underline{\theta}_1|\underline{\theta}_2) \equiv \tilde{D}(\underline{\mu}_1|\underline{\theta}_2) = A(\underline{\theta}_2) + A^*(\underline{\mu}_1) - \langle \underline{\mu}_1, \underline{\theta}_2 \rangle. \tag{8.12}$$

The representation given in Equation (8.12) is known as the *mixed form* of the Kullback-Leibler divergence. Recall the definition of A^* given by

$$A^*(\underline{\mu}) := \sup_{\underline{\theta} \in \tilde{\Theta}}\{\langle \underline{\mu}, \underline{\theta} \rangle - A(\underline{\theta})\}$$

and recall Equation (8.6) from theorem 8.1,

$$A(\underline{\theta}) = \sup_{\underline{\mu} \in M}\{\langle \underline{\theta}, \underline{\mu} \rangle - A^*(\underline{\mu})\}.$$

Equation (8.6) may be rewritten as

$$\inf_{\underline{\mu} \in M}\{A(\underline{\theta}) + A^*(\underline{\mu}) - \langle \underline{\theta}, \underline{\mu} \rangle\} = 0.$$

It follows that $\inf_{\underline{\mu} \in M} \tilde{D}(\underline{\mu}|\underline{\theta}) = 0$.

Finally, taking $\underline{\mu}_2 = \Lambda(\underline{\theta}_2)$ and applying Equation (8.10) once again to Equation (8.12) yields the so-called *dual form* of the Kullback-Leibler divergence:

$$\tilde{D}(\underline{\mu}_1|\underline{\mu}_2) \equiv D(\underline{\theta}_1|\underline{\theta}_2) = A^*(\underline{\mu}_1) - A^*(\underline{\mu}_2) - \langle \underline{\theta}_2, \underline{\mu}_1 - \underline{\mu}_2 \rangle. \tag{8.13}$$

8.8 Mean field theory

In this section, probability distributions of the form

$$p_{\underline{X}}(\underline{x}|\underline{\theta}) = \exp\left\{\sum_\alpha \theta_\alpha \phi_\alpha(\underline{x}) - A(\underline{\theta})\right\} h(\underline{x})$$

are considered. Mean field theory techniques are discussed and it is shown how they may be used to obtain estimates of the log partition function $A(\underline{\theta})$. This is equivalent to the problem of finding an appropriate normalizing constant to make a function into a probability density, a problem that often arises when updating using Bayes' rule.

Mean field theory is based on the variational principle of Equation (8.6). The two fundamental difficulties associated with the variational problem are the nature of the constraint set $\mathcal{M}$ and the lack of an explicit form for the dual function A^*. Mean field theory entails limiting the optimization to a subset of distributions for which A^* is relatively easy to characterize.

More specifically, the discussion in this chapter is restricted to the case where the functions ϕ_α are either linear or quadratic. The problem therefore reduces to considering a graph $\mathcal{G} = (V, E)$, where the node set V denotes the variables and the edge set E denotes a direct association between the variables. For this discussion, the edges in E are assumed to be undirected. As usual, $V = \{X_1, \ldots, X_d\}$. Let $\tilde{V} = \{1, \ldots, d\}$ denote the indexing set and let $\tilde{E} = \{(s, t)|(X_s, X_t) \in E\}$. Specifically, the probability distributions under consideration are of the form

$$p_{\underline{X}}(\underline{x}|\underline{\theta}) = \exp\left\{\sum_{s\in\tilde{V}} \theta_s x_s + \sum_{(s,t)\in\tilde{E}} \theta_{(s,t)} x_s x_t - A(\underline{\theta})\right\}.$$

Let H denote a sub-graph of $\mathcal{G}$ over which it is feasible to perform exact calculations. In an exponential formulation, the set of all distributions that respect the structure of H can be represented by a linear subspace of the exponential parameters. Let $\mathcal{I}(H)$ denote the subset of indices associated with cliques in H. Then the set of exponential parameters corresponding to distributions structured according to H is given by

$$\mathcal{E}(H) := \left\{\underline{\theta} \in \tilde{\Theta} \mid \theta_\alpha = 0, \ \alpha \in \mathcal{I}\backslash\mathcal{I}(H)\right\}.$$

The simplest example is to consider the completely disconnected graph $H = (V, \phi)$. Then

$$\mathcal{E}(H) = \left\{\underline{\theta} \in \tilde{\Theta} \mid \theta(s, t) = 0, \ (s, t) \in E\right\}.$$

The associated distributions are of the product form

$$p_{\underline{X}}(\underline{x}|\underline{\theta}) = \prod_{s\in\tilde{V}} p_{X_s}(x_s|\theta_s).$$

Optimization and lower bounds Let $p_{\underline{X}}(\underline{x}|\underline{\theta})$ denote the *target distribution* that is to be approximated. The basis of mean field approximation is the following: any valid mean parameter specifies a lower bound on the log partition function, established using Jensen's inequality.

Proposition 8.1 (Mean Field Lower Bound)

$$A(\underline{\theta}) \geq \sup_{\mu \in \mathcal{M}} \left\{ \langle \underline{\theta}, \underline{\mu} \rangle - A^*(\underline{\mu}) \right\}$$

Proof of Proposition 8.1 The proof is given for discrete variables; the proof for continuous variables is exactly the same, replacing the sum with an integral.

$$A(\underline{\theta}) = \log \sum_{\underline{x} \in \mathcal{X}} \exp\{\langle \underline{\theta}, \Phi(\underline{x}) \rangle\}$$

$$= \log \sum_{\underline{x} \in \mathcal{X}} p_{\underline{X}}(\underline{x}|\underline{\theta}) \exp\{\langle \underline{\theta}, \Phi(\underline{X}) \rangle - \log p_{\underline{X}}(\underline{x}|\underline{\theta})\}$$

$$= \log E_{\underline{\theta}}[\exp\{\langle \underline{\theta}, \Phi(\underline{X}) \rangle - \log p_{\underline{X}}(\underline{X}|\underline{\theta})\}]$$

$$\overset{(a)}{\geq} \langle \underline{\theta}, E_{\underline{\theta}}[\Phi(\underline{X})] \rangle - E_{\underline{\theta}}[\log p_{\underline{X}}(\underline{X}|\underline{\theta})\}]$$

$$= \langle \underline{\theta}, \underline{\mu} \rangle - A^*(\underline{\mu}).$$

The inequality (a) follows from Jensen's inequality; the last line follows from Theorem 8.1. □

There are difficulties in computing the lower bound in cases where there is not an explicit form for $A^*(\underline{\mu})$. The mean field approach circumvents this difficulty by restricting to

$$\mathcal{M}(G; H) := \left\{ \mu \in \mathbf{R}^d \mid \mu = E_{\underline{\theta}}[\Phi(\underline{X})], \theta \in \mathcal{E}(H) \right\}.$$

Note that $\mathcal{M}(G; H) \subset \mathcal{M}$, hence

$$A(\theta) \geq \sup_{\mu \in \mathcal{M}} \left\{ \langle \underline{\theta}, \underline{\mu} \rangle - A^*(\underline{\mu}) \right\} \geq \sup_{\mu \in \mathcal{M}(G;H)} \left\{ \langle \underline{\theta}, \underline{\mu} \rangle - A^*(\underline{\mu}) \right\}.$$

This lower bound is the best that can be obtained by restricting to H.

Let $\underline{\mu}^{(n)}$ denote a sequence such that for each n, $\underline{\mu}^{(n)} \in \mathcal{M}(G, H)$, such that $\underline{\mu}^{(n)} \overset{n \to +\infty}{\longrightarrow} \mu$ and such that

$$\langle \underline{\theta}, \underline{\mu}^{(n)} \rangle - A^*(\underline{\mu}^{(n)}) \overset{n \to +\infty}{\longrightarrow} \sup_{\mu \in \mathcal{M}(G;H)} \left\{ \langle \underline{\theta}, \underline{\mu} \rangle - A^*(\underline{\mu}) \right\}.$$

Note that $\mu \in \overline{\mathcal{M}(G; H)}$. Since $\theta \in \tilde{\Theta}$, it follows that $\mu \in \mathcal{M}$. The distribution associated with $\underline{\mu}$ minimizes the Kullback-Leibler divergence between the approximating distribution and the target distribution, subject to the constraint that $\mu \in \overline{\mathcal{M}(G; H)}$. Recall the mixed form of the Kullback-Leibler divergence; namely, Equation (8.12).

$$\tilde{D}(\underline{\mu}|\theta) = A(\theta) - A^*(\underline{\mu}) - \langle \underline{\mu}, \underline{\theta} \rangle.$$

Naive mean field updates In the *naive mean field* approach, a fully factorised distribution is chosen. This is equivalent to the approximation obtained by taking an empty edge set to approximate the original distribution. The naive mean field updates are a set of recursions for finding a stationary point of the resulting optimization problem.

Example 8.5 Consider the sigmoid belief network model. Here $\underline{X} = (X_1, \ldots, X_d)$ and $\mathcal{X} = \{0, 1\}^d$. Suppose that the distribution may be factorized along a graph $\mathcal{G} = (V, E)$. Let $\tilde{E} = \{(i, j) | (X_i, X_j) \in E\}$. The probability function is given by

$$p_{\underline{X}}(\underline{x}|\underline{\theta}) = \exp\left\{\sum_{j=1}^{d} \theta_j x_j + \sum_{(i,j)\in\tilde{E}} \theta_{(i,j)} x_i x_j - A(\underline{\theta})\right\}.$$

The *naive mean field* approach involves considering the graph with no edges. In this restricted class,

$$p_{\underline{X}}(\underline{x}|\underline{\theta}) = \exp\left\{\sum_{j=1}^{d} \theta_j x_j - A(\underline{\theta}^{(H)})\right\},$$

where $\underline{\theta}^{(H)}$ is the collection of parameters $\theta_s^{(H)} = \theta_s$, $s = 1, \ldots, d$ and $\theta^{(H)}(s, t) \equiv 0$. Note that

$$\mu_s = E_{\underline{\theta}}[\phi_s(\underline{X})] = E_{\underline{\theta}}[X_s]$$

and

$$\mu_{(s,t)} = E_{\underline{\theta}}[\phi_{s,t}(\underline{X})] = E_{\underline{\theta}}[X_s X_t].$$

When $\underline{\theta} \in H$, it follows that $(X_s)_{s=1}^{d}$ are independent, so that

$$\mu_{(s,t)} = E_{\underline{\theta}}[X_s X_t] = \mu_s \mu_t.$$

The optimization is therefore restricted to the set of parameters

$$\mathcal{M}(G; H) = \{(\mu_s)_{s=1}^{d}, (\mu_{(s,t)})_{(s,t)\in\{1,\ldots,d\}^2} | 0 \leq \mu_s \leq 1, \mu_{(s,t)} = \mu_s \mu_t.\}$$

With the restriction to product form distributions, $(X_s)_{s=1}^{d}$ are independent Bernoulli variables and hence

$$A_H^*(\underline{\mu}) = \sum_{s=1}^{d} \{\mu_s \log \mu_s + (1 - \mu_s) \log(1 - \mu_s).$$

Set

$$F(\underline{\mu}; \underline{\theta}) = \sum_{s=1}^{d} \theta_s \mu_s + \sum_{(s,t)\in\tilde{E}} \theta_{(s,t)} \mu_s \mu_t - \sum_{s=1}^{d} (\mu_s \log \mu_s + (1 - \mu_s) \log(1 - \mu_s)),$$

then the lower bound is given by

$$A(\underline{\theta}) \geq \sup_{(\mu_s)_{s=1}^{d}\in[0,1]^d} F(\underline{\mu}; \underline{\theta}).$$

Note that, for each μ_s, the function F is strictly convex. It is easy to see that the maximum is attained when, for all $1 \leq s \leq t$, $(\mu_t)_{t=1}^d$ satisfies

$$\theta_s + \sum_{t:(s,t)\in\tilde{E}} \theta_{(s,t)}\mu_t - \log \frac{\mu_s}{1 - \mu_s} = 0,$$

or

$$\log \frac{\mu_s}{1 - \mu_s} = \theta_s + \sum_{t\in\mathcal{N}(s)} \theta_{(s,t)}\mu_t.$$

Note that if

$$\log \frac{y}{1 - y} = x,$$

then

$$y = \sigma(x),$$

where

$$\sigma(x) = \frac{1}{1 + e^{-x}}.$$

The algorithm then proceeds by setting

$$\mu_s^{(j+1)} = \sigma(\theta_s + \sum_{t\in\mathcal{N}(s)} \theta_{(s,t)}\mu_t^{(j)}).$$

As discussed in [114] (p. 222), the lower bound thus computed seems to provide a good approximation to the true value.

8.9   Conditional Gaussian distributions

One very important family of distributions is the family of *conditional Gaussian distributions*. These are not of themselves of exponential type, but the conditional distributions are of exponential type. An example of the situation in view here is found in Section 10.9, where some of the variables are continuous and, conditioned on the discrete variables, have Gaussian distribution.

Let V be a set of variables, where $V = D \cup C$; D is a set of discrete variables and C is a set of continuous variables. Let $\tilde{V}$ denote the indexing set, where $\tilde{V}$ is decomposed into $\tilde{V} = \Delta \cup \Gamma$; Δ is the indexing set for the discrete variables and Γ is the indexing set for the continuous variables. Let $|\Delta|$ denote the number of variables in Δ and let $|\Gamma|$ denote the number of variables in Γ. The random vector $\underline{X}$ of variables in V will be written

$$\underline{X} = (\underline{X}_\Delta, \underline{X}_\Gamma),$$

where $\underline{X}_\Delta$ denotes the vector of discrete variables and $\underline{X}_\Gamma$ denotes the vector of continuous variables. Random vectors will be *row* vectors. The state space is

$$\mathcal{X} = \mathcal{X}_1 \times \ldots \times \mathcal{X}_{|\Delta|} \times \mathcal{X}_{|\Delta|+1} \times \ldots \times \mathcal{X}_{|\Delta|+|\Gamma|},$$

where $\mathcal{X}_j$ denotes the state space for variables j. The following notation will also be used:

$$\mathcal{X}_\Delta = \mathcal{X}_1 \times \ldots \times \mathcal{X}_{|\Delta|}, \quad \mathcal{X}_\Gamma = \mathcal{X}_{|\Delta|+1} \times \ldots \times \mathcal{X}_{|\Delta|+|\Gamma|},$$
$$\mathcal{X} = \mathcal{X}_\Delta \times \mathcal{X}_\Gamma.$$

Attention is restricted to the case where the continuous variables have Gaussian distribution, conditional on the discrete variables, so $\mathcal{X}_\Gamma = \mathbf{R}^{|\Gamma|}$. For the discrete variables,

$$\mathcal{X}_j = \{i_j^{(1)}, \ldots, i_j^{(k_j)}\}.$$

A particular configuration $\underline{i} \in \mathcal{X}_\Delta$ is called a *cell*.

The following notation will be used to indicate that a random vector $\underline{X}_1$ conditioned on $\underline{X}_2 = \underline{x}_2$ has distribution F:

$$\underline{X}_1 \mid \underline{X}_2 = \underline{x}_2 \sim F.$$

The *moment generating function* is necessary to define a multivariate normal distribution.

Definition 8.6 (Moment Generating Function) *Let* $\underline{X} = (X^1, \ldots, X^d)$ *be a random vector. Its* moment generating function *is the function* $M_{\underline{X}} : \mathbf{R}^d \to \overline{\mathbf{R}}$, *where* $\overline{\mathbf{R}} \cup \{+\infty\} \cup \{-\infty\}$, *is defined as*

$$M_{\underline{X}}(p_1, \ldots, p_d) = E\left[\exp\left\{\sum_{j=1}^d p_j X_j\right\}\right].$$

The moment generating function is useful, because it *uniquely determines the distribution* of a random vector $\underline{X}$. That is, a joint probability determines a unique moment generating function, and the moment generating function uniquely determines a corresponding joint probability. The moment generating function is essentially a Laplace transform.

A *multivariate normal distribution* is defined as follows:

Definition 8.7 (Multivariate Normal Distribution) *A random vector* $\underline{X} = (X^1, \ldots, X^d)$ *is said to have a* multivariate normal *distribution, written* $\underline{X} \sim N(\underline{\mu}, \mathbf{C})$, *if its moment generating function is of the form*

$$\phi(p_1, \ldots, p_d) = \exp\left\{\sum_{j=1}^d p_j \mu_j + \frac{1}{2}\sum_{jk} p_j p_k C_{jk}\right\}, \quad \underline{p} \in \mathbf{R}^d.$$

If a random vector $\underline{X} \sim N(\underline{\mu}, \mathbf{C})$, then $E[X^i] = \mu^i$ for each $i = 1, \ldots, d$ and $\text{Cov}(X_i, X_j) = C_{ij}$ for each (i, j). If $\mathbf{C}$ is positive definite, then the joint density function of $\underline{X} = (X^1, \ldots, X^d)$ is given by

$$\pi_{X_1,\ldots,X_d}(x_1, \ldots, x_d) = \frac{1}{(2\pi)^{d/2}|\mathbf{C}|^{1/2}} \exp\left\{-\frac{1}{2}(\underline{x} - \underline{\mu})\mathbf{C}^{-1}(\underline{x} - \underline{\mu})\right\}, \quad \underline{x} \in \mathbf{R}^d,$$

where $\underline{x} = (x_1, \ldots, x_d)$ and $\underline{\mu} = (\mu_1, \ldots, \mu_d)$ are row vectors and $|\mathbf{C}|$ denotes the determinant of $\mathbf{C}$.

The *conditional Gaussian* distribution, or CG distribution, may now be defined.

Definition 8.8 (CG Distribution) *A collection of random variables $\underline{X} = \left(\underline{X}_\Delta, \underline{X}_\Gamma\right)$ is said to follow a CG distribution if for each $\underline{i} \in \mathcal{X}_\Delta$,*

$$\underline{X}_\Gamma | \underline{X}_\Delta = \underline{i} \sim N\left(\underline{\mu}(\underline{i}), \mathbf{C}(\underline{i})\right). \tag{8.14}$$

The notation for such a conditional Gaussian distribution is

$$\underline{X} \sim \mathrm{CG}(|\Delta|, |\Gamma|).$$

If the numbers of discrete and continuous random variables are, respectively, $|\Delta| = p$ and $|\Gamma| = q$, then $X \sim \mathrm{CG}(p, q)$.

If $\mathbf{C}^{-1}$ is well defined, then the conditional density function of $\underline{X}_\Gamma$ conditioned on $\underline{X}_\Delta = \underline{i}$ is

$$\pi_{\underline{X}_\Gamma | \underline{X}_\Delta}\left(\underline{x} | \underline{i}\right) = \frac{1}{(2\pi)^{q/2}\sqrt{\det\mathbf{C}(\underline{i})}} e^{-\frac{1}{2}(\underline{x} - \underline{\mu}(\underline{i}))\mathbf{C}(\underline{i})^{-1}(\underline{x} - \underline{\mu}(\underline{i}))^t}, \tag{8.15}$$

for all $\underline{i} \in \mathcal{X}_\Delta$ such that

$$p_{\underline{X}_\Delta}(\underline{i}) > 0.$$

For this discussion, it is assumed that $p_{\underline{X}_\Delta}(\underline{i}) > 0$ for each $\underline{i} \in \mathcal{X}_\Delta$.

Directly from Equation (8.15),

$$p_{\underline{X}_\Delta}(\underline{i})\pi_{\underline{X}_\Gamma | \underline{X}_\Delta}\left(\underline{x} | \underline{i}\right) = \chi(\underline{i}) e^{g(\underline{i}) + \underline{x}h(\underline{i}) - \frac{1}{2}\underline{x}K(\underline{i})\underline{x}^t} \tag{8.16}$$

where $\chi(\underline{i}) = 1$ if $p_{\underline{X}_\Delta}(\underline{i}) > 0$ and 0 if $p_{\underline{X}_\Delta}(\underline{i}) = 0$, $h(\underline{i}) = \mathbf{C}(\underline{i})^{-1}\underline{\mu}(\underline{i})$, $K(\underline{i}) = \mathbf{C}(\underline{i})^{-1}$ and

$$g(\underline{i}) = \log p_{\underline{X}_\Delta}(\underline{i}) + \frac{1}{2}\left(\log \det K(\underline{i}) - |\Gamma| \log 2\pi - \underline{x}K(\underline{i})\underline{x}^t\right).$$

The family of *joint* distributions is *not* an exponential family, but from Equation (8.15), it is clear that conditioning on the discrete variables gives a family of multivariate normal distributions, which is an exponential family. The *canonical parameters* of this exponential family are $\left(h(\underline{i}), K(\underline{i})\right)$ and it is easy to see that the *mean parameters* of the CG distribution are $\left(\underline{\mu}(\underline{i}), \mathbf{C}(\underline{i})\right)$ since, conditioned on $\underline{X}_\Delta = \underline{i}$, it follows that $E[\underline{X}_\Gamma] = \underline{\mu}(\underline{i})$ and $E[\underline{X}_\Gamma^t \underline{X}_\Gamma] = \mathbf{C}(\underline{i})$ (recall that random vectors are taken to be row vectors).

Parametrization of the CG distribution   The CG distribution may be parametrized in terms of the exponential family to which the conditional distributions belong. The *canonical parameters* for the *joint* distribution, defined by the pair of functions $\left(p_{\underline{X}_\Delta}, \pi_{\underline{X}_\Gamma | \underline{X}_\Delta}\right)$ are *defined* as (g, h, K), where the parameters $(h(\underline{i}), K(\underline{i}))$ are the canonical parameters of the conditional distribution and $g(\underline{i})$ is the log partition function of the conditional distribution, conditioned on $\underline{X}_\Delta = \underline{i}$.

Similarly, the *mean parameters* are *defined* as $(p, \mu, \mathbf{C})$, where $(\mu(\underline{i}), \mathbf{C}(\underline{i}))$ are the mean parameters of the conditional distribution and $p(\underline{i})$ is the appropriate multiplier obtained from Equation (8.15).

Proposition 8.2 *Let $\underline{X}$ have a conditional Gaussian distribution. Let V denote the set of variables and $\tilde{V}$ the indexing set. Let A and B be two disjoint sets such that $\tilde{V} = A \cup B$, then the conditional distribution of $\underline{X}_A$ given $\underline{X}_B = x_B$ is conditional Gaussian.*

Proof of Proposition 8.2 The following calculation shows that $\underline{X}_{A \cap \Gamma} \mid \{\underline{X}_B = x_B\} \cup \{\underline{X}_{A \cap \Delta} = \underline{x}_{A \cap \Delta}\}$ has a multivariate Gaussian distribution. Firstly, it is clear that

$$\pi_{\underline{X}_{A \cap \Gamma} \mid \underline{X}_{A \cap \Delta} \cdot \underline{X}_B} \left(\underline{x}_{A \cap \Gamma} \mid \underline{x}_{A \cap \Delta}, \underline{x}_B \right) = \pi_{\underline{X}_{A \cap \Gamma} \mid \underline{X}_\Delta \cdot \underline{X}_{B \cap \Gamma}} \left(\underline{x}_{A \cap \Gamma} \mid \underline{x}_\Delta, \underline{x}_{B \cap \Gamma} \right);$$

this is obtained simply by reorganising the sets of variables.

The conditional density function on the right hand side is obtained by conditioning the distribution of $\underline{X}_{A \cap \Gamma} \mid \underline{X}_\Delta = \underline{x}_\Delta$ on $\underline{X}_{B \cap \Gamma} = \underline{x}_{B \cap \Gamma}$. Since $(\underline{X}_{A \cap \Gamma}, \underline{X}_{B \cap \Gamma}) \mid \underline{X}_\Delta = \underline{x}_\Delta$ has a multivariate Gaussian distribution, and the conditional distribution of a multivariate Gaussian, conditioning on some of its component variables is again multivariate Gaussian, it follows that the conditional distribution is multivariate. The proof is complete. □

8.9.1 CG potentials

Definition 8.9 (Conditional Gaussian Potential) *A CG potential $\phi(\underline{x})$, for $\underline{x} = (\underline{i}, \underline{x}_\Gamma) \in \mathcal{I} \times \mathbf{R}^q$, for a discrete space $\mathcal{I}$ (where $\underline{x}$ is a row vector) is any function of the form*

$$\phi(\underline{x}) = \chi(\underline{i}) e^{g(\underline{i}) + \underline{x}_\Gamma h(\underline{i}) - \frac{1}{2} \underline{x}_\Gamma K(\underline{i}) \underline{x}_\Gamma^t},$$

where χ takes the values 1 or 0 and $K(\underline{i})$ is symmetric for each $\underline{i} \in \mathcal{I}$.

A CG potential is said to have canonical parameters (g, h, K).

If, furthermore, $K(i)$ is positive definite for each $i \sim \mathcal{I}$, then the mean parameters *are defined by*

$$\mu(\underline{i}) = K(\underline{i})^{-1} h(\underline{i}), \qquad C(\underline{i}) = K(\underline{i})^{-1}$$

and

$$p_{\underline{X}_\Delta}(\underline{i}) = \chi(\underline{i}) \exp \left\{ g(\underline{i}) - \frac{1}{2} \left(\log \det K(\underline{i}) - q \log 2\pi - \underline{x}_\Gamma K(\underline{i}) \underline{x}_\Gamma^t \right) \right\},$$

where $q = |\Gamma|$.

The margins of a CG potential with respect to a subset of the continuous variables may be computed in the standard way for multivariate Gaussian distributions, but for a *potential*, the integral may not be 1.

8.9.2 Some results on marginalization

If the variables to be integrated out are discrete, then complicated mixture distributions arise. The following proposition gives a special case.

Proposition 8.3 *Let $A \subseteq \tilde{V}$ denote a subset of the indexing set for the variables. If $\underline{X}$ is CG and $B = \tilde{V} \setminus A$ (namely, B is the set of all indices in $\tilde{V}$ that are not in A) and $B \subseteq \Delta$ and*

$$\underline{X}_B \perp \underline{X}_\Gamma \mid \underline{X}_{\Delta \setminus B},$$

then $\underline{X}_A \sim CG$.

Proof of Proposition 8.3 Clearly, from the definition of a CG distribution, it is necessary and sufficient to show that

$$\underline{X}_{A \cap \Gamma} \mid \underline{X}_{\Delta \setminus B} \sim N_{|A \cap \Gamma|}.$$

(multivariate normal, with dimension $\mid A \cap \Gamma \mid$). The proof requires the following identity: If $\underline{X}_B \perp \underline{X}_\Gamma \mid \underline{X}_{\Delta \setminus B}$, then

$$\pi_{\underline{X}_\Gamma \mid \underline{X}_\Delta} (\underline{x}_\Gamma \mid \underline{X}_\Delta) = \pi_{\underline{X}_\Gamma \mid \underline{X}_B, \underline{X}_{\Delta \setminus B}} (\underline{x}_\Gamma \mid \underline{x}_B, \underline{x}_{\Delta \setminus B}) = \pi_{\underline{X}_\Gamma \mid \underline{X}_{\Delta \setminus B}} (\underline{x}_\Gamma \mid \underline{x}_{\Delta \setminus B}).$$

This follows almost directly from the first characterization of conditional independence from Theorem 2.1. The result in Theorem 2.1 was stated for discrete variables; it is straightforward to verify that it holds for conditional density functions. Recall that, from the definition of a CG distribution, $\pi_{\underline{X}_\Gamma \mid \underline{X}_\Delta} \left(\underline{x}_\Gamma \mid \underline{x}_\Delta \right)$ is a multivariate normal distribution. Therefore the conditional distribution of $\underline{X}_\Gamma$ conditioned on $\underline{X}_{\Delta \setminus B}$ is multivariate Gaussian, therefore the conditional distribution of $\underline{X}_{\Gamma \cap A}$ conditioned on $\underline{X}_{\Delta \setminus B}$ is multivariate Gaussian. The proof is complete. $\square$

8.9.3 CG regression

Definition 8.10 (CG Regression) *Let $\underline{Z} = (Z_1, \ldots, Z_s)$ be a continuous random (row) vector and let $\underline{I}$ be a discrete random (row) vector with probability function p_I. Let $\mathcal{I}$ denote the state space for I. If a random (row) vector $\underline{Y} = (Y_1, \ldots, Y_r)$ has the property that*

$$\underline{Y} \mid (\underline{I} = \underline{i}, \underline{Z} = \underline{z}) \sim N_r \left(\underline{A}(\underline{i}) + \underline{z} \underline{B}(\underline{i}), \underline{C}(\underline{i}) \right),$$

where for each $i \in \mathcal{I}$

- *$\underline{A}(i)$ is a $1 \times r$ row vector for each $i \in \mathcal{I}$,*
- *$\mathbf{B}(i)$ is an $s \times r$ matrix,*
- *$\mathbf{C}(i)$ is a positive semi-definite symmetric matrix,*

then $\underline{Y}$ is said to follow a CG regression.

Let V denote a set of variables, containing both discrete and continuous variables, which have been ordered so that the probability distribution may be factorized along a directed acyclic graph $\mathcal{G} = (V, E)$. Let X_γ be a continuous variable, with parent set $\Pi(\gamma)$. Suppose that $\underline{X}$ has a conditional Gaussian distribution. Then the conditional distribution for X_γ, conditioned on its parent nodes $\Pi(\gamma)$ is the CG regression

$$X_\gamma \mid \Pi(\gamma) \sim N \left(\alpha(\underline{i}) + \underline{z} \underline{\beta}, \sigma(\underline{i}) \right),$$

where the discrete variables of $\Pi(\gamma)$ take values $\underline{i}$ and the continuous variables of $\Pi(\gamma)$ take values $\underline{z}$. Here $\alpha(\underline{i})$ is a number, $\sigma(\underline{i}) > 0$ and β is a column vector with dimension equal to the dimension of the continuous component $\underline{z}$ so that $\underline{z}\beta$ is a well defined inner product. Thus, the conditional density corresponds to a CG potential $\phi(\underline{i}, \underline{z}, x_\gamma)$, equal to

$$\phi(\underline{i}, \underline{z}, x_\gamma) = \frac{1}{\sqrt{2\pi\sigma(\underline{i})}} \exp\left\{ -\frac{(x_\gamma - \alpha(\underline{i}) + \underline{z}\beta)^2}{2\sigma(\underline{i})} \right\} \tag{8.17}$$

The canonical characteristics of this potential are easily found manipulating the expression in Equation (8.17).

Notes   The material for Chapter 8 is taken mostly from M.J. Wainright and M.I. Jordan [115]. It is developed further in [114]. Possible improvements to the lower bound are proposed by K. Humphreys and D.M. Titterington in [116]. The book by O. Barndorff-Nielsen [113] is the standard treatise of exponential families and the required convex analysis. Conditional Gaussian distributions and their applications to Bayesian networks are discussed in [67].

8.10   Exercises: Graphical models and exponential families

1. Which of the following families of distributions are exponential families? Obtain the minimal sufficient statistics for those which are.

   (a)
   $$p(x|\theta) = \begin{cases} \frac{1}{10} & x = 0.1n + \theta, \quad n \in \{0, 1, \dots, 9\} \\ 0 & \text{otherwise} \end{cases}$$

   (b) The family of $N(\mu, \mu)$ distributions, where $\mu > 0$ (that is mean and variance both μ)

   (c) The family $X|X \neq 0$, where $X \sim Bi(n, p)$.

2. Prove Lemma 8.1.

3. Consider a one parameter exponential family, in canonical parameters, with probability function, or density function,

   $$p(x|\theta) = \exp\{\theta\phi(x) - A(\theta)\}h(x).$$

   Show that

   $$E[\phi(X)] = A'(\theta)$$

   and

   $$\text{Var}(\phi(X)) = A''(\theta).$$

4. Let (X_1, X_2, X_3) be random variables, with joint probability function

   $$p(x_1, x_2, x_3|\eta) = \frac{n!}{x_1!x_2!x_3!} \prod_{j=1}^{3} p_i^{x_i}, \quad x_1 + x_2 + x_3 = n,$$

   where $p_1 = \eta^2$, $p_2 = 2\eta(1 - \eta)$ and $p_3 = (1 - \eta)^2$ and $0 \leq \eta \leq 1$.

   (a) Is this an exponential family?

   (b) Obtain the minimal sufficient statistic for θ.

   (c) Compute the mean parameter in terms of η.

   (d) Compute the Fenchel Legendre conjugate of the log partition function.

   (e) Prove that the Kullback-Leibler divergence is given by

   $$D(\theta_1|\theta_2) = A(\theta_2) - A(\theta_1) - \langle \mu_1, \theta_2 - \theta_1 \rangle.$$
   $$\tilde{D}(\mu_1|\theta_2) = A(\theta_2) + A^*(\mu_1) - \langle \mu_1, \theta_2 \rangle$$
   $$\tilde{\tilde{D}}(\mu_1|\mu_2) = A^*(\mu_1) - A^*(\mu_2) - \langle \theta_2, \mu_1 - \mu_2 \rangle.$$

   State the definitions of the terms used in this equation.

(f) Compute the primal form of the Kullback-Leibler divergence $D(\theta_1 \| \theta_2)$, where θ_1 and θ_2 are the canonical parameters. Compute the dual form, expressed in terms of the mean parameters.

5. **(Propositional Logic)** Consider a treatment for high blood pressure which, under 'normal' circumstances is effective in 9 cases out of 10. The patient may have additional conditions, which cause the treatment to fail. If the patient has condition R, it causes the treatment to fail with probability $\frac{1}{7}$. If the patient has condition W, it causes the treatment to fail with probability $\frac{1}{4}$. If the patient has condition C, it causes the treatment to fail with probability $\frac{1}{3}$. If the patient has condition B, it causes the treatment to fail with probability $\frac{1}{2}$. Assume that all these factors act independently on the probability that the treatment fails. Compute the probabilities for success and failure for all possible combinations of the factors listed above.

6. **(Mean Field Update)** Consider a probability function, given by

$$p_{\underline{X}}(x|\underline{\theta}) = \exp\left\{ \sum_{j=1}^{n} \theta(j)x(j) + \sum_{(i,j)\in E} \theta(i,j)x(j) - A(\underline{\theta}) \right\},$$

where $\underline{\theta} = \{(\theta(j))_{j=1}^{n}, (\theta(j,k)), (j,k) \in E\}$, E denotes the edge set and $\mathbf{x} \in \{0, 1\}^n$. Let q denote the probability function

$$q_{\underline{X}}(x|\underline{\theta}) = \exp\left\{ \sum_{j=1}^{n} \theta(j)x(j) - A_H(\underline{\theta}) \right\}.$$

Let

$$A_H^*(\underline{\mu}) = \sup_{\underline{\theta}}\{\langle \mu, \theta\rangle - A_H(\underline{\theta}).$$

(a) Prove that

$$A_H^*(\underline{\mu}) = \sum_{j=1}^{n} \{\mu(j)\log\mu(j) + (1 - \mu(j))\log\mu(j)\}.$$

(b) Prove that

$$A(\underline{\theta}) \geq \sup_{\underline{\mu}}\left\{ \sum_{j=1}^{n} \theta(j)\mu(j) + \sum_{(j,k)\in E} \theta(j,k)\mu(j)\mu(k) - A_H^*(\underline{\mu}) \right\}.$$

(c) Consider the probability distribution

$$p(x_1, x_2, x_3; \theta) = \exp\left\{ \sum_{j=1}^{3} \theta(j)x_j + \theta(1,2)x_1x_2 + \theta(1,3)x_1x_3 - A(\theta) \right\}.$$

Show that the expression in the previous part is maximized for $(\mu(1), \mu(2), \mu(3))$ that satisfy

$$\log \frac{\mu(1)}{1 - \mu(1)} = \theta(1) + \theta(1, 2)\mu(2) + \theta(1, 3)\mu(3)$$

$$\log \frac{\mu(2)}{1 - \mu(2)} = \theta(2) + \theta(1, 2)\mu(1)$$

$$\log \frac{\mu(3)}{1 - \mu(3)} = \theta(3) + \theta(1, 3)\mu(1).$$

(d) Write a MATLAB code to compute numerical approximations to the values $(\mu(1), \mu(2), \mu(3))$ that give the naive mean field approximation to the log partition function $A(\theta)$.

7. Let

$$\underline{X} = (\underline{X}_\Delta, X_\Gamma) \sim \text{CG}(|\Delta|, 1).$$

Let $\mathcal{I}$ denote the state space for $\underline{X}_\Delta$ and let p denote the probability function for the random vector $\underline{X}_\Delta$. Prove that

$$E[X_\Gamma] = \sum_{i \in \mathcal{I}} p(i)\mu(i)$$

and

$$\text{Var}(X_\Gamma) = \sum_{i \in \mathcal{I}} p(i)\sigma(i)^2 + \sum_{i \in \mathcal{I}} p(i)(\mu(i) - E[X_\Gamma])^2,$$

stating clearly any results about multivariate normal random variables that you are using.

8. Let $X \sim \text{CG}(2, 2)$ and let I_1 and I_2 be binary variables. Find the canonical parameters for the distribution.

9. Prove that if a conditional Gaussian distribution is marginalized over a subset of the continuous variables, the resulting distribution is again a CG distribution. Find the canonical characteristics of the marginal distribution in terms of the original canonical characteristics, stating clearly any results about multivariate normal random variables that you are using.

10. Suppose that hard evidence is entered into a subset of the continuous variables of a CG distribution. Show that the updated distribution is again a CG distribution and express the canonical characteristics of the updated distribution in terms of the canonical characteristics of the original distribution, stating clearly any results about multivariate normal random variables that you are using.

9

Causality and intervention calculus

9.1   Introduction

Causality is a notion with a manipulative component. Wold R.H. Strotz and H.O.A. [117] state:

> '... in common scientific usage, (causality) has the following general mean-
> ing: z is the cause of y if, by the hypothesis that it is, or would be, possible,
> by controlling z indirectly, to control y, at least stochastically'.

Hence, causal inference in this chapter is meant to answer to predictive queries about the effect of a hypothetical or pondered manipulation or *intervention*.[1] Causal predictive inference requires a machinery to signify intervention, i.e. when one actively changes the value of one or more of the variables. Examples of manipulations are medical treatments and manual interceptions in an automatic controller. These change the states in a data-generating mechanism by upsetting the normal forces working on it. This is the basic principle of a 'controlled experiment'. In order to assess whether or not a particular variable has a causal effect on another, the values of that variable are assigned purely at random, by the controller, without reference to any other factors.

---

[1] There are also *counter-factual* causal inferences of the form of an explanation *If and event A had not occurred, then C would not have occurred*, which are not explicitly covered here.

One argument against making causal inferences has been that statistics lacks the language to model effects of intervention. In order to discuss effects of intervention in the language of DAGs, an edge between two nodes in a Bayesian network is expressly interpreted as a causal link. In other words, the states of a parent variable are said to be direct causes, stochastically, of the states of a child variable. This is the sense in which the discussion about causal Bayesian networks is to be understood. This does not perhaps contribute to a deeper understanding of the generic concept of causality, but it is sufficient for predictive causal inference, and in this setting, the modelling assumptions are transparent.

The present chapter introduces the 'do-calculus' of probability updating by intervention: some of the variables in a causal Bayesian network are *forced* to take certain values, rather than simply being *observed* to have these values. If the variable is *forced* into a certain value, and then the conditional probabilities computed, this is known as 'do' conditioning; while if the variable is simply *observed*, this is known as 'see' conditioning. The 'do-calculus' is due to Judea Pearl in [118] and [1]. It enables conclusions to be drawn about the effects of *active* interventions, based on passive observations.

The updating of probabilities based on 'see' conditioning (Bayes' rule, Jeffrey's rule) are defined without reference to any causal structure. The conditional probabilities are estimated directly from observations and it is important that the variables are merely observed and that they have not been forced. The updating associated with 'do' conditioning depends upon the causal structure. It is the 'do' conditional probabilities that are estimated in a controlled randomized experiment. The calculus described in this chapter enables one to compute intervention probabilities from the conditional probabilities in causal Bayesian networks.

Pearl's 'do-calculus' describes how to treat a *perfect intervention*; namely, an intervention which essentially 'cuts off' the influence of the parents to a node. Some real world interventions may be modelled in this way; for example, gene knock outs. Many interventions, however, are not so precise in their effects and Pearl's calculus has been extended by Eaton and Murphy in [119] to take this into account.

In times past, only a few expositions of statistics considered causal inference. A notable exception is H.O.A. Wold [120]. Nowadays, there are discussions of causality in statistical textbooks; for example, in D. Edwards [121], Chapter 9. Even though probabilistic reasoning and statistical studies avoid claims that they have established causal links and restrict claims to 'correlation' and 'association', statistical methods are routinely used incorrectly to justify causal inference from data, see D. Freedman [122]. D.V. Lindley [123] quotes a statement by Terry Speed: 'considerations of causality should be treated as they have always been treated in statistics: preferably not at all (but if necessary, then with great care).' It should thus come as no surprise that the methodologies in Pearl [1] and Spirtes *et al.* [48] have met with critical reviews, for example, [20] (Freedman and Humphreys), which should be taken seriously by every statistician studying 'do-calculus'.

One practical result of the theory of 'do-calculus' is that sets of *confounding* variables (any common ancestor to two nodes in a causal DAG is a confounder) may be characterized. That is, one may locate a *sufficient* set of variables that, if they were known, one could establish the correct causal effect between variables of interest.

It will be shown that a sufficient set for estimating the causal effect of X on Y is any set of non-descendants of X that d-separate X from Y after all the arrows emanating from X have been removed. This criterion, known as the 'back door criterion', provides a mathematical definition of confounding and helps to identify accessible sets of variables that ought to be investigated.

Another effect of the theory of 'do-calculus' is to give a precise meaning to the well known phrase *learning by doing* in the sense that there can be perfect interventions that can resolve the question of which structure to use from a Markov equivalence class. In practise, a controlled experiment provides a perfect intervention.

D.V. Lindley [123] notes that the do-calculus is well adapted for Bayesian use; the separate assessments of $p(Y \mid see(x))$ and $p(Y \mid do(x))$ do not violate coherence. Lindley also states that 'the "do"-calculus is an extremely sensible method of developing a calculus for controlled experiments.' 'Do'-calculus also has applications to situations other than controlled experiments.

9.2   Conditioning by observation and by intervention

Suppose that $X = x$ is *observed*. Then the conditional probability of $Y = y$ can be expressed, using Bayes' rule, as

$$p_{Y|X}(y|x) = \frac{p_{X|Y}(x|y)p_Y(y)}{p_X(x)}.$$

This formula describes the way that the probability distribution of the random variable Y changes after $X = x$ is *observed*. This is denoted by

$$p(Y = y|\text{see}(x)).$$

The *causal* probability calculus, *also* describes how to modify the probability distribution of the random variable Y in the presence of an active *intervention* to *force* the random variable X to take the value x, where the intervention is independent of the state of the system. This is denoted by $X \leftarrow x$. The conditional distribution of Y, after an intervention has forced the value x on the variable X, is denoted by

$$p(Y = y|\text{do}(x)) = p(y\|x) = p(Y = x|X \leftarrow x).$$

There is a distinction between these two types of conditioning. As P. Spirtes *et al.* [48] put it:

> 'How can an observed distribution p be used to obtain reliable predictions of the effects of alternative policies that would impose a new marginal distribution on some set of variables? The very idea of imposing a policy that would directly change the distribution of some variable ... necessitates that the resulting distribution p_{MAN} will be different from p. p alone cannot be used to predict p_{MAN}, but p and a causal structure can be.'

Here 'MAN' is an abbreviation of 'manipulate'.

This chapter shows how the causal structure of the Bayesian network may be used to construct an intervention calculus.

9.3   The intervention calculus for a Bayesian network

Notations   Consider a directed acyclic graph (DAG) $\mathcal{G} = (V, E)$ where the node set V is finite. Let $\tilde{V}$ denote the indexing set for the nodes. For each $v \in \tilde{V} = \{1, \dots, d\}$, the random variable X_v takes its values in a finite state space $\mathcal{X}_v = \{x_v^{(1)}, \dots, x_v^{(k_v)}\}$. As usual, Π_j will denote the set of parent variables of variable X_j. The notation $\tilde{\Pi}_j$ will be used to denote the indexing set of the variables in Π_j.

For a set of indices A, the space $\mathcal{X}_A = \times_{v \in A} \mathcal{X}_v$.

The edges E in the graph represent the *causal relationships* between the variables, so that a *parent* of X_v is a *direct cause* of X_v.

$$\underline{X}_{\tilde{V}} = \times_{v \in \tilde{V}} X_v = (X_1, \dots, X_d),$$

while $\underline{x}_{\tilde{V}}$ will denote a value in $\mathcal{X}_1 \times \dots \times \mathcal{X}_d$. That is,

$$\underline{x}_{\tilde{V}} = \times_{v=1}^{d} x_v^{(i_v)} = (x_1^{(i_1)}, \dots, x_d^{(i_d)});$$

that is,

$$\underline{X}_{\tilde{V}} = \underline{x}_{\tilde{V}} \Leftrightarrow (X_1 = x_1^{(i_1)}, \dots, X_d = x_d^{(i_d)}).$$

For any $A \subset \tilde{V}$, $\underline{X}_A$ is defined as

$$\underline{X}_A = \times_{v \in A} X_v,$$

while $\underline{x}_A$ will denote a value in $\times_{v \in A} \mathcal{X}_v$. That is,

$$\underline{x}_A = \times_{v \in A} x_v^{(i_v)}.$$

Since the superscripts are implied, they will often be *omitted* in the text.

The *set difference*, written

$$\tilde{V} \backslash A,$$

is defined as all the indices in $\tilde{V}$ which are not included in A. The following notation will be used in situations where the ordering of the variables is not important:

$$\underline{x}_{\tilde{V}} = \times_{v=1}^{d} x_v = \times_{v \in \tilde{V} \backslash A} x_v \times_{v \in A} x_v = \underline{x}_{\tilde{V} \backslash A} \cdot \underline{x}_A.$$

Let ϕ be a function defined on $\mathcal{X}_1 \times \dots \times \mathcal{X}_d$. Then the quantity $\sum_{\tilde{V} \backslash A} \phi$ is defined as

$$\left[\sum_{\tilde{V} \backslash A} \phi \right] (\underline{x}_{\tilde{V}}) = \sum_{\underline{x}_{\tilde{V} \backslash A}} \phi(\underline{x}_{\tilde{V} \backslash A} \cdot \underline{x}_A).$$

Definition 9.1 (The Intervention Formula)   *The conditional probability of $\underline{X}_{\tilde{V} \backslash A} = \underline{x}_{\tilde{V} \backslash A}$, given that the variables $\underline{X}_A$ were forced to take the values $\underline{x}_A$ independently of all else, is written*

$$p_{\underline{X}_{\tilde{V}} \| \underline{X}_A}(\underline{x}_{\tilde{V}} | \underline{X}_A \leftarrow \underline{x}_A) \quad \text{or} \quad p_{\underline{X}_{\tilde{V} \backslash A} \| \underline{X}_A}(\underline{x}_{\tilde{V}} \| \underline{x}_A)$$

and defined as

$$p_{\underline{X}_{\tilde{V}\backslash A}\|\underline{X}_A}(\underline{x}_{\tilde{V}}|\underline{X}_A \leftarrow \underline{x}_A) = p_{\underline{X}_{\tilde{V}\backslash A}\|\underline{X}_A}(\underline{x}_{\tilde{V}}\|\underline{x}_A)$$

$$= \frac{p_{\underline{X}_{\tilde{V}}}(\underline{x}_{\tilde{V}})}{\prod_{v\in A} p_{X_v|\Pi_v}(x_v^{(i_v)}|\underline{x}_{\tilde{\Pi}_v})}$$

$$= \prod_{v\in\tilde{V}\backslash A} p_{X_v|\Pi_v}(x_v^{(i_v)}|\underline{x}_{\tilde{\Pi}_v}). \tag{9.1}$$

The function $p_{\underline{X}_{\tilde{V}\backslash A}\|\underline{X}_A}(\cdot\|\underline{x}_A)$ *defines a probability distribution over the space* $\mathcal{X}_{\tilde{V}\backslash A}$, *the space where the variables on which no intervention has been made take their values.*

This means *instantiation of the variables indexed by the set A* and *elimination of those edges in E which lead from the parents of the nodes indexed by A to the nodes indexed by A*. The terminology 'local surgery' is used to describe such an elimination. A local surgery is performed and the conditional probabilities on the remaining edges are multiplied. This yields a factorization along a mutilated graph. The *direct causes* of the manipulated variable are put out of effect.

The idea of deletion of connections (in terms of wiping out equations in a multivariate model) is found in R.H. Strotz and H.O.A. Wold [117].

The quantity $p_{\underline{X}_{\tilde{V}\backslash A}\|\underline{X}_A}(\cdot\|\underline{x}_A)$ from Definition 9.1 defines a family of probability measures over $\mathcal{X}_{\tilde{V}\backslash A}$, which depending on the parameter $\underline{x}_A$, the values forced on the variables indexed by A. This family includes original probability measure; if $A = \phi$, then

$$p_{\underline{X}_{\tilde{V}\backslash A}\|\underline{X}_A}(\cdot\|\underline{x}_A) = p_{\underline{X}_{\tilde{V}}}(\cdot).$$

The family of probability measures defined in Definition 9.1 is the *intervention measure*. In addition, the final expression on the right hand side of Equation (9.1) is called the *intervention formula*. This formula is due to Pearl, but is also given independently in the first edition of P. Spirtes *et al.* [48]. See also C. Meek and C. Glymour [124], p. 1010.

Intervention   An 'intervention' is an action taken to force a variable into a certain state, without reference to its own current state, or the states of any of the other variables. It may be thought of as choosing the values $\underline{x}_A^*$ for the variables $\underline{X}_A$ by using a random generator independent of the variables $\underline{X}_{\tilde{V}}$.

Remark   In the same style of notation, *conditioning by observation* is

$$p_{\underline{X}_{\tilde{V}\backslash A}|\underline{X}_A}(\underline{x}_{\tilde{V}\backslash A}|\text{see}(\underline{x}_A)) = p_{\underline{X}_{\tilde{V}\backslash A}|\underline{X}_A}(\underline{x}_{\tilde{V}\backslash A}|\underline{x}_A). \tag{9.2}$$

where, by the standard definition of conditional probability,

$$p_{\underline{X}_{\tilde{V}\backslash A}|\underline{X}_A}(\underline{x}_{\tilde{V}\backslash A}|\underline{x}_A) = \frac{p_{\underline{X}_{\tilde{V}}}(\underline{x}_{\tilde{V}})}{\sum_{\underline{y}_{\tilde{V}}:\underline{y}_A=\underline{x}_A} p(X_1 = y_{i_1}, \dots X_d = y_{i_n})}. \tag{9.3}$$

Comparing Equation (9.1) with Equation (9.3) gives the following corollary.

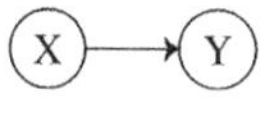

Figure 9.1 A DAG for X having causal effect on Y.

Corollary 9.1 *If X_i has no parents, i.e. if $\Pi_i = \phi$, then for all $\underline{x}_{\tilde{V}\backslash\{i\}} \in \times_{j\in\tilde{V}\backslash\{i\}} \mathcal{X}_j$ and all $y \in \mathcal{X}_i$,*

$$p_{\underline{X}_{\tilde{V}\backslash\{i\}} \| X_i}(\underline{x}_{\tilde{V}\backslash\{i\}} \| y) = p_{\underline{X}_{\tilde{V}} | X_i}(\underline{x}_{\tilde{V}} | y).$$

Example 9.1 Consider the DAG given in Figure 9.1, for 'X having causal effect on Y'. The factorization of $p_{X,Y}$ along the DAG in Figure 9.1 is

$$p_{X,Y}(x, y) = p_{Y|X}(y|x)p_X(x)$$

and the intervention formula gives

$$p_{Y\|X}(y\|x) = p_{Y|X}(y|x).$$

Since X is a parent of Y, intervening to force $X = x$ transforms the distribution over Y in exactly the same way as *observing* $X = x$. But if instead Y is forced, the intervention formula yields

$$p_{X\|Y}(x\|y) = p_X(x).$$

Clearly, $p_{X\|Y}(x\|y) \neq p_{X|Y}(x|y)$ as functions unless X and Y are independent.

The causal graph in Figure 9.1 may be used to illustrate the following situation, where X denotes 'rain', with values 'yes' or 'no', while Y denotes 'barometer reading', with values 'high' or 'low'.

- The reading in a barometer is useful to predict rain:

$$p \text{ (rain | barometer reading} = \text{high)} > P \text{ (rain | barometer reading} = \text{low)}$$

- But forcing the barometer will not cause rain:

$$P \text{ (rain | barometer} \leftarrow \text{high)} = P \text{ (rain | barometer} \leftarrow \text{low)}$$

Example 9.2: The DAG for a wet pavement Consider the DAG given in Figure 9.2, which represents a causal model for a wet pavement.

The *season* A has four states: spring, summer autumn, winter. Rain B has two states: yes/no. Sprinkler C has two states: on/off. Wet pavement D has two states: yes/no. Slippery pavement has two states: yes/no.

The joint probability distribution is factorized as

$$p_{A,B,C,D,E} = p_A p_{B|A} p_{C|A} p_{D|B,C} p_{E|D}.$$

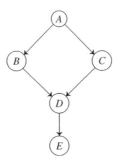

Figure 9.2   DAG for wet pavement, no intervention.

Suppose, without reference to the values of any of the other variables and without reference to the current state of the sprinkler, 'sprinkler on' is now *enforced*. Then

$$p_{A,B,C,D,E}(\cdot|C \leftarrow 1) = \frac{p_{A,B,C,D,E}(\cdot,\cdot,1,\cdot,\cdot)}{p_{C|A}(1|\cdot)}$$

$$= p_A p_{B|A} p_{D|B,C}(\cdot|\cdot,1) p_{E|D}.$$

After *observing* that the sprinkler is on, it may be inferred that the season is dry and that it probably did not rain and so on. If 'sprinkler on' is enforced, without reference to the state of the system when the action is taken, then no such inference should be drawn in evaluating the effects of the intervention. The resulting DAG is given in Figure 9.3. It is the same as before, except that $C = 1$ is fixed and the edge between C and A disappears.

The deletion of the factor $p_{C|A}$ represents the understanding that whatever relationships existed between sprinklers and seasons prior to the action, found from

$$p_{A,B,D,E|C}(\cdot,\cdot,\cdot,\cdot|1)$$

are no longer in effect when the state of the variable is *forced*, as in a controlled experiment, without reference to the state of the system. This is an example of the difference between *seeing* and *doing*. After *observing* that the sprinkler is on, it may be inferred

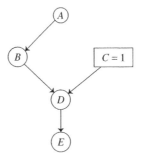

Figure 9.3   Sprinkler 'on' is forced.

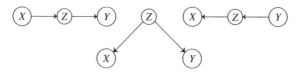

Figure 9.4   Three Markov equivalent graphs.

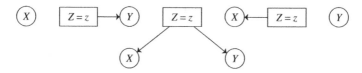

Figure 9.5   Graphs from Figure 9.4 with intervention $Z \leftarrow z$ applied.

that the season is dry, that it probably did not rain and so on. No such inferences may be drawn in evaluating the effects of the intervention 'ensure that the sprinkler is on'.

9.3.1   Establishing the model via a controlled experiment

The three graphs in Figure 9.4 are Markov equivalent. The chains $X \to Z \to Y$ and $X \leftarrow Z \leftarrow Y$ and the fork $X \leftarrow Z \to Y$ are all Markov equivalent, with conditional independence structure $X \perp Y|Z$. If there are *causal* relations between the variables, then it is not possible to distinguish which of the models is appropriate from the data alone.

If one of the graphs in Figure 9.4 represents an appropriate causal structure between the variables, and if it is possible to intervene by controlling the variables Z, then it is possible to distinguish which of the models is appropriate. Figure 9.5 shows the associated structural model when the control $Z \leftarrow z$ has been applied, forcing Z to be independent of its ancestors. A controlled experiment, where the direct causal links between Z and its parent variables have been eliminated, will exhibit independence structure $X \perp (Y, Z)$ in the first case, $X \perp Y|Z$ in the second $(X, Z) \perp Y$ in the third. The original experiment determines the equivalence; the additional controlled experiment, if it is possible to carry it out, will determine which graph within the equivalence class is appropriate.

9.4   Properties of intervention calculus

The following propositions summarize some basic properties of intervention calculus.

Proposition 9.1   *If X_i has no parents, then for all $\underline{x}_{\tilde{V}} \in \mathcal{X}_{\tilde{V}}$*

$$p_{\underline{X}_{\tilde{V}\backslash\{i\}} \| X_i}(\underline{x}_{\tilde{V}\backslash\{i\}} \| x_i) = p_{\underline{X}_{\tilde{V}\backslash\{i\}} | X_i}(\underline{x}_{\tilde{V}\backslash\{i\}} | x_i). \qquad \Box$$

Proposition 9.2 introduces the term *exogeneity*. A variable is exogenous to a model if it is not determined by other parameters and variables in the model, but is set externally and any changes to it come from external forces. In this context, it simply means that 'do' conditioning on a variable affects the offspring in the same way that it would if the variable took that value without the external intervention.

Proposition 9.2 (Exogeneity and Invariance) *For each $j = 1, \ldots, d$, let Π_j denote the set of parent variables for variable X_j, $\tilde{\Pi}_j$ the indexing set for the variables in Π_j, so that $\mathcal{X}_{\tilde{\Pi}_j}$ denotes the state space for Π_j. For each variable in the variable set $V = \{X_1, \ldots, X_d\}$, exogeneity holds, where exogeneity for variable X_j is defined as follows: for each $(x, \pi) \in \mathcal{X}_j \times \mathcal{X}_{\tilde{\Pi}_j}$,*

$$p_{X_j \| \Pi_j}(x \| \pi) = p_{X_j | \Pi_j}(x | \pi). \tag{9.4}$$

For all $j = 1, \ldots, d$ and each $S \subseteq \tilde{V}$ such that $S \cap (\{j\} \cup \{\tilde{\Pi}_j\}) = \phi$, modularity/invariance holds. This is defined as follows for each $(x, \pi, \underline{x}_S) \in \mathcal{X}_j \times \mathcal{X}_{\tilde{\Pi}_j} \times \mathcal{X}_S$,

$$p_{X_j \| \Pi_j, \underline{X}_S}(x \| \pi, \underline{x}_S) = p_{X_j | \Pi_j}(x | \pi). \tag{9.5}$$

Proof of Proposition 9.2 Equation (9.4) is established first. $p_{X_j \| \Pi_j}(. \| \pi)$ is a marginal distribution which depends on the enforced value $\Pi_j \leftarrow \pi$. For all $(x, \pi) \in \mathcal{X}_j \times \mathcal{X}_{\tilde{\Pi}_j}$,

$$p_{X_j \| \Pi_j}(x \| \pi) = \sum_{\underline{x}_{\tilde{V} \backslash \tilde{\Pi}_j} | x_j = x} p_{\underline{X}_{\tilde{V} \backslash \tilde{\Pi}_j} \| \Pi_j}(\underline{x}_{\tilde{V} \backslash \tilde{\Pi}_j} \| \pi).$$

An application of the intervention formula (9.1) yields

$$p_{X_j \| \Pi_j}(x \| \pi) = \sum_{\underline{x}_{\tilde{V} \backslash \tilde{\Pi}_j} | x_j = x} \left(\prod_{v \in \tilde{V} \backslash \tilde{\Pi}_j} p_{X_v | \Pi_v}(x_v | \underline{x}_{\tilde{\Pi}_v}) \right) \Bigg|_{(x_j, \underline{x}_{\tilde{\Pi}_j}) = (x, \pi)}.$$

Successive application of the distributive law, together with

$$\sum_{x \in \mathcal{X}_v} p_{X_v | \Pi_v}(x | \pi_v) = 1$$

for any $\pi_v \in \mathcal{X}_{\tilde{\Pi}_v}$ gives

$$p_{X_j \| \Pi_j}(x \| \pi) = p_{X_j | \Pi_j}(x | \pi)$$

for all $(x, \pi) \in \mathcal{X}_j \times \mathcal{X}_{\tilde{\Pi}_j}$ as required. The proof of Equation (9.5) may be carried out by a similar marginalization. $\qquad \square$

The property described by Equation (9.5) expresses the notion of invariance, or *modularity* found in Woodward [21]. Once all the *direct* causes of a variable X_j are controlled, no other interventions will affect the probability of X_j.

The following property is another straightforward consequence of the definition.

Proposition 9.3 *For any $(x, \pi) \in \mathcal{X}_j \times \mathcal{X}_{\tilde{\Pi}_j}$,*

$$p_{\Pi_j \| X_j}(\pi \| x) = p_{\Pi_j}(\pi).$$

Proof of Proposition 9.3 By marginalization, followed by an application of the intervention formula (9.1), for each $(x, \pi) \in \mathcal{X}_j \times \mathcal{X}_{\tilde{\Pi}_j}$,

$$p_{\Pi_j \| X_j}(\pi \| x) = \sum_{\underline{x}_{\tilde{V} \backslash (\{j\} \cup \{\Pi_j\})}} p_{\underline{X}_{\tilde{V} \backslash (\{j\} \cup \tilde{\Pi}_j)}, \Pi \| X}(\underline{x}_{\tilde{V} \backslash (\{j\} \cup \tilde{\Pi}_j)}, \pi \| x)$$

$$= \sum_{\underline{x}_{\tilde{V} \backslash (\{j\} \cup \tilde{\Pi}_j)}} \left. \frac{p_{\underline{X}_{\tilde{V}}}(\underline{x}_V)}{p_{X_j | \Pi_j}(x | \pi)} \right|_{(x_j, \underline{x}_{\tilde{\Pi}_j}) = (x, \pi)}$$

$$= \left. \frac{\sum_{\underline{x}_{\tilde{V} \backslash (\{j\} \cup \tilde{\Pi}_j)}} p_{X_{\tilde{V}}}(\underline{x}_{\tilde{V}})}{p_{X_j | \Pi_j}(x | \pi)} \right|_{(x_j, \underline{x}_{\tilde{\Pi}_j}) = (x, \pi)}$$

$$= \frac{p_{X_j, \Pi_j}(x, \pi)}{p_{X_j | \Pi_j}(x | \pi)}$$

$$= p_{\Pi_j}(\pi).$$

$\square$

The probability measure *after* intervention is factorized along the mutilated graph. The following proposition determines the probabilities on the mutilated graph.

Proposition 9.4 *Let $A \subset V$ and let $\tilde{A}$ denote the indexing set for the variables in A. Then, for $X_j \notin A$ and any $(x, \underline{x}_{\tilde{\Pi}_j \backslash \tilde{A}}, \underline{x}_A) \in \mathcal{X}_j \times \mathcal{X}_{\tilde{\Pi}_j} \times \mathcal{X}_A$,*

$$p_{X_j | \Pi_j \backslash A \| A}(x | \underline{x}_{\tilde{\Pi}_j \backslash \tilde{A}} \| \underline{x}_{\tilde{A}}) = p_{X_j | \Pi_j \backslash A, \Pi_j \cap A}(x | \underline{x}_{\tilde{\Pi}_j \backslash \tilde{A}}, \underline{x}_{\tilde{\Pi}_j \cap \tilde{A}}),$$

where the conditioning is taken in the sense of: first *the 'do' conditioning $\underline{X}_{\tilde{A}} \leftarrow \underline{x}_{\tilde{A}}$ is applied and* then *the set of variables $\Pi_j \backslash A$ is observed.*

The *causality calculus* means here that the conditional specifications are unchanged for variables which are not used for the intervention.

Proof of Proposition 9.4 By definition of conditional probability,

$$p_{X_j | \Pi_j \backslash A \| A}(x | \underline{x}_{\tilde{\Pi}_j \backslash \tilde{A}} \| \underline{x}_{\tilde{A}}) = \frac{p_{X_j, \Pi_j \backslash A \| A}(x, \underline{x}_{\tilde{\Pi}_j \backslash \tilde{A}} \| \underline{x}_{\tilde{A}})}{p_{\Pi_j \backslash A \| A}(\underline{x}_{\tilde{\Pi}_j \backslash \tilde{A}} \| \underline{x}_{\tilde{A}})}.$$

An application of the intervention formula to the numerator gives

$$p_{X_j, \Pi_j \backslash A \| A}(x, \underline{x}_{\tilde{\Pi}_j \backslash \tilde{A}} \| \underline{x}_{\tilde{A}}) = \prod_{v \in \{j\} \cup \tilde{\Pi}_j \backslash \tilde{A}} p_{X_v | \Pi_v}(x_v | \underline{x}_{\tilde{\Pi}_v})$$

and to the denominator gives

$$p_{\Pi_j \backslash A \| A}(\underline{x}_{\tilde{\Pi}_j \backslash \tilde{A}} \| \underline{x}_{\tilde{A}}) = \prod_{v \in \tilde{\Pi}_j \backslash \tilde{A}} p_{X_v | \Pi_v}(x_v | \underline{x}_{\tilde{\Pi}_v}).$$

Putting these together clearly gives

$$p_{X_j|\Pi_j\backslash A\|A}(x|\underline{x}_{\tilde{\Pi}_j\backslash\tilde{A}}\|\underline{x}_{\tilde{A}}) = p_{X_j|\Pi_j\backslash A,\Pi_j\cap A}(x|\underline{x}_{\tilde{\Pi}_j\backslash\tilde{A}},\underline{x}_{\tilde{\Pi}_j\cap\tilde{A}}),$$

as claimed. The proof is complete. □

As Lauritzen [125] states, the intervention calculus is shown to have the property that conditional specifications are unchanged for variables that are not used in the intervention.

Example 9.3: Wet pavement revisited Consider the conditional probability $p_{D|B\|C}(.|.\|1)$. Here B and C are parents of D, and in the notation of the preceding proposition (using A in the sense of the previous proposition), $D = X_j$, $\Pi_j\backslash A = B$, $X_{\tilde{A}} = C$ and $x_{\tilde{A}} = x_{\tilde{\Pi}_j\cap\tilde{A}} = 1$. Plugging into the formula in the preceding proposition,

$$p_{D|B\|C}(.|.\|1) = p_{D|B,C}(.|.,1).$$

The right hand side may be thought of as a *pre-intervention probability*, which can be estimated from the data *before* the intervention $C \leftarrow 1$ is made. In this case, an estimate of the pre-intervention probability $p_{D|B,C}(.|., 1)$ is also an estimate of the *post-intervention* probability $p_{D|B\|C}(.|.\|1)$.

9.5   Transformations of probability

The following proposition is almost a direct consequence of the definition. It presents a simple rearrangement of the intervention formula in a special case.

Proposition 9.5

$$p_{\underline{X}_{\tilde{V}\backslash\{j\}\|X_j}}(\underline{x}_{\tilde{V}\backslash\{j\}}\|x) = p_{\underline{X}_{\tilde{V}\backslash(\{j\}\cup\tilde{\Pi}_j)}|X_j,\Pi_j}(\underline{x}_{\tilde{V}\backslash(\{j\}\cup\tilde{\Pi}_j)}|x,\underline{x}_{\tilde{\Pi}_j})p_{\Pi_j}(\underline{x}_{\tilde{\Pi}_j})$$

Proof of Proposition 9.5 An application of the definition gives

$$p_{\underline{X}_{\tilde{V}\backslash\{j\}\|X_j}}(\underline{x}_{\tilde{V}\backslash\{j\}}\|x) = \prod_{v\in\tilde{V}\backslash\{j\}} p_{X_v|\Pi_v}(x_v|\underline{x}_{\tilde{\Pi}_j}).$$

One term has been removed in the product, namely, $p_{X_j|\Pi_j}(x|\underline{x}_{\tilde{\Pi}_j})$, so that (with $x_j = x$)

$$\prod_{v\in\tilde{V}\backslash\{j\}} p_{X_v|\Pi_v}(x_v|\underline{x}_{\tilde{\Pi}_j}) = \frac{p_{\underline{X}_{\tilde{V}}}(\underline{x}_{\tilde{V}})}{p_{X_j|\Pi_j}(x|\underline{x}_{\tilde{\Pi}_j})}$$

$$= \frac{p_{\underline{X}_{\tilde{V}}}(\underline{x}_{\tilde{V}})p_{\Pi_j}(\underline{x}_{\tilde{\Pi}_j})}{p_{X_j,\Pi_j}(x,\underline{x}_{\tilde{\Pi}_j})}$$

$$= p_{\underline{X}_{\tilde{V}\backslash(\{j\}\cup\tilde{\Pi}_j)}|X_j,\Pi_j}(\underline{x}_{\tilde{V}\backslash(\{j\}\cup\tilde{\Pi}_j)}|x,\underline{x}_{\tilde{\Pi}_j})p_{\Pi_j}(\underline{x}_{\tilde{\Pi}_j})$$

as required. □

The effect of the intervention may be viewed as follows: all the conditional probability potentials remain the same except for $p_{X_j|\Pi_j}$. After the intervention $X_j \leftarrow x$, this is replaced by the potential:

$$p_{X_j|\Pi_j}(x|\pi) = 1 \qquad \forall \pi \in \mathcal{X}_{\tilde{\Pi}_j},$$

and

$$p_{X_j|\Pi_j}(y|\pi) = 0 \qquad \forall y \in \mathcal{X}_j\backslash\{x\}, \pi \in \mathcal{X}_{\tilde{\Pi}_j}.$$

Consider the probability over the remaining variables in $V\backslash\{X_j\}$ after conditioning on $\{X_j = x\}$. The following two equations illustrate the differences in the way the probability mass is distributed following 'see' conditioning on the one hand and 'do' conditioning on the other.

$$p_{\underline{X}_{\tilde{V}\backslash A}|\underline{X}_A}(\underline{x}_{\tilde{V}\backslash A}|\mathrm{see}(\underline{x}_A)) = p_{\underline{X}_{\tilde{V}\backslash A}|\underline{X}_A}(\underline{x}_{\tilde{V}\backslash A}|\underline{x}_A) = \frac{p_{\underline{X}_{\tilde{V}}}(\underline{x}_{\tilde{V}})}{p_{\underline{X}_A}(\underline{x}_A)}.$$

It follows that for any set of variables

$$p_{\underline{X}_{\tilde{V}\backslash A}\|\underline{X}_A}(\underline{x}_{\tilde{V}\backslash A}\|\mathrm{do}(\underline{x}_A)) = \prod_{v\in\tilde{V}\backslash A} p_{X_v|\Pi_v}(x_v|\underline{x}_{\tilde{\Pi}_v}).$$

Proposition 9.6 (Adjustment for Direct Causes) *Let $B \subset V$ be a set of random variables in a DAG that are disjoint from $\{X_j\} \cup \Pi_j$ and let $\tilde{B}$ denote the indexing set for the variables in B. Then for any $(x, \underline{x}_{\tilde{B}}) \in \mathcal{X}_j \times \mathcal{X}_{\tilde{B}}$,*

$$p_{B\|X_j}(\underline{x}_{\tilde{B}}\|x) = \sum_{\mathcal{X}_{\tilde{\Pi}_j}} p_{B|X_j,\Pi_j}(\underline{x}_{\tilde{B}}|x, \underline{x}_{\tilde{\Pi}_j})p_{\Pi_j}(\underline{x}_{\tilde{\Pi}_j}). \tag{9.6}$$

Proof of Proposition 9.6 Firstly,

$$p_{B\|X_j}(\underline{x}_{\tilde{B}}\|x) = \sum_{\mathcal{X}_{\tilde{V}\backslash(\tilde{B}\cup\{j\})}} p_{V\backslash\{j\}}(\underline{x}_{\tilde{V}\backslash\{j\}}|x).$$

By Proposition 9.5, this may be written as

$$p_{B\|X_j}(\underline{x}_{\tilde{B}}\|x) = \sum_{\mathcal{X}_{\tilde{V}\backslash(\tilde{B}\cup\{j\})}} p_{V\backslash(\{X_j\}\cup\Pi_j)}(\underline{x}_{\tilde{V}\backslash(\{j\}\cup\tilde{\Pi}_j)}|x, \underline{x}_{\tilde{\Pi}_j})p_{\Pi_j}(\underline{x}_{\tilde{\Pi}_j}).$$

A marginalization over $\mathcal{X}_{\tilde{V}\backslash(\tilde{B}\cup\{j\}\cup\tilde{\Pi}_j)}$ gives

$$p_{B|X_j}(\underline{X}_{\tilde{B}}|x) = \sum_{\mathcal{X}_{\tilde{\Pi}_j}} p_{B|X_j,\Pi_j}(\underline{x}_{\tilde{B}}|x, \underline{x}_{\tilde{\Pi}_j})p_{\Pi_j}(\underline{x}_{\tilde{\Pi}_j})$$

as required. The proof is complete. □

In Proposition 9.6, the 'do' probability is computed by first 'see' conditioning on the direct causes Π_j of X_j and then averaging over them.

9.6   A note on the order of 'see' and 'do' conditioning

The causal calculus is well defined for a causal network if *firstly* some of the variables are forced, irrespective of the state of the network, and *then* some other variables are observed. Consider a set of variables $V = \{X_1, \ldots, X_d\}$. Let A, B and C denote three disjoint subsets of variables and $\tilde{A}, \tilde{B}, \tilde{C}$ their indexing sets. Conditioning is to be read from right to left. The quantity

$$p_{A|B\|C}(x_{\tilde{A}}|x_{\tilde{B}}\|x_{\tilde{C}})$$

may be computed quite easily using the calculus developed so far by first modifying the network according to the 'do' conditioning $C \leftarrow x_{\tilde{C}}$, and then applying Bayes' rule to condition on $\{B = x_{\tilde{B}}\}$ in the usual way.

In terms of the 'causal' interpretation defined above, conditioning the other way round, first *see* $B = x_{\tilde{B}}$ and then *force* $C = x_{\tilde{C}}$ does not appear to be so well defined. It is difficult to give it a 'causal' interpretation, particularly if there are variables in C that are ancestors of variables in B. Mathematically, the 'see' followed by 'do' conditioning, for a causal network, can be defined, although the result may not have a practical value. If the 'see' conditioning is carried out first, then these variables can no longer be influenced by the 'do' conditioned variables, even though the 'do' conditioned variables may be ancestors.

If $B = x_{\tilde{B}}$ is *observed*, these variables are now fixed, so the states of the variables in C can no longer have any influence. More precisely, to define 'see' followed by 'do' conditioning, one has to assume that the variables $V = \{X_1, \ldots, X_d\}$ have a *causal order*, $(X_{\sigma(1)}, \ldots, X_{\sigma(d)})$, where $X_{\sigma(1)}$ has no ancestors and, for each $j \in \{2, \ldots, d\}$, the parent set is either empty, or is chosen from $(X_{\sigma(1)}, \ldots, X_{\sigma(j-1)})$. Furthermore, it is assumed that the same causal order holds for the remaining random variables *after* a 'see' conditioning of the network. After the 'see' conditioning $C = x_{\tilde{C}}$, the conditional probability potentials $p_{X_{\sigma(j)}|\Pi_{\sigma(j)}}$ are replaced, for $\sigma(j) \notin \tilde{C}$, by

$$\tilde{p}_{X_{\sigma(j)}|\Pi_{\sigma(j)}\backslash C} = p_{X_{\sigma(j)}|\Pi_{\sigma(j)}\backslash C, C}(.|., x_{\tilde{C}}).$$

The 'do' conditioning is then applied to this new network.

Example 9.4   Consider three variables (X_1, X_2, X_3), where the probability distribution may be factorized along the DAG given in Figure 9.6. Consider $p_{X_1|X_3\|X_2}(x_1|x_3\|x_2)$ and $p_{X_1\|X_2|X_3}(x_1\|x_2|x_3)$, where the conditioning is taken from right to left.

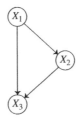

Figure 9.6   DAG for 'see' and 'do' example.

$$X_1 \longrightarrow X_2$$

Figure 9.7   DAG for 'see' and 'do' example, after 'see' conditioning.

In terms of the original tables,

$$p_{X_1|X_3\|X_2}(x_1|x_3\|x_2) = \frac{p_{X_1}(x_1)p_{X_3|X_1,X_2}(x_3|x_1,x_2)}{\sum_y p_{X_1}(y)p_{X_3|X_1,X_2}(x_3|y,x_2)}. \tag{9.7}$$

If the 'see' conditioning $X_3 = x_3$ is applied first, the network after this conditioning is given by Figure 9.7. The conditional probability potentials, after conditioning on $X_3 = x_3$, are given by $\tilde{p}_{X_1}$ and $\tilde{p}_{X_2|X_1}$, which are given in terms of the original potentials in Equations (9.8) and (9.9).

$$\tilde{p}_{X_1}(.) = p_{X_1|X_3}(.|x_3) = \frac{\sum_z p_{X_1}(.)p_{X_3|X_1,X_2}(x_3|.,z)p_{X_2|X_1}(z|.)}{\sum_{y,z} p_{X_1}(y)p_{X_3|X_1,X_2}(.|y,z)p_{X_2|X_1}(z|y)} \tag{9.8}$$

$$\tilde{p}(X_2|X_1) = p(X_2|X_1, X_3 = x_3) = \frac{p(X_1)p(X_2|X_1)p(X_3 = x_3|X_1, X_2)}{\sum_y p(X_1)p(X_2 = y|X_1)p(X_3 = x_3|X_1, X_2 = y)}. \tag{9.9}$$

Now, the application of the 'do $X_2 \leftarrow x_2$' conditioning breaks the causal link between X_1 and X_2, so that

$$p_{X_1\|X_2|X_3}(x_1\|x_2|x_3) = \tilde{p}_{X_1}(x_1) = \frac{\sum_z p_{X_1}(x_1)p_{X_3|X_1,X_2}(x_3|x_1,z)p_{X_2|X_1}(z|x_1)}{\sum_{y,z} p_{X_1}(y)p_{X_3|X_1,X_2}(x_3|y,z)p_{X_2|X_1}(z|y)}. \tag{9.10}$$

The formula is clearly different from the formula for $p_{X_1|X_3\|X_2}(x_1|x_3\|x_2)$, which is given by Equation (9.7).

Although a mathematical definition may be given to applying a 'see' conditioning first, it is difficult to see how to make sense of this in terms of causality.

9.7   The 'Sure Thing' principle

The following result is taken from [1], p. 181, where Pearl refers to it as the 'Sure Thing' principle.

Proposition 9.7 *Consider three binary variables A, B, C with the network given in Figure 9.10. If*

$$p_{B|C\|A}(1|1\|1) < p_{B|C\|A}(1|1\|0)$$

and

$$p_{B|C\|A}(1|0\|1) < p_{B|C\|A}(1|0\|0)$$

then

$$p_{B\|A}(1\|1) < p_{B\|A}(1\|0).$$

The notation means: first A is forced, then C is observed.

Proof of Proposition 9.7 Firstly,

$$p_{B\|A}(1\|1) = p_{B|C\|A}(1|1\|1)p_{C\|A}(1\|1) + p_{B|C\|A}(1|0\|1)p_{C\|A}(0\|1).$$

Since C is a parent of A,

$$p_{C\|A}(.\|1) = p_C(.).$$

It follows that

$$p_{B\|A}(1\|1) = \sum_{x=0}^{1} p_{B|C\|A}(1|x\|1)p_{C\|A}(x\|1) = \sum_{x=0}^{1} p_{B|C\|A}(1|x\|1)p_C(x).$$

Similarly,

$$p_{B\|A}(1\|A0) = \sum_{x=0}^{1} p_{B|C\|A}(1|x\|0)p_C(x).$$

It now follows directly from the assumptions that

$$p_{B\|A}(1\|1) < p_{B\|A}(1\|0),$$

as advertized.   ☐

Simpson's paradox resolved by a controlled experiment   Consider three binary variables, A, B and C. Simpson's paradox is the observation that there are situations where

$$\frac{p_{B|C,A}(1|1,1)/p_{B|C,A}(0|1,1)}{p_{B|C,A}(1|1,0)/p_{B|C,A}(0|1,0)} > 1 \quad \text{and} \quad \frac{p_{B|C,A}(1|0,1)/p_{B|C,A}(0|0,1)}{p_{B|C,A}(1|0,0)/p_{B|C,A}(0|0,0)} > 1,$$

but $\frac{p_{B|A}(1|1)/p_{B|A}(0|1)}{p_{B|A}(1|0)p_{B|A}(0|0)} < 1$. For example, suppose that A denotes 'treatment', B denotes 'recovery' and C denotes 'gender'. With no information about the causal relations between the variables, Simpson's paradox states that even if the 'treatment' may improve the chances of recovery for both men and women, it may nevertheless be bad for the population as a whole.

If the model is that shown in Figure 9.8, where A denotes 'treatment', $(1 = \text{applied}, 0 = \text{no treatment})$, C 'gender' (male/female), B recovery (yes/no) and the value of A is randomly assigned to each individual so that a proper controlled experiment is carried out, then the causal link between C and A is broken. The 'sure thing' principle may then be applied, which states that if the treatment improves the chances of recovery for both men and women, it is good for the population as a whole.

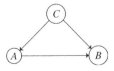

Figure 9.8   A = treatment, B = recovery, C = gender.

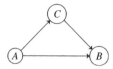

Figure 9.9   A = treatment/B = recovery/C = blood pressure.

Now consider the DAG given in Figure 9.9, where A denotes treatment, B recovery and C blood pressure.

Suppose that 'treatment' is randomly assigned to individuals in an appropriate manner so that a controlled experiment is carried out. In this model, blood pressure is on the causal pathway influencing recovery. The controlled experiment has not removed the direct causal link between A and B and hence the possibility of Simpson's paradox cannot be excluded. It could be that although the treatment is comparatively good within the group where high blood pressure is observed after treatment and also comparatively good within the group where low blood pressure is observed after treatment, it may be bad for the population as a whole. This could happen if 'treatment' increases blood pressure and increased blood pressure reduces the chances of recovery.

9.8   Back door criterion, confounding and identifiability

Given a causal Bayesian network and observational data, where the values of the variables in a set A have been forced (due, for example, to a controlled experiment), the task is to estimate the conditional probability distribution over the remaining variables; $p_{V \backslash A \| A} \left(x_{\bar{V} \backslash \bar{A}} \| x_{\bar{A}} \right)$. The problem is simplified if $p_{V \backslash A \| A} \left(x_{\bar{V} \backslash \bar{A}} \| x_{\bar{A}} \right)$ can be modified until no 'do' operations appear, so that the required conditional probability potentials may be estimated using observational data, thus reducing a causal query to a probabilistic query.

Confounding   Consider the DAG given in Figure 9.10.
The factorization is

$$p_{A,B,C} = p_{B|A,C} \, p_{A|C} \, p_C.$$

Consider $p_{B \| A}(.\|a)$. Note that

$$p_{B \| A}(.\|a) = \sum_{c \in \mathcal{X}_C} p_{B,C \| A}(., c \| a)$$

and that

$$p_{B,C \| A}(., . \| a) = p_{B|C \| A}(.|.\|a) p_{C \| A}(.\|a) = p_{B|A,C}(.|a, .) p_C,$$

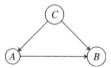

Figure 9.10   Illustration for Confounding.

where in the second term, the 'do' conditioning of $A \leftarrow a$ is applied first, and then C is observed. It follows that

$$p_{B\|A}(.\|a) = \sum_{c \in \mathcal{X}_C} p_{B|A,C}(.|a, c) p_C(c).$$

This shows that to estimate $p_{B\|A}(.\|a)$ from data, it is necessary to be able to estimate the potentials $p_{B|A,C}$ and p_C. If C is *observable*, then the effect on the probability potentials of B of manipulating A may be estimated. But if C is a *hidden* random variable (sometimes the term *latent* is used) in the sense that no *direct* sample of the outcomes of C may be obtained, it will not be possible to estimate the probabilities used on the right hand side and hence it will not be possible to predict the effect on B of manipulating A. This is known as *confounding*.

Semi-Markovian model   The model described above is an example of a *semi-Markovian model*. Let $V = \{X_1, \ldots, X_d\}$ and suppose that the probability distribution over the variables V may be factorized along a directed acyclic graph $\mathcal{G} = (V, E)$. Now suppose that $V = U \cup Z$, where U is a set of *unobserved* variables, while Z is the set of observed variables. Assume that in $\mathcal{G}$, no variable in U is a descendant of any variable in Z. Such a model is known as a *semi-Markovian model*, following [126]. Let $\tilde{Z}$ and $\tilde{U}$ denote the indexing sets for the variables in Z and U respectively. The joint distribution of the observed variables becomes a mixture of products of conditional probabilities,

$$p_Z\left(\underline{x}_{\tilde{Z}}\right) = \sum_{\mathcal{X}_{\tilde{U}}} \prod_{v \in \tilde{Z}} p_{X_v|\Pi_v}\left(x_v \mid \underline{x}_{\tilde{\Pi}_v}\right) p_U\left(\underline{x}_{\tilde{U}}\right). \tag{9.11}$$

Now consider an intervention $Z_T \leftarrow \underline{x}_{Z_T}$, where $Z_T \subset Z$. Then the 'do' conditional probability for the remaining variables in Z is

$$p_Z\left(\underline{x}_{\tilde{Z}}\right) = \sum_{\mathcal{X}_{\tilde{U}}} \prod_{v \in \tilde{Z} \backslash \tilde{Z}_T} p_{X_v|\Pi_v}\left(x_v \mid \underline{x}_{\tilde{\Pi}_v}\right) p_U\left(\underline{x}_{\tilde{U}}\right). \tag{9.12}$$

The question of *identifiability*, considered next, is whether it is possible to express

$$p_{Z \backslash Z_T \| Z_T}\left(\underline{x}_{\tilde{Z} \backslash \tilde{Z}_T} \| \underline{x}_{\tilde{Z}_T}\right)$$

uniquely as a function of the observed distribution $p_Z\left(\underline{x}_{\tilde{Z}}\right)$; that is, without involving either the unknown conditional probability tables $P_{Z_v|\Pi_v}$ for $v \in \{1, \ldots, d\}$ such that $\Pi_v \cup U \neq \phi$ or the unknown distribution $p_U\left(\underline{x}_{\tilde{U}}\right)$.

Back door criterion   Recall that there are three basic types of connection in a DAG: chain, collider and fork. Any sequence of nodes with edges between successive nodes, regardless of direction, is known as a *trail*. Two subsets of nodes A and B are *d*-separated by a set of nodes C if on all trails between a node in A and B there is an intermediate node X such that

- **either** the connection is a chain or a fork and $X \in C$

- **or** the connection is a collider and neither X *nor any of its descendants* are in C.

The notation

$$A \perp B \|_{\mathcal{G}} C$$

denotes that A and B are d-separated by C. That is, C *blocks* every trail from a node in A to a node in B; if all the nodes of C are instantiated, then there are no active trails between A and B. Theorem 2.2 states that $A \perp B\|_{\mathcal{G}} C$ implies that $A \perp B|C$. That is, if the probability distribution factorizes along the graph $\mathcal{G}$, then d-separation implies conditional independence.

Definition 9.2 (Back Door Criterion) *A set of nodes C satisfies the* back door criterion *relative to an ordered pair of nodes $(X_i, X_j) \in V \times V$ if*

1. *no node of C is a descendant of X_i **and***

2. *C blocks every trail (in the sense of d-separation) between X_i and X_j which contains an edge pointing to X_i.*

If A and B are two disjoint subsets of nodes, C is said to satisfy the back door criterion relative to (A, B) if it satisfies the back door criterion relative to any pair $(X_i, X_j) \in A \times B$.

The name 'back door criterion' reflects the fact that the second condition requires that only trails with nodes pointing at X_i be blocked. The remaining trails can be seen as entering X_i through a back door.

Example 9.5   Consider the back door criterion DAG, given in Figure 9.11. The sets of variables $C_1 = \{X_3, X_4\}$ and $C_2 = \{X_4, X_5\}$ satisfy the back door criterion relative to the ordered pair of nodes (X_i, X_j), whereas $C_3 = \{X_4\}$ does *not* satisfy the criterion relative to the ordered pair of nodes (X_i, X_j); if X_4 is instantiated, the Bayes ball may pass through the *collider* connection from X_1 to X_2.

Identifiability   Suppose that $X_1, \ldots, X_n$ and Z are sets of variables in a Bayesian network and that Z satisfies the back door criterion with respect to (X_i, X_j). The aim is to show that the set of variables Z plays a similar role to the variable C in the discussion on confounding. Firstly, since no variables of Z are descendants of X_i, it follows that

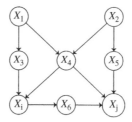

Figure 9.11   Back door criterion.

$p_{Z\|X_i}(.|x_i) = p_Z(.)$. This is seen as follows: first, marginalise over all variables that are descendants of X_i. Now consider the resulting reduced DAG, where all descendants of X_i have been eliminated. The result follows directly, by noting that since no variables of the set Z are descendants of X_i, and the conditional probability tables for the Bayesian network remain unchanged except that the causal links between the set of variables $\Pi_i \cup \{X_i\}$ removed. It follows that

$$p_{X_j\|X_i}(.\|x_i) = \sum_{z\in\mathcal{X}_Z} p_{X_j,Z\|X_i}(., z\|x_i)$$

$$= \sum_{z\in\mathcal{X}_Z} p_{X_j|Z\|X_i}(.|z\|x_i)p_{Z\|X_i}(z\|x_i)$$

where the conditioning is to be taken that *first* $X_i \to x_i$ is imposed and *then* Z is observed. Since Z d-separates X_i from X_j via any trail with arrow pointing into X_i, it follows that the same probability tables are used for the computation of $p_{X_j|Z\|X_i}(.|.\|x_i)$ and $p_{X_j|Z,X_i}(.|., x_i)$, from which

$$p_{X_j|Z\|X_i}(.|.\|x_i) = p(X_j|Z, X_i = x_i),$$

so that

$$p_{X_j\|X_i}(.\|x_i) = \sum_{z\in\mathcal{X}_Z} p_{X_j|Z,X_i}(.|z, x_i)p_Z(z). \tag{9.13}$$

If a set of variables Z satisfying the back door criterion with respect to (X_i, X_j) can be chosen such that p_Z and $p_{X_j|Z,X_i}$ can be estimated from the observed data, then the distribution $p_{X_j\|X_i}$ can also be estimated from the observed data.

Definition 9.3 (Identifiability) *If a set of variables Z satisfies the back door criterion relative to (X, Y), then the causal effect of X to Y is given by the formula*

$$p_{Y\|X}(.\|x) = \sum_{z\in\mathcal{X}_Z} p_{Y|X,Z}(.|x, z)p_Z(z) \tag{9.14}$$

and the causal effect of X on Y is said to be identifiable.

The formula given in Equation (9.14) is named *adjustment for concomitants*. The word *identifiability* refers to the fact that the existence of the concomitants Z satisfying the back door criterion makes it possible to compute, or identify, $p_{Y\|X}(y\|x)$ uniquely from any p which is strictly positive over V.

Notes   In the main, Chapter 9 presents material found in [125], [1] and [121]. The paper [127] summarizes the recent developments in the problem of identifiability and presents an algorithmic solution. The identifiability question asks whether it is possible to compute the probability of some set of effect variables given an intervention on another set of variables in the presence of non-observable variables, using data that is not obtained from a controlled experiment. The results by Y. Huang, M. Valtorta in [127] show that the do-calculus rules of J. Pearl [118] and [128] are complete in the sense that if a causal effect is identifiable, then there exists a sequence of applications of the 'do' rules that transforms the causal effect formula to a formula that only contains observational quantities. The philosophical paper [129] argues for Bayesian networks as the proper representation of stochastic causality.

9.9   Exercises: Causality and intervention calculus

1. Consider a Bayesian network with the following DAG

Figure 9.12   Directed acyclic graph.

where A, B and C are binary variables (i.e. taking values either 0 or 1), together with probabilities

$$p_A = \left(\frac{1}{2}, \frac{1}{2} \right)$$

$p_{B|A} =$

A\B	1	0
1	0.75	0.25
0	0.25	0.75

$p_{C|B} =$

B\C	1	0
1	0.125	0.875
0	0.675	0.375

    (a) Compute $p_{B\|C}(1\|1)$.

    (b) Compute $p_{A\|C}(1\|1)$.

    (c) Compute $p_{A\|B}(1\|1)$

    (d) Compute $p_{C\|B}(1\|1)$

2. Consider the *joint* probability table for the three variables A, B and C.

$p_{A,B,C} =$

A	B	C 1	C 0
1	1	0.15	0.22
	0	0.04	0.09
0	1	0.1	0.03
	0	0.26	0.11

Compute $p_{A,B}$, p_B, $p_{B,C}$ and $p_{C|A,B}(1|.,.)$. Compare your answers with $p_{C|B}(1|.)$. Show that this is an example of *Simpson's paradox*.

    Suppose that the values in the table for $p_{A,B,C}$ are the 'empirical' probabilities obtained from 400 observations. The following model is considered to be appropriate.

Figure 9.13   Directed acyclic graph.

(a) Which conditional probabilities do we need for this Bayesian network?

(b) What are the estimates of $p_{A,B,C}$ obtained using these estimates for the conditional probability?

(c) How many parameters does this model have?

(d) Let $\mathbf{x}$ denote the empirical probabilities and $\mathbf{y}$ denote the fitted probabilities. Calculate

$$d_K(\mathbf{x}, \mathbf{y}).$$

If the fitted model holds, then one would expect

$$400 d_K(\mathbf{x}, \mathbf{y}) \sim \chi_1^2.$$

The reason: the empirical distribution has seven parameters, while the fitted model has six parameters. For large numbers of observations, the Kullback-Leibler distance is approximately $\sum_j \frac{(x_j - y_j)^2}{x_j}$. Multiply by the number of observations to obtain the χ^2 statistic. Degrees of freedom is the difference in number of parameters.

(e) Assuming that the probabilities given in the table for $p(A, B, C)$ are exact, and the distribution factorizes according to the DAG given in Figure 9.14, compute

$$p_{C|A\|B}(1|1\|1)$$

$$p_{C|A\|B}(1|1\|0)$$

$$p_{C|A\|B}(1|0\|1)$$

$$p_{C|A\|B}(1|0\|0)$$

and compare with $p_{C\|B}(1\|1)$ and $p_{C\|B}(1\|0)$.

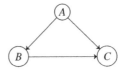

Figure 9.14   DAG for A, B, C.

3. Let A, B, C, W be disjoint sets of nodes in a Bayesian network. Let $\mathcal{G}$ denote the directed acyclic graph describing the causal network, and let $\mathcal{G}^{-C}$ denote the graph with all edges between C and parents of C removed.

Prove that if A and B are d-separated by (C, W) on the graph $\mathcal{G}^{-C}$, then

$$p_{A|W,B\|C}(x_A|x_W, x_B, \|x_C) = p_{A|W\|C}(x_A|x_W\|x_C),$$

where the conditioning is performed from right to left.

4. Let $\mathcal{G}$ be a directed acyclic graph, and suppose that a probability distribution p may be factorized along $\mathcal{G}$. Let $\mathcal{G}^{-X}$ denote the graph obtained by deleting from $\mathcal{G}$ all arrows pointing towards X (that is, all links between X and its parents are deleted). Prove that if Y and Z are d-separated in $\mathcal{G}^{-X}$ by X, then

$$p_{Y|Z\|X}(\cdot|\cdot\|x) = p_{Y\|X}(\cdot\|x),$$

where the conditioning is taken from right to left.

5. Suppose the causal relations between the variables $(X_1, X_2, X_3, X_4, X_5, X_6, Y, Z)$ may be expressed by the DAG given in Figure 9.15. Prove that $C_1 = \{X_1, X_2\}$ and $C_2 = \{X_4, X_5\}$ satisfy the back door criterion relative to the ordered pair of nodes (Y, Z), while $C_3 = \{X_4\}$ does not. Which sets of nodes satisfy the back door criterion with respect to the ordered set of nodes (Z, Y)?

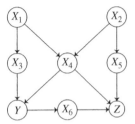

Figure 9.15   Causal relations between variables.

6. Let a set of variables C satisfy the back door criterion relative to (X, Y). Prove that

$$p_{Y\|X}(y\|x) = \sum_c p_{Y|C\|X}(y|c\|x)p_{C\|X}(c\|x).$$

7. Let C be a set of variables in a Bayesian network and let X be a variable such that C contains no descendants of X. Prove, from the definition, that

$$p_{C\|X}(c\|x) = p_C(c).$$

8. Let $V = \{X_1, \dots, X_d\}$ denote a set of variables. Let $V = Z \cup U$, where the variables in Z are observable and the variables in U are unobservable. Assume that the probability distribution over the variables in V may be factorized along a directed acyclic graph $\mathcal{G} = (V, E)$, where no variable in U is a descendant of any variable in Z; that is, the model is semi-Markovian. Consider a single variable, say $X_j \in Z$. Assume that there exists no fork of the form $\{(X_i, X_j), (X_i, X_k)\}$, where $X_k \in Z$ and $j \neq k$, and $X_i \in U$ is an unobservable variable. That is, there are no confounders between X_j and the rest of the observable variables. Then show that

$$p_{Z\setminus\{X_j\}\|X_j}(\underline{x}_{\bar{Z}\setminus\{j\}}\|x_j) = p_{Z\setminus(\{x_j\}\cup\Pi_j)|X_j.\Pi_j}\left(\underline{x}_{\bar{Z}\setminus(\{j\}\cup\bar{\Pi}_j)}|x_j, \underline{x}_{\bar{\Pi}_j}\right) p_{\Pi_j}(\underline{x}_{\bar{\Pi}_j}).$$

10

The junction tree and probability updating

A Bayesian network presents a factorization of a probability distribution according to a directed acyclic graph. Chapter 4 introduced the basic material about decomposable graphs that enabled the construction of a junction tree from a directed acyclic graph. Trees often provide the basis of efficient algorithms and in this chapter, the junction tree is used as the basis for developing a method of updating the probability distribution.

The *probability updating task* is the following: let $\underline{X} = (X_1, \ldots, X_d)$ denote a set of variables, with a probability function $p_{\underline{X}}$ that factorizes along a directed acyclic graph to form a Bayesian network. The task is to compute the conditional probability distribution of $\underline{X}$ given some evidence (Definition 3.1) that is entered into the network. Chapter 10 develops a method based on the junction tree and applies it for hard evidence received in a network where all the variables are discrete. Section 10.9 extends the technique to hard evidence received for a conditional Gaussian (CG) distribution (defined in Section 8.9) and Section 10.10 to virtual and soft evidence.

10.1 Probability updating using a junction tree

Let $\mathbf{e} = (e_1, \ldots, e_m)$ denote hard evidence potential (Definition 3.1); that is, a collection of hard findings. A hard finding is defined as an instantiation of a variable in the network; for each $j \in \{1, \ldots, m\}$, e_j is a potential containing 0s and 1s corresponding to the instantiation. The task is to compute the conditional distribution $p_{\underline{X}|\mathbf{e}}$ given the hard

Bayesian Networks: An Introduction   T. Koski, J. Noble
© 2009 John Wiley & Sons, Ltd

evidence **e**. Following Equation (3.3), the conditional distribution is defined as

$$p_{\underline{X}|\mathbf{e}} = \frac{p_{\underline{X};\mathbf{e}}}{p(\mathbf{e})},$$

where

$$p(\mathbf{e}) = \sum_{\underline{x} \in \mathcal{X}} p_{\underline{X};\mathbf{e}}(\underline{x})$$

and

$$p_{\underline{X};\mathbf{e}} = p_{\underline{X}} \prod_{j=1}^{m} e_j.$$

The main object of this chapter is to illustrate how this may be carried out effectively using the following procedure:

1. Moralize the DAG of the Bayesian network.

2. Triangulate the moralized graph.

3. Let the cliques of the triangulated graph be the nodes of a tree, which is a *junction tree*.

4. Use the Markov properties of the probability distribution to associate potentials to the separators and nodes of the junction tree.

5. Propagate (i.e. send messages to update the potentials on the separators and nodes of the junction tree) through the junction tree.

For producing a tree that is computationally efficient, the most important step of the construction is the triangulation of the moralized graph. There are many ways to add edges to triangulate a graph and it is important, with large networks, to find a method that is optimal. An efficient triangulation will lead to a junction tree that produces the smallest possible largest clique size, and should find the triangulation within polynomial time.

An algorithm for constructing a triangulation that is close to optimal, when the state spaces $(\mathcal{X}_v)_{v=1}^{d}$ are the same size for each variable in $(X_1, \ldots, X_d)$ is found in [130].

In general, probabilistic inference is an NP-hard problem. The complexity of the inference techniques are not discussed in this text; an analysis of the complexity is found in [82].

10.2   Potentials and the distributive law

Notations   The following paragraph repeats various notations. Let $\tilde{V} = \{1, \ldots, d\}$ denote the indexing set for the d nodes for a graph $\mathcal{G} = (V, E)$. To each node is associated a *random variable*. To each node $j \in \tilde{V}$, is associated a finite state space $\mathcal{X}_j = (x_j^{(1)}, \ldots, x_j^{(k_j)})$, the set of possible states for the random variable X_j. The state space of $\underline{X} = (X_1, \ldots, X_d)$ is denoted by $\mathcal{X} = \times_{j=1}^{d} \mathcal{X}_j$. Let $D \subset \tilde{V}$ denote a subset of the nodes. The notation

$$\underline{x}_D = \times_{v \in D} x_v$$

is used to denote a configuration (or a collection of outcomes) on the nodes in D. The state space of the variables in D is denoted by

$$\mathcal{X}_D = \times_{v \in D} \mathcal{X}_v.$$

Suppose $D \subseteq W \subseteq \tilde{V}$ and that $\underline{x}_W \in \mathcal{X}_W$. That is, $\underline{x}_W = \times_{v \in W} x_v$. Then, ordering the variables of W so that $\mathcal{X}_W = \mathcal{X}_D \times \mathcal{X}_{W \setminus D}$, the projection of $\underline{x}_W$ onto D is defined as the variable $\underline{x}_D$ that satisfies

$$\underline{x}_W = (\underline{x}_D, \underline{x}_{W \setminus D}),$$

where the meaning of the notation '(,)' is clear from the context. Here $A \backslash B$ denotes the set difference, i.e. the elements in the set A not included in B.

Definition of a potential and charge   Let

$$\Phi = \{\phi_1, \dots, \phi_m\}$$

be a set of non-negative real valued functions on $\mathcal{X}$. The functions $\phi_j \in \Phi$ are called *potentials*. The set of potentials Φ is known as a *charge*. For each $j = 1, \dots, m$, $\mathcal{X}_{D_j}$ will denote the state space for potential ϕ_j, while the set $D_j \subset \tilde{V}$ denotes the set of indices for the argument variables of ϕ_j. The set $\mathcal{X}_{D_j}$ is called the *domain* of ϕ_j.

The *joint probability* function $p_{X_1, \dots, X_d}$ is itself a potential, with domain $\mathcal{X}$. If the joint probability function may be factorized according to a DAG $\mathcal{G} = (V, E)$, the decomposition is written as

$$p_{X_1, \dots, X_d} = \prod_{j=1}^{d} p_{X_j | \Pi_j}.$$

Then for each $j = 1, \dots, d$, ϕ_j defined by $\phi_j(x_v, \underline{x}_{\tilde{\Pi}_j}) = p_{X_j | \Pi_j}(x_j | \underline{x}_{\tilde{\Pi}_j})$ is a potential with domain $\mathcal{X}_{D_j} = \mathcal{X}_j \times \mathcal{X}_{\tilde{\Pi}_j}$ and $D_j = \{j\} \cup \tilde{\Pi}_j$.

Example 10.1   Consider a probability function over six variables that may be factorized along the directed acyclic graph in Figure 10.1.

The potentials corresponding to the conditional probabilities are

$$\phi_1 = p_{X_1}, \phi_2 = p_{X_2 | X_1}, \phi_3 = p_{X_3 | X_1},$$

$$\phi_4 = p_{X_4 | X_2}, \phi_5 = p_{X_5 | X_2, X_3}, \phi_6 = p_{X_6 | X_3}.$$

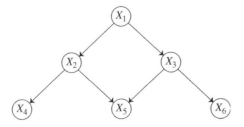

Figure 10.1   A Bayesian network on six variables.

The corresponding *domains* are

$$\mathcal{X}_{D_1} = \mathcal{X}_1$$
$$\mathcal{X}_{D_2} = \mathcal{X}_2 \times \mathcal{X}_1$$
$$\mathcal{X}_{D_3} = \mathcal{X}_3 \times \mathcal{X}_1$$
$$\mathcal{X}_{D_4} = \mathcal{X}_4 \times \mathcal{X}_2$$
$$\mathcal{X}_{D_5} = \mathcal{X}_5 \times \mathcal{X}_2 \times \mathcal{X}_3$$
$$\mathcal{X}_{D_6} = \mathcal{X}_6 \times \mathcal{X}_3.$$

Definition 10.1 (Contraction) *Recall Definition 2.26 for multiplication of potentials. A contraction of a charge, or set of potentials is an operation of multiplication and division of potentials, after extending them to $\mathcal{X}$, that returns a function over $\mathcal{X}$.*

For example,

$$\Phi(\underline{x}) = \prod_{j=1}^{m} \phi_j(\underline{x})$$

is a contraction, when the ϕ_j have first been expanded to the domain $\mathcal{X}_D = \mathcal{X}_{\bigcup_{j=1}^m D_j}$.

In other settings, the charge may contain potentials of two types;

$$\Phi = \{\phi_1, \ldots, \phi_{m_1}, \psi_1, \ldots, \psi_{m_2}\}$$

and

$$\Phi(\underline{x}) = \frac{\prod_{j=1}^{m_1} \phi_j(\underline{x})}{\prod_{j=1}^{m_2} \psi_j(\underline{x})}$$

is a contraction, where the domains of all the potentials have been extended to $\mathcal{X}$ before the operations of multiplication and division were applied.

The same notation is often used to denote the *contraction* of a charge and of the *set of potentials* (the charge). The context makes it clear which is intended.

Evidence potentials   Suppose that a random variable X_v is known to be instantiated with the value y. This is a piece of hard evidence and may be expressed as an evidence potential, which is a table containing 1s and 0s. Using notation $\underline{x} = (x_1, \ldots, x_d)$, this evidence may be expressed as a potential $e_v^{(y)}$ over the domain $\mathcal{X}$ defined as

$$e_v^{(y)}(\underline{x}) = \begin{cases} 1 & x_v = y \\ 0 & x_v \neq y. \end{cases} \tag{10.1}$$

Let $U \subseteq V$ and let $\tilde{U}$ denote the indexing set for the variables in U. An *evidence potential*, denoted by e_U, is defined for $\underline{y} \in \mathcal{X}_U, \underline{x} \in \mathcal{X}$, where the components of $\underline{y}$ are indexed by $\tilde{U}$, as

$$e_U^{(\underline{y})}(\underline{x}) = \prod_{v \in \tilde{U}} e_v^{(y_v)}(\underline{x}),$$

where e_v is defined by Equation (10.1). The interpretation is that e_U is a finding in the sense of seeing; not in the sense of doing.

For a Bayesian network with the nodes indexed by $\tilde{V} = \{1, \ldots, d\}$ and joint probability distribution factorized recursively as

$$p_{\underline{X}} = \prod_{v=1}^{d} p_{X_v|\Pi_v},$$

the quantity $p_{\underline{X};e_{\tilde{U}}^{(y)}} = p_{\underline{X}}.e_{\tilde{U}}^{(y)}$, is obtained by multiplication of tables to give

$$p_{\underline{X};e_{\tilde{U}}^{(y)}}(\underline{x}) = \prod_{v=1}^{d} p_{X_v|\Pi_v}(x_v|\underline{x}_{\tilde{\Pi}_v}) \prod_{v \in U} e_v^{(y_v)}(\underline{x}).$$

The product is zero for all $\underline{x} \in \mathcal{X}$ such that $\underline{x}_U \neq \underline{y}$; that is, if $x_v \neq y_v$ for some $v \in \tilde{U}$. It follows that, for fixed $\underline{y} \in \mathcal{X}_U$, $p_{\underline{X};e_{\tilde{U}}^{(y)}}$ may be considered as a potential with domain $\mathcal{X}_{V \setminus U}$.

10.2.1 Marginalization and the distributive law

Recall the discussion of marginalization in Section 2.5. The distributive law is used when marginalizing a product of potentials. It can be written as follows: let ϕ_1 be a potential with domain $\mathcal{X}_{D_1}$ and let ϕ_2 be a potential with domain $\mathcal{X}_{D_2}$. Suppose that $A \subset D_1 \cup D_2$ and the product $\phi_1\phi_2$ (Definition 2.26) is to be marginalized over $\mathcal{X}_A$. If $A \cap D_1 = \phi$ (the empty set), then

$$\sum_{\mathcal{X}_A} \phi_1\phi_2 = \phi_1 \sum_{\mathcal{X}_A} \phi_2.$$

More particularly, suppose that ϕ_1 has domain $\mathcal{X}_{D_1 \cup D_3}$ and ϕ_2 has domain $\mathcal{X}_{D_2 \cup D_3 \cup D_4}$, where D_1, D_2, D_3 and D_4 are disjoint. In coordinates, the distributive law may be written as

$$\sum_{x_2 \in \mathcal{X}_{D_2}} \phi_1(\underline{x}_1, \underline{x}_3)\phi_2(\underline{x}_2, \underline{x}_3, \underline{x}_4) = \phi_1(\underline{x}_1, \underline{x}_3) \sum_{x_2 \in \mathcal{X}_{D_2}} \phi_2(\underline{x}_2, \underline{x}_3, \underline{x}_4).$$

The effect of the distributive law is that the potential over $\mathcal{X}_{D_1} \times \mathcal{X}_{D_3} \times \mathcal{X}_{D_4}$ is first marginalized down to a function over $\mathcal{X}_{D_3} \times \mathcal{X}_{D_4}$. The function is transmitted to the function over $\mathcal{X}_{D_2} \times \mathcal{X}_{D_3}$, to which it is multiplied. The domains of the two functions to be multiplied have to be extended to $\mathcal{X}_{D_1} \times \mathcal{X}_{D_3} \times \mathcal{X}_{D_4}$. Using X_1, X_2, X_3, X_4 to denote the associated domains $\mathcal{X}_{D_1}, \mathcal{X}_{D_2}, \mathcal{X}_{D_3}$ and $\mathcal{X}_{D_4}$, the domains under consideration for the operations are illustrated in Figure 10.2. First, the potential ϕ_2, defined over (X_2, X_3, X_4) is considered. This is marginalized to a potential over (X_3, X_4) and is then extended, by multiplying with ϕ_1, to a potential over (X_1, X_3, X_4).

Example 10.2: A marginalization   Consider the computation for marginalizing a contraction of a charge Φ defined over a state space $\mathcal{X} = \mathcal{X}_1 \times \mathcal{X}_2 \times \mathcal{X}_3 \times \mathcal{X}_4 \times \mathcal{X}_5$ where

$$\Phi(\underline{x}) = \phi_1(x_1, x_3, x_5)\phi_2(x_1, x_2)\phi_3(x_3, x_4)\phi_4(x_5, x_6).$$

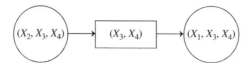

Figure 10.2   The distributive law.

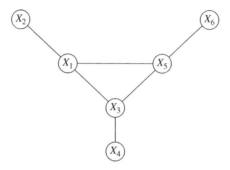

Figure 10.3   Associations of variables.

More particularly, consider the computation of

$$\Phi^{\downarrow 0} = \sum_{\underline{x} \in \mathcal{X}} \Phi(\underline{x}),$$

where the notation $\Phi^{\downarrow U}$ is defined in the discussion of marginalization in Section 2.5. The bucket elimination method may be used, with (for example) the order of summation: $x_2, x_4, x_6, x_5, x_3, x_1$. The sum may be written as

$$\sum_{x_1 \in \mathcal{X}_1} \sum_{x_3 \in \mathcal{X}_3} \sum_{x_5 \in \mathcal{X}_5} \phi_1(x_1, x_3, x_5) \sum_{x_6 \in \mathcal{X}_6} \phi_4(x_5, x_6) \sum_{x_4 \in \mathcal{X}_4} \phi_3(x_3, x_4) \sum_{x_2 \in \mathcal{X}_2} \phi_2(x_1, x_2).$$

The computation, carried out in this order, may be represented by the graph in Figure 10.3. A computational tree, according to the distributive law, is given in Figure 10.4.

10.3   Elimination and domain graphs

The notations described at the beginning of Section 2.5 will be used. Let U denote a subset of V and set $W = V \backslash U$. Let $\tilde{U}$, $\tilde{W}$ and $\tilde{V}$ denote the indexing sets of U, W and V respectively. Consider the computation of

$$\Phi^{\downarrow U}(\underline{x}) = \left(\sum_{\underline{x}_{V \backslash U} \in \mathcal{X}_{V \backslash U}} \Phi(\underline{x}_{V \backslash U}, \underline{x}_U) \right)$$

for a contraction

$$\Phi(\underline{x}) = \prod_{j=1}^{m} \phi_j(\underline{x}_{D_j}).$$

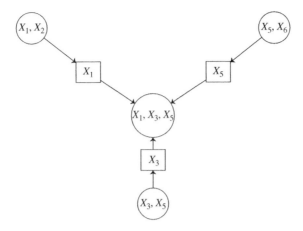

Figure 10.4   A computational tree for the marginalization.

Recall that the operation $\Phi^{\downarrow U}(\underline{x})$ means marginalizing Φ over all variables not in the set U. The variable x_v, with index $v \in \tilde{W} = \tilde{V}\backslash\tilde{U}$ is *eliminated* from $\sum_{\underline{x}_{V\backslash U} \in \mathcal{X}_{V\backslash U}} \Phi(\underline{x}_{V\backslash U}, \underline{x}_U)$ by the following procedure, where *contraction* means multiplying together all the potentials in the charge.

1. Let Φ_v (or Φ_{X_v}) denote the *contraction* of the potentials in Φ that have X_v in their domain; that is,

$$\Phi_v = \prod_{j|v \in D_j} \phi_j.$$

2. Let $\phi^{(v)}$ (or $\phi^{(X_v)}$) denote the function $\sum_{x_v \in \mathcal{X}_v} \Phi_v$.

3. Find a new set of potentials Φ^{-v} (or Φ^{-X_v}) by setting

$$\Phi^{-v} = (\Phi \cup \{\phi^{(v)}\})\backslash\Phi_v.$$

This is the definition of Φ^{-v}, also denoted by Φ^{-X_v}. Those potentials that do *not* contain X_v in their domain have been retained; the others have been multiplied together and then marginalized over $\mathcal{X}_v$ (thus eliminating the variable) to give $\phi^{(v)}$. This potential has been added to the collection, and all those containing X_v (other than $\phi^{(v)}$) have been removed.

Note that the notation Φ^{-X_v} has two meanings: it is used to denote the collection of potentials, and it is also used to denote the contraction of the charge obtained by multiplying together the potentials in the collection. The meaning is determined by the context. Having removed X_v, it remains to compute

$$\sum_{\underline{x}_{W\backslash\{X_v\}}} \Phi^{-X_v}(\underline{x}_U, \underline{x}_{W\backslash\{X_v\}}).$$

Proposition 10.1 *Let Φ be a contraction over a domain $\mathcal{X}_W$ and let $U \subset W$. The quantity*

$$\Phi^{\downarrow U}(\underline{x}_U) = \sum_{\underline{x}_{W\backslash U} \in \mathcal{X}_{W\backslash U}} \Phi(\underline{x}_{W\backslash U}, \underline{x}_U)$$

can be computed through successive elimination of the variables

$$X_v \in W \backslash U.$$

Proof of Proposition 10.1 By the commutative laws for multiplication and adding, together with the distributive law, the elimination of X_v gives

$$\Phi^{\downarrow U}(\underline{x}_U) = \sum_{\underline{x}_{W \backslash \{X_v\}} \in \mathcal{X}_{W \backslash \{X_v\}}} \Phi^{-X_v}(\underline{x}_U, \underline{x}_{W \backslash \{X_v\}}).$$

By the argument given above, Φ^{-X_v} is the contraction of a potential over $\mathcal{X}_{W \backslash \{X_v\}}$ and it is clear, by induction, that the marginal can be computed through successive elimination. □

The problem, of course, is to find an elimination sequence which gives as small *elimination domains* as possible. Elimination domains were defined in Definition 4.17 and, in this context, the elimination domain is the union of the domains of potentials in Φ having X_v in their domain. This is facilitated by considering the *domain graph*.

Definition 10.2 (Domain Graph) *The domain graph for the set of potentials in Φ is an undirected graph with the variables as nodes and the links between any pair of variables which are members of the same domain.*

Figure 10.5 illustrates the domain graph associated with Figure 10.1. Figure 10.3 illustrates the domain graph associated with Figure 10.4. Note here that the domain graph is the *moral graph*. It is clear that the domain graph for *any* Bayesian network is the moral graph, since by definition all the parents are connected to each other and to the variable.

Eliminating a node   Let $\mathcal{G} = (V, E)$ be an undirected graph, where $V = \{X_1, \ldots, X_d\}$. Recall Definition 4.15; eliminating a node. When a node X_v is eliminated from the graph $\mathcal{G}$, the resulting graph is denoted by $\mathcal{G}^{-X_v}$. If $\mathcal{G}$ is the domain graph for a set of potentials Φ, then it is clear from Definition 4.15 that the graph $\mathcal{G}^{-X_v}$ is the *domain graph* for the set of potentials Φ^{-X_v}.

Junction trees   Let

$$\Phi = \{\phi_1, \ldots, \phi_m\}$$

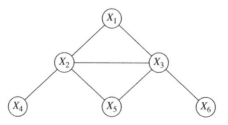

Figure 10.5   Domain graph of Bayesian network in Figure 10.1.

be a set of potentials on $\mathcal{X}$. Let $\mathcal{G}$ be a triangulated graph. A *junction tree for* Φ is a junction tree for $\mathcal{G}$ such that

- Each ϕ_j is associated with a clique $C_j = \{X_{j_1}, \ldots X_{j_k}\}$ such that $\mathcal{X}_{C_j}$, the domain of ϕ_j is the state space of the variables $\{X_{j_1}, \ldots, X_{j_k}\}$.

- Each edge is labelled by a separator consisting of the nodes remaining after elimination in a clique by an elimination sequence.

Figure 4.9 shows a directed acyclic graph, Figure 4.10 shows its moralized version, Figure 4.11 shows a triangulation of the moral graph and finally, Figure 4.13 shows a junction tree constructed from the directed acyclic graph in Figure 4.9.

Summary: Constructing an inference algorithm[1]

- Take a Bayesian network and find its domain graph $\mathcal{G}$.

- Triangulate the domain graph: $\mathcal{G} \rightarrow \mathcal{G}'$.

- Find an elimination sequence in $\mathcal{G}'$.

- The elimination sequence determines the elimination domains.

- The elimination domains are cliques.

- Organize the cliques into a junction tree (which is possible, following the results of Chapter 4).

- Associate the potentials to the junction tree.

It remains to describe a scheme of message passing (propagation) for the task of marginalization to compute

$$p_{\underline{X}|e_U}(\underline{x})^{\downarrow A} = \left(\sum_{\underline{x}_{\tilde{V}\backslash A} \in \mathcal{X}_{\tilde{V}\backslash A}} p_{\underline{X}|e_U}(\underline{x}_A, \underline{x}_{\tilde{V}\backslash A}) \right).$$

The task may be performed computationally using 'HUGIN'.[2] HUGIN propagation for the task is based on representing joint distribution of a Bayesian network using the so-called *Aalborg formula*

$$p_{\underline{X}}(x_1, \ldots, x_d) = \frac{\prod_{C \in \mathcal{C}} \phi_C(\underline{x}_C)}{\prod_{S \in \mathcal{S}} \phi_S(\underline{x}_S)},$$

where

$$\mathcal{C} = \text{cliques of the triangulated moral graph}$$

---

[1] This algorithm is sometimes referred to as an *inference engine*.
[2] http://www.hugin.com/
HUGIN is a product from the HUGIN EXPERT A/S, a software company, with head office in the town of Aalborg in Denmark, developing intelligent solutions in areas such as information management, data mining, decision analysis, troubleshooting, decision support, prediction, diagnosis, risk management, safety assessment, control systems, all based on Bayesian networks.

and

$$S = \text{separators of the junction tree}$$

and each ϕ_C and ϕ_S is the potential over the respective clique C and separator S. The propagation presented is the approach of Lauritzen and Spiegelhalter, discussed in [131]; the technicalities differ slightly from the implementation in HUGIN.

10.4   Factorization along an undirected graph

Let $\mathcal{G} = (V, E)$ be an undirected graph, where $V = \{X_1, \ldots X_d\}$ is a set of discrete variables.

Definition 10.3 *A joint probability* $p_{\underline{X}}$ *over a random vector* $\underline{X} = (X_1, \ldots, X_d)$ *is said to be* factorized *according to* $\mathcal{G}$ *if there exist* potentials *or* factors, ϕ_A *defined on* $\times_{v \in \tilde{A}} \mathcal{X}_v$ *where A is a complete set of nodes in* $\mathcal{G}$, *and* $\tilde{A}$ *is the index set for A, such that*

$$p_{\underline{X}}(\underline{x}) = \prod_A \phi_A(\underline{x}_A)$$

where the notation is clear (see Section 2.5); the product is over all the potentials.

Joint probability distributions factorized along a graph are known as *Markov probability distributions* and a corresponding Markov property will be established later on. If the factorization holds, then clearly, for any $W \subset V$,

$$p_{\underline{X}_W}(\underline{x}_W) = \sum_{\underline{x}_{V\setminus W} \in \mathcal{X}_{V\setminus W}} \prod_A \phi_A(\underline{x}_W, \underline{x}_{V\setminus W}),$$

where, for each A, the domain of ϕ_A has first been extended to $\mathcal{X}$.

Recall Definition 4.5 of a separator and Definition 4.12 of a decomposition. In the definition, A, B or S may be the empty set, ϕ.

Proposition 10.2 *Let* $\mathcal{G}$ *be a decomposable undirected graph and let* (A, B, S) *decompose* $\mathcal{G}$. *Then the following two statements are equivalent:*

1. *p factorizes along* $\mathcal{G}$ *and*

2. *both* $p_{A \cup S}$ *and* $p_{B \cup S}$ *factorize along* $\mathcal{G}_{A \cup S}$ *and* $\mathcal{G}_{B \cup S}$ *respectively and*

$$p(\underline{x}) = \frac{p_{A \cup S}(\underline{x}_{A \cup S}) p_{B \cup S}(\underline{x}_{B \cup S})}{p_S(\underline{x}_S)}.$$

Proof of Proposition 10.2, 1) $\implies$ **2)** Since the graph is decomposable, its cliques can be organized as a junction tree. Hence, without loss of generality, the factorization can be taken to be of the form

$$p(\underline{x}) = \prod_{K \in \mathcal{C}} \phi_K(\underline{x}_K),$$

where the product is over the cliques of $\mathcal{G}$. Since (A, B, S) decomposes $\mathcal{G}$, any clique of $\mathcal{G}$ can either be taken as a subset of $A \cup S$ or as a subset of $B \cup S$. Furthermore, S is a *strict* subset of any clique of $A \cup S$ containing S and S is a *strict* subset of any clique of $B \cup S$ containing S. Letting K denote a clique, it follows that

$$p(\underline{x}) = \prod_{K \subseteq A \cup S} \phi_K(\underline{x}_K) \prod_{K \subseteq B \cup S} \phi_K(\underline{x}_K).$$

Since S is itself complete, it is a subset of any clique containing S, so that no clique in the decomposition will appear in both $A \cup S$ and $B \cup S$. Set

$$h(\underline{x}_{A \cup S}) = \prod_{K \subseteq A \cup S} \phi_K(\underline{x}_K)$$

and

$$k(\underline{x}_{B \cup S}) = \prod_{K \subseteq B \cup S} \phi_K(\underline{x}_K),$$

then

$$p(\underline{x}) = h(\underline{x}_{A \cup S}) k(\underline{x}_{B \cup S})$$

and the marginal distribution is given by

$$p_{A \cup S}(\underline{x}_{A \cup S}) = \sum_{\mathcal{X}_B} h(\underline{x}_{A \cup S}) k(\underline{x}_{B \cup S}).$$

The distributive law now yields

$$p_{A \cup S}(\underline{x}_{A \cup S}) = h(\underline{x}_{A \cup S}) \sum_{\mathcal{X}_B} k(\underline{x}_{B \cup S}) = h(\underline{x}_{A \cup C}) k(\underline{x}_S),$$

where

$$k(\underline{x}_S) := \sum_{\mathcal{X}_B} k(\underline{x}_{B \cup S}).$$

Similarly,

$$p_{B \cup S}(\underline{x}_{B \cup S}) = k(\underline{x}_{B \cup S}) h(\underline{x}_S),$$

where

$$h(\underline{x}_S) := \sum_{\mathcal{X}_A} h(\underline{x}_{A \cup S}).$$

It follows that

$$p(\underline{x}) = h(\underline{x}_{A \cup S}) k(\underline{x}_{B \cup S}) = \frac{p(\underline{x}_{A \cup S}) p(\underline{x}_{B \cup S})}{k(\underline{x}_S) h(\underline{x}_S)}.$$

Furthermore,

$$p_S(\underline{x}_S) = \sum_{\mathcal{X}_{A \cup S}} h(\underline{x}_{A \cup S}) k(\underline{x}_{B \cup S}) = \sum_{\mathcal{X}_A} h(\underline{x}_{A \cup S}) \sum_{\mathcal{X}_B} k(\underline{x}_{B \cup S}) = h(\underline{x}_S) k(\underline{x}_S).$$

It follows that

$$p(\underline{x}) = \frac{p_{A \cup S}(\underline{x}_{A \cup S}) p_{B \cup S}(\underline{x}_{B \cup S})}{p_S(\underline{x}_S)}.$$

In the course of the proof, it has also been shown that p_{AUS} and p_{BUS} are factorizable along the corresponding graphs. This establishes the proof of 1) $\implies$ 2).   □

Proof of Proposition 10.2, 2) $\implies$ **1)** This is clear.   □

The Markov property By a recursive application of the proposition, together with

$$p(\underline{x}) = \prod_{C \subseteq A \cup S} \phi_C(\underline{x}_C) \prod_{C \subseteq B \cup S} \phi_C(\underline{x}_C),$$

it follows that

$$p(\underline{x}) = \frac{\prod_{C \in \mathcal{C}} p_C(\underline{x}_C)}{\prod_{S \in \mathcal{S}} p_S(\underline{x}_S)},$$

where $\mathcal{C}$ denotes the set of cliques and $\mathcal{S}$ denotes the set of separators, and thus the desired Markov property has been established.   □

This result may be extended to more general undirected graphs and the following proposition for factorization of probability distributions on undirected graphs has been proved by Lauritzen in [132]. The generality is unnecessary for the scope of this text and therefore the proof is omitted.

Proposition 10.3 *Let $\mathcal{G} = (V, E)$ be an undirected graph and let p be a probability distribution over the variable set V. If $p(\underline{x}) > 0$ for all $\underline{x} \in \mathcal{X}$, then p may be factorized along $\mathcal{G}$ if and only if for any sets of variables $A, B, S \subset V$ such that A and B are separated by S,*

$$A \perp B | S$$

(that for any sets A and B separated by S, the sets of variables A and B are conditionally independent given S).   □

This property is known as the *undirected global Markov property*.

10.5   Factorizing along a junction tree

Let p be a probability distribution that factorizes along a directed acyclic graph $\mathcal{G} = (V, E)$. The factorization is given by

$$p_{\underline{X}}(\underline{x}) = \prod_{v=1}^{d} p_{X_v | \Pi_v}(x_v | \underline{x}_{\tilde{\Pi}_v}),$$

where $\tilde{\Pi}_j$ denotes the indexing set for the parent set Π_j. It is clear that this may be expressed as a factorization according to the moralized graph $\mathcal{G}^{mor}$, which is undirected:

$$p_{\underline{X}}(\underline{x}) = \prod_{v=1}^{d} \phi_{A_v}(\underline{x}_{A_v})$$

where $A = \{X_v\} \cup \Pi_v$ and

$$\phi_{A_v}(\underline{x}_{A_v}) = p_{X_v|\Pi_v}(x_v | \underline{x}_{\bar{\Pi}_v}).$$

Hence a probability distribution factorized along the DAG is also factorized along the moral graph $\mathcal{G}^{mor}$ and the global Markov property is seen to hold on $\mathcal{G}^{mor}$. For implementing algorithms, the problem is that it may not be possible to represent the sets $(A_v)_{v=1}^d$ on a tree. To enable this, $\mathcal{G}^{mor}$ is *triangulated* to give $(\mathcal{G}^{mor})^t$. Recall that $(\mathcal{G}^{mor})^t$ is decomposable and its cliques can be organized into a junction tree $\mathcal{T}$. Then

$$p_{\underline{X}}(\underline{x}) = \prod_{C \in \mathcal{C}} \phi_C(\underline{x}_C),$$

where $\phi_C(\underline{x}_C)$ is the product of all those $p(x_v | \underline{x}_{\bar{\Pi}_v})$, *all* of whose arguments belong to C. By moralization, there is always one such clique. Note that this factorization is not necessarily unique. It follows that

$$p_{\underline{X}}(\underline{x}) = \frac{\prod_{C \in \mathcal{C}} p_C(\underline{x}_C)}{\prod_{S \in \mathcal{S}} p_S(\underline{x}_S)}, \tag{10.2}$$

where $\mathcal{C}$ denotes the set of cliques and $\mathcal{S}$ denotes the set of separators of $(\mathcal{G}^{mor})^t$. Furthermore, the cliques of the expression in Equation (10.2) may be organized according to a junction tree. This is the definition of a factorization along a junction tree.

Definition 10.4 (Factorization along a Junction Tree, Marginal Charge) *Let $p_{\underline{X}}$ be a probability distribution over a random vector $\underline{X} = (X_1, \ldots, X_d)$. Suppose that the variables can be organized as a junction tree, with cliques $\mathcal{C}$ and separators $\mathcal{S}$ such that $p_{\underline{X}}$ has representation given in Equation (10.2), where p_C and p_S denote the marginal probability functions over the clique variables $C \in \mathcal{C}$ and separator variables $S \in \mathcal{S}$ respectively. The representation in Equation (10.2) is known as the* factorization along the junction tree, *and the charge*

$$\Phi = \{p_S : S \in \mathcal{S}, \ p_C : C \in \mathcal{C}\}$$

is known as the marginal charge.

From the foregoing discussion, it is clear that Definition 10.4 is a special case of Definition 10.3, where the potentials are appropriately defined.

Entering evidence   Equation (10.2) expresses the prior distribution in terms of potentials over the cliques and separators of $(\mathcal{G}^{mor})^t$, or the junction tree. Suppose that new hard evidence e_U is obtained on the variables U; namely, that for $U \subseteq V$, $\{\underline{X}_U = \underline{y}_U\}$ and the probability over the variables $V \backslash U$ has to be updated accordingly. Then

$$p_{\underline{X};e_U}(\underline{x}) = \begin{cases} \frac{\prod_{C \in \mathcal{C}} p_C(\underline{x}_C)}{\prod_{S \in \mathcal{S}} p_S(\underline{x}_S)} & \underline{x}_U = \underline{y}_U \\ 0 & \underline{x}_U \neq \underline{y}_U. \end{cases}$$

Now, set $\phi_C(\underline{x}_C) = p_C(\underline{x}_{C \backslash U}, \underline{y}_{C \cap U})$ and $\phi_S(\underline{x}_S) = p_S(\underline{x}_{S \backslash U}, \underline{y}_{S \cap U})$. Then

$$p_{\underline{X}_{V \backslash U}, \underline{X}_U}(\underline{x}_{V \backslash U}, \underline{y}_U) = \frac{\prod_{C \in \mathcal{C}} \phi_C(\underline{x}_C)}{\prod_{S \in \mathcal{S}} \phi_S(\underline{x}_S)}. \tag{10.3}$$

The posterior distribution

$$p_{\underline{X}_{\tilde{V}\setminus U} \mid \underline{X}_U}(\cdot \mid \underline{y}_U) = \frac{p_{\underline{X}_{\tilde{V}\setminus U}, \underline{X}_U}(\cdot, \underline{y}_U)}{p_{\underline{X}_U}(\underline{y}_U)}$$

may then be computed by finding a representation of the function $p_{\underline{X}_{V\setminus U}, \underline{X}_U}(\cdot, \underline{y}_U)$ over domain $\mathcal{X}_{V\setminus U}$ using the algorithm defined below. The algorithm such that for any function $f : \mathcal{X} \to \mathbf{R}_+$ (not necessarily a probability function) that is expressed as

$$f(\underline{x}) = \frac{\prod_{C \in \mathcal{C}} \phi_C(\underline{x}_C)}{\prod_{S \in \mathcal{S}} \phi_S(\underline{x}_S)}, \tag{10.4}$$

for a collection of potentials $\Phi = \{\phi_C, \ C \in \mathcal{C}, \ \phi_S, \ S \in \mathcal{S}\}$ where $\mathcal{C}$ and $\mathcal{S}$ are the cliques and separators of a junction tree, the algorithm updates Φ to a collection of potentials $\Phi^* = \{\phi_C^*, \ C \in \mathcal{C}, \ \phi_S^*, \ S \in \mathcal{S}\}$ that satisfy

$$\phi_C^*(\underline{x}_C) = \sum_{\underline{z} \in \mathcal{X}_{V \setminus C}} f(\underline{z}, \underline{x}_C)$$

and

$$\phi_S^*(\underline{x}_S) = \sum_{\underline{z} \in \mathcal{X}_{V \setminus S}} f(\underline{z}, \underline{x}_S)$$

for each $C \in \mathcal{C}$ and each $S \in \mathcal{S}$. It follows that the probability of the evidence is

$$p_{\underline{X}_U}(\underline{y}_U) = \sum_{\underline{z} \in \mathcal{X}_{C \setminus (U \cap C)}} \phi_C^*(\underline{z}, \underline{y}_{U \cap C}) = \sum_{\underline{z} \in \mathcal{X}_{S \setminus (U \cap S)}} \phi_S^*(\underline{z}, \underline{y}_{U \cap S})$$

for all $S \in \mathcal{S}$ and all $C \in \mathcal{C}$. The conditional probability distribution over the remaining variables $V\setminus U$ may therefore be computed by marginalizing the clique or separator with the smallest domain, giving a representation of the conditional distribution in terms of marginal distributions over the cliques and separators.

10.5.1   Flow of messages initial illustration

Consider a non-negative function with domain $X \times Y \times Z$, $F : X \times Y \times Z \to \mathbf{R}_+$, which may be written as

$$F(x, y, z) = \frac{f(x, z)g(y, z)}{h(z)}, \tag{10.5}$$

for potentials $f : X \times Z \to \mathbf{R}_+$, $g : Y \times Z \to \mathbf{R}_+$ and $h : Z \to \mathbf{R}_+$.

The decomposition shown in Equation (10.5) for the function F is of the form given in Equation (10.4). The graph illustrating the associations between variables is given in Figure 10.6, with cliques $C_1 = \{X, Z\}$, $C_2 = \{Z, Y\}$ and separator $S = \{Z\}$ arranged according to the junction tree in Figure 10.7.

The following algorithm returns a representation $F(x, y, z) = \frac{F_1(x,z)F_2(y,z)}{F_3(z)}$, where

$$F_1(x, z) = \sum_{y \in Y} F(x, y, z), \quad F_2(y, z) = \sum_{x \in X} F(x, y, z), \quad F_3(z) = \sum_{(x,y) \in X \times Y} F(x, y, z).$$

Figure 10.6   Undirected graph for the three variables.

Figure 10.7   Junction tree for message passing.

Firstly,

$$F_1(x, z) = \sum_{y \in Y} F(x, y, z) = \sum_y \frac{f(x, z)g(y, z)}{h(z)} = \frac{f(x, z)}{h(z)} \sum_{y \in Y} g(y, z).$$

Define the auxiliary function $h^*(z) = \sum_y g(y, z)$, and the update $f^*(x, y) = f(x, y)\frac{h^*(z)}{h(z)}$, then clearly

$$f^*(x, z) = f(x, z)\frac{h^*(z)}{h(z)} = F_1(x, z).$$

The calculation of the marginal function $F_1(x, z)$ by means of the auxiliary function $h^*(z)$ may be described as being done by passing a *local message flow* from ZY to XZ through their separator Z. The factor

$$\frac{h^*(z)}{h(z)}$$

is called the *update ratio*. It follows that

$$F(x, y, z) = \frac{f(x, z)g(y, z)}{h(z)} = \frac{f(x, z)g(y, z)h^*(z)}{h^*(z)h(z)} = F_1(x, z)\frac{1}{h^*(z)}g(y, z).$$

The passage of the flow has resulted in a new representation of $F(x, y, z)$ similar to the original, but where one of the factors is a marginal function.

Similarly, a message can be passed in the other direction, i.e. from XZ to ZY Using the same procedure, set

$$\tilde{h}(z) = \sum_{x \in X} F_1(x, z) = \sum_{(x,y) \in X \times Y} F(x, y, z) = F_3(z).$$

Next, set

$$\tilde{g}(y, z) = g(y, z)\frac{\tilde{h}(z)}{h^*(z)}.$$

It then follows that $\tilde{g}(y, z) = F_2(y, z)$.

This follows because

$$F(x, y, z) = F_1(x, z)\frac{1}{\tilde{h}(z)}\tilde{g}(y, z) = F_1(x, z)\frac{1}{F_3(z)}\tilde{g}(y, z)$$

and hence, since $F_3(z) = \sum_{x \in X} F_1(x, z)$, it follows that

$$F_2(y, z) = \sum_{x \in X} F(x, y, z) = \tilde{g}(y, z) \sum_{x \in X} F_1(x, z) \frac{1}{F_3(z)} = \tilde{g}(y, z).$$

□

Passing messages in both directions results in a new overall representation of the function $F(x, y, z);$.

$$F(x, y, z) = f^*(x, z) \frac{1}{h^*(z)} g(y, z) = f^*(x, z) \frac{1}{h^*(z)} \frac{h^*(z)}{\tilde{h}(z)} \tilde{g}(y, z)$$

$$= f^*(x, z) \frac{1}{\tilde{h}(z)} \tilde{g}(y, z)$$

$$= F_1(x, z) \frac{1}{F_3(z)} F_2(y, z).$$

The original representation using potentials has been transformed into a new representation where all the potentials are marginal functions.

The idea is now extended to arbitrary non-negative functions represented on junction trees.

10.6   Local computation on junction trees

Consider a junction tree T with nodes C and separators S and let Φ be a charge

$$\Phi = \{\phi_C : C \in C, \phi_S : S \in S\}, \tag{10.6}$$

that is, a collection of potentials such that $\phi_C : X_C \to \mathbb{R}_+$ and $\phi_S : X_S \to \mathbb{R}_+$ for each $C \in C$ and each $S \in S$.

Definition 10.5 (Contraction of a Charge on a Junction Tree) *The contraction of a charge (Equation (10.6)) over a junction tree is defined as*

$$f(x) = \frac{\prod_{C \in C} \phi_C(x_C)}{\prod_{S \in S} \phi_S(x_S)}. \tag{10.7}$$

Local message passing   Let C_1 and C_2 be two adjacent neighbouring nodes in T separated by S^0. Set

$$\phi_{S^0}^*(x_{S^0}) = \sum_{\underline{z} \in X_{C_1 \backslash S^0}} \phi_{C_1}(\underline{z}, \underline{x}_{S^0}) \tag{10.8}$$

and set

$$\lambda_{S^0} = \frac{\phi_{S^0}^*(x_{S^0})}{\phi_{S^0}(\underline{x}_{S^0})} \tag{10.9}$$

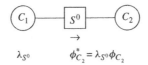

Figure 10.8   Flow from C_1 to C_2.

Note that, directly from the definition of division of potentials (Definition 2.24), $\lambda_{S^0} = 0$ for $\phi_{S^0} = 0$. The *update ratio* is defined as the quantity λ_{S^0}. The 'message passing' is defined as the operation of updating ϕ_{S^0} to $\phi_{S^0}^*$ and ϕ_{C_2} to

$$\phi_{C_2}^* = \lambda_{S^0}\phi_{C_2}. \tag{10.10}$$

All other potentials remain unchanged. The scheme of local message passing is illustrated in Figure 10.8.

Lemma 10.1 Let $f : \mathcal{X} \to \mathbf{R}_+$ be the contraction of a a charge $\Phi = \{\phi_S, S \in \mathcal{S}, \phi_C, C \in \mathcal{C}\}$ on a junction tree (Definition 10.5), where $\mathcal{C}$ is the collection of cliques and $\mathcal{S}$ the collection of separators.

A flow does not change the contraction of the charge.

Proof of Lemma 10.1 The initial contraction is given by

$$f(\underline{x}) = \frac{\prod_{C \in \mathcal{C}} \phi_C(\underline{x}_C)}{\prod_{S \in \mathcal{S}} \phi_S(\underline{x}_S)}. \tag{10.11}$$

Firstly, recall Definition 2.24. By the definition of division of potentials, $f(\underline{x}) = 0$ for all $\underline{x}$ such that $\phi_S(\underline{x}_S) = 0$ for some $S \in \mathcal{S}$. This is part of the definition in the hypothesis that f has a representation of the form given in Equation (10.11). Let the charge, after the flow from C_1 to C_2, be denoted by

$$\Phi^* = \{\phi_C^* : C \in \mathcal{C}, \phi_S^* : S \in \mathcal{S}\}$$

and the contraction

$$f^*(\underline{x}) := \frac{\prod_{C \in \mathcal{C}} \phi_C^*(\underline{x}_C)}{\prod_{S \in \mathcal{S}} \phi_S^*(\underline{x}_S)}. \tag{10.12}$$

Note that

$$f^*(\underline{x}) = \frac{\phi_{C_2}^*(\underline{x}_{C_2}) \prod_{C \in \mathcal{C}, C \neq C_2} \phi_C(\underline{x}_C)}{\phi_{S^0}^*(\underline{x}_{S^0}) \prod_{S \in \mathcal{S}, S \neq S^0} \phi_S(\underline{x}_S)}. \tag{10.13}$$

There are three cases to consider.

- For $\underline{x}$ such that $\phi_{S^0}(\underline{x}_{S^0}) > 0$ and $\phi_{S^0}^*(\underline{x}_{S^0}) > 0$,

$$\frac{\phi_{C_2}^*}{\phi_{S^0}^*} = \frac{\phi_{C_2}\lambda_{S^0}}{\phi_{S^0}^*} = \frac{\phi_{C_2}\left(\frac{\phi_{S^0}^*}{\phi_{S^0}}\right)}{\phi_{S^0}^*} = \frac{\phi_{C_2}}{\phi_{S^0}}$$

and the result is proved.

- For second case and the third case, the results follow from the convenient arrangement of the definitions for division by zero in the sense of tables. For x such that $\phi_{S^0}(x_{S^0}) = 0$, $f(x) = 0$, from the definition. Furthermore, from Equation (10.9), $\lambda_{S^0} = 0$ (this follows from division by zero in the sense of division of potentials) and hence, by the definition of $\phi_{C_2}^*$, it follows that that $\phi_{C_2}^* = 0$. It therefore follows from Equation (10.12) that $f^*(x) = 0$, so that $0 = f^*(x) = f(x)$.

- For x such that $\phi_{S^0}(x_{S^0}) > 0$, but $\phi_{S^0}^*(x_{S^0}) = 0$, it follows directly from Equation (10.13) using the definition of division by zero in the sense of potentials) that $f^*(x) = 0$. It remains to show that $f(x) = 0$. From the definition,

$$0 = \phi_{S^0}^*(x_{S^0}) = \sum_{z \in \mathcal{X}_{C_1 \backslash S^0}} \phi_{C_1}(z, x_{S^0}).$$

Since $\phi_{C_1}(x_{C_1}) \geq 0$ for all $x_{C_1} \in \mathcal{X}_{C_1}$, it follows that $\phi_{C_1}(z, x_{S^0}) = 0$ for all $z \in \mathcal{X}_{C_1 \backslash S^0}$. Since

$$f(x) = \phi_{C_1}(x_{C_1}) \frac{\phi_{C_2}(x_{C_2})}{\phi_{S^0}(x_{S^0})} \frac{\prod_{C \in \mathcal{C}, C \neq C_1, C_2} \phi_C(x_C)}{\prod_{S \in \mathcal{S}, S \neq S^0} \phi_S(x_S)},$$

it follows directly from the facts that the domains of the cliques other than C_1 and C_2 and separators other than S^0 do not include $\mathcal{X}_{S^0}$, and that $\frac{\phi_{C_2}(x_{C_2})}{\phi_{S^0}(x_{S^0})} < +\infty$ that $f(x) = 0$, hence $f(x) = f^*(x)$.

In all cases, it follows that a flow does not change the contraction of a charge.   □

Having shown how to update for a simple example, and having shown that message passing does not alter the contraction of a charge, it remains to schedule the message transmissions in an efficient way for a general junction tree to update the potentials over the cliques and separators.

10.7   Schedules

The aim of this section is to describe how to construct a series of transmissions between the various cliques of a junction tree, to update a set of potentials, whose contraction is a probability distribution, to the posterior probability distributions over the cliques and separators. First, some definitions and notations are established.

Definition 10.6 (Sub-tree, Neighbouring Clique) *A sub-tree T' of a junction tree T is a connected set of nodes of T together with the edges in T between them.*

A clique C of a junction tree T is a neighbour of a sub-tree T if the corresponding node of T is not a node of T' but is connected to T' by an edge of T.

The following definition gives the technical terms that will be used.

Definition 10.7 (Schedule, Active Flow, Fully Active Schedule) *A schedule is an ordered list of directed edges of T specifying which flows are to be passed and in which order.*

A flow is said to be active *relative to a schedule if before it is sent the source has already received active flows from all its neighbours in T, with the exception of the* sink; *namely, the node to which it is sending its flow. It follows that the first active flow must originate in a leaf of T. This leaf serves as a* root *for the junction tree. A schedule is* full *if it contains an active flow in each direction along every edge of the tree T. A schedule is* active *if it contains only* active flows. *It is* fully active *if it is both full and active.*

Example 10.3 Figures 10.9 and 10.10 depict a DAG and the corresponding junction tree.

A fully active schedule for the junction tree given in Figure 10.10 would be:

$$AT \rightarrow ELT, BLS \rightarrow BEL, BDE \rightarrow BEL, EK \rightarrow ELT, ELT \rightarrow BEL$$

$$BEL \rightarrow ELT, ELT \rightarrow EK, ELT \rightarrow AT, BEL \rightarrow BLS, BEL \rightarrow BDE.$$

Definition 10.8 (Lazy Propagation) *The method of updating a distribution that factorizes over a junction tree is known as* lazy propagation.

Proposition 10.4 *For any tree T, there exists a fully active schedule.*

Proof of Proposition 10.4 If there is only one clique, the proposition is clear; no transmissions are necessary. Assume that there is more than one clique. Let C_0 denote a leaf in T. Let T_0 be a sub-tree of T obtained by removing C_0 and the corresponding edge S_0. Assume that the proposition is true for T_0. Adding the edge

$$C_0 \rightarrow S_0 \rightarrow T_0$$

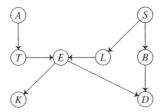

Figure 10.9   Example of a DAG.

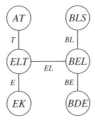

Figure 10.10   Corresponding junction tree.

to the beginning of the schedule and

$$C_0 \leftarrow S_0 \leftarrow T_0$$

to the end of the schedule provides a fully active schedule for T. □

The aim of a substantial part of the remainder of the chapter is to show that if the contraction of a charge Φ on a junction tree (Definition 10.5) is a probability distribution, then after the passage of a fully active schedule of flows over a junction tree, the resulting charge is the *marginal charge*. That is, all the potentials of the charge are probability functions over the respective cliques and separators. Furthermore, there is *global* consistency after the passage of a fully active schedule of flows over a junction tree. This will be defined later, but loosely speaking, it means that if there are several apparent ways to compute a probability distribution over a set of variables using the potentials of the marginal charge, they will all give the same answer.

Definition 10.9 (The Base of a Sub-tree, Restriction of a Charge, Live Sub-tree) *Let T' be a sub-tree of T, with nodes $C' \subseteq C$ and edges $S' \subseteq S$. The base of T' is defined as the set of variables*

$$U' := \cup_{C \in C'} C.$$

Let

$$\Phi = \{\phi_C : C \in C, \phi_S : S \in S\}$$

be a charge for T. Its restriction *to T' is defined as*

$$\Phi_{T'} = \{\phi_C : C \in C', \phi_S : S \in S'\}.$$

Recall Definition 10.5. The contraction *of $\Phi_{T'}$ is defined as*

$$\frac{\prod_{C \in C'} \phi_C(\underline{x}_C)}{\prod_{S \in S'} \phi_S(\underline{x}_S)}.$$

A sub-tree T' is said to be live *with respect to the schedule of flows if it has already received active flows from all its neighbours.*

Proposition 10.5 *Let*

$$\Phi^0 = \{\phi_C^0 : C \in C, \phi_S^0 : S \in S\}$$

denote an initial charge for a function f that has factorization

$$f(\underline{x}) = \frac{\prod_{C \in C} \phi_C^0(\underline{x}_C)}{\prod_{S \in S} \phi_S^0(\underline{x}_S)}$$

where C and S are the sets of cliques and separators for a junction tree T. Suppose that Φ^0 is modified by a sequence of flows according to some schedule. Then, whenever T' is live, the contraction of the charge for T' is the margin of the contraction f of the charge for T on U'.

Proof of Proposition 10.5 Assume that $T' \subset T$ and that T' is live. Let C^* denote that last neighbour to have passed a flow into T'. Let T^* be the sub-tree obtained by adding C^* and the associated edge S^* to T'. Let C^*, S^* and U^* be the cliques, separators and the base of T^*. By the junction tree property of T, the separator associated with the edge S^* joining C^* to T' is

$$S^* = C^* \cap U'.$$

Furthermore,

$$C^* = C' \cup \{C^*\} \qquad \text{and} \qquad S^* = S' \cup \{S^*\}.$$

and the set of base variables for T^* is

$$U^* = U' \cup C^*.$$

The induction hypothesis   The assertion holds for the contraction of the charge on T^*. Set

$$f_{U^*}(\underline{x}_{U^*}) = \sum_{U \backslash U^*} f(\underline{x}).$$

Then the inductive hypothesis states that

$$f_{U^*}(\underline{x}_{U^*}) = \frac{\prod_{C \in C^*} \phi_C(\underline{x}_C)}{\prod_{S \in S^*} \phi_S(\underline{x}_S)}. \tag{10.14}$$

Let

$$\Phi = \{\phi_C : C \in C, \phi_S : S \in S\}$$

denote the charge *just before the last flow from C^* into T'*. Lemma 10.1 states that a flow does not change the contraction of a charge. This is applied to the contraction of the charge restricted to T^* which, from Equation (10.14) is given by

$$f_{U^*}(\underline{x}_{U^*}) = \frac{\phi_{C^*}(\underline{x}_{C^*}) \prod_{C \in C'} \phi_C(\underline{x}_C)}{\phi_{S^*}(\underline{x}_{S^*}) \prod_{S \in S'} \phi_S(\underline{x}_S)}. \tag{10.15}$$

Set

$$\alpha_{U'} = \frac{\prod_{C \in C'} \phi_C(\underline{x}_C)}{\prod_{S \in S'} \phi_S(\underline{x}_S)},$$

then Equation (10.15) may be rewritten as

$$f_{U^*}(\underline{x}_{U^*}) = \frac{\phi_{C^*}(\underline{x}_{C^*})}{\phi_{S^*}(\underline{x}_{S^*})} \alpha_{U'}. \tag{10.16}$$

The aim is to find the margin $f_{U'}$ of f on U' and to show that after the flow,

$$f_{U'}(\underline{x}_{U'}) = \frac{\prod_{C \in C'} \phi_C(\underline{x}_C)}{\prod_{S \in S'} \phi_S(\underline{x}_S)}.$$

Note that

$$U' \subset U^* \subset U,$$

so that

$$U \backslash U' = (U \backslash U^*) \cup (U^* \backslash U').$$

Since the variables may be summed in any order,

$$\sum_{X_{U \backslash U'}} f = \sum_{X_{(U \backslash U^*)}} \sum_{X_{(U^* \backslash U')}} f = \sum_{X_{(U^* \backslash U')}} \left(\sum_{X_{(U \backslash U^*)}} f \right) = \sum_{X_{(U^* \backslash U')}} f_{U^*}. \tag{10.17}$$

It follows that

$$\sum_{X_{U^* \backslash U'}} f_{U^*} = \alpha_{U'} \sum_{X_{U^* \backslash U'}} \frac{\phi_{C^*}(\underline{x}_{C^*})}{\phi_{S^*}(\underline{x}_{S^*})}. \tag{10.18}$$

Since $S^* = C^* \cap U'$ and $U^* = U' \cup C^*$, it follows that

$$U^* \backslash U' = C^* \backslash S^*,$$

so that

$$\sum_{X_{U^* \backslash U'}} \frac{\phi_{C^*}(\underline{x}_{C^*})}{\phi_{S^*}(\underline{x}_{S^*})} = \sum_{C^* \backslash S^*} \frac{\phi_{C^*}(\underline{x}_{C^*})}{\phi_{S^*}(\underline{x}_{S^*})} = \frac{1}{\phi_{S^*}(\underline{x}_{S^*})} \sum_{C^* \backslash S^*} \phi_{C^*}(\underline{x}_{C^*}) = \frac{\phi^*_{S^*}(\underline{x}_{S^*})}{\phi_{S^*}(\underline{x}_{S^*})} \overset{(def)}{=} \lambda_{S^*}.$$

Recall Equation (10.9); λ_{S^*} is the *update ratio*. This, together with Equation (10.18), may be applied to Equation (10.17) to give

$$\sum_{X_{U \backslash U'}} f = \alpha_{U'} . \lambda_{S^*}.$$

But *after* the flow into T', $\alpha_{U'}$ is *updated* as

$$\alpha^*_{U'} = \lambda_{S^*} \alpha_{U'},$$

because the potential ϕ_{C^*} over the nearest neighbour C^* in T' is updated to $\lambda_{S^*} \phi_{C^*}$ (the update defined in Equation (10.10)). Hence, using $\phi^*_{C^*}$ to denote the update of ϕ_{C^*} and $\phi^*_{S^*}$ to denote the update of ϕ_{S^*} and using the inductive hypothesis that a flow does not alter the contraction of a charge on T^*, it follows that

$$f_{U^*}(\underline{x}_{U^*}) = \frac{\phi_{C^*}(\underline{x}_{C^*})}{\phi_{S^*}(\underline{x}_{S^*})} \alpha_{U'} = \frac{\phi^*_{C^*}(\underline{x}_{C^*})}{\phi^*_{S^*}(\underline{x}_{S^*})} \alpha^*_{U'}.$$

Recall Equation (10.8); the potential over the separator $\phi_{S^*}(\underline{x}_{S^*})$ is updated to

$$\phi^*_{S^*}(\underline{x}_{S^*}) = \sum_{z \in X_{C^* \backslash S^*}} \phi_{C^*}(\underline{z}, \underline{x}_{S^*}).$$

Since the flow is *from* C^* to T', $\phi^*_{C^*} = \phi_{C^*}$. It follows that

$$\sum_{X_{U^* \backslash U'}} f_{U^*}(\underline{x}_{U^*}) = \sum_{\underline{z} \in X_{C^* \backslash S^*}} \frac{\phi_{C^*}(\underline{x}_{C^*})}{\phi^*_{S^*}(\underline{x}_{S^*})} \alpha^*_{U'} = \alpha^*_{U'},$$

from which it follows that after the flow,

$$\sum_{U\setminus U'} f = \frac{\prod_{C\in C'} \phi_C(\underline{x}_S)}{\prod_{S\in S'} \phi_S(\underline{x}_S)},$$

which is the definition of the contraction of the charge on T'. It follows that after the flow, the contraction of the charge on T' is $\sum_{U\setminus U'} f$, as required. The proof is complete.

Corollary 10.1 *Let* $\{\phi_C, C \in C, \; \phi_S, S \in S\}$ *denote the current potentials over the cliques and separators. For any set* $A \subseteq V$, *let* $f_A = \sum_{\underline{x}_{V\setminus A}} f$; *the marginal over* A. *Whenever a clique* C *is live, its potential is* $\phi_C = f_C = \sum_{\underline{x}_{V\setminus C}} f$.

Proof of Corollary 10.1 A single clique is a sub-tree. The result is immediate from Proposition 10.5. □

Corollary 10.2 *Using the notation of Corollary 10.1, whenever active flows have passed in both directions across an edge in* T, *the potential for the associated separator is* $\phi_S = f_S = \sum_{\underline{x}_{V\setminus S}} f$.

Proof of Corollary 10.2 The potential ϕ_S for the associated separator is, by definition of the update,

$$\phi_S = \sum_{\underline{x}_{C\setminus S}} \phi_C,$$

so that

$$\sum_{\underline{x}_{C\setminus S}} \phi_C = \sum_{\underline{x}_{C\setminus S}} f_C = f_S,$$

because ϕ_C is f_C by the previous corollary. □

Proposition 10.6 (The Main Result) *After passage of a fully active schedule of flows, the resulting charge is the charge consisting of the marginals over the cliques and separators and its contraction represents* f. *In other words, the following formula, known as the Aalborg formula* (see [52]);

$$f(\underline{x}) = \frac{\prod_{C\in C} f_C(\underline{x}_C)}{\prod_{S\in S} f_S(\underline{x}_S)}.$$

Proof of Proposition 10.6 This follows from the previous two corollaries and Lemma 10.1, stating that the contraction is unaltered by the flows. □

Thin junction trees Sometimes the cliques in a junction tree may contain rather many variables, which may be problematic if one is using numerical methods to compute marginal distributions. This problem is addressed by R.G. Cowell, A.P. Dawid, S.L. Lauritzen and D.J. Spiegelhalter in [67], who discuss methods of breaking the probability distributions of the cliques into smaller marginal distributions. The paper [133] develops an algorithm that tries to keep the maximum size of cliques below a prescribed bound. Such junction trees are known as *thin junction trees*. The thin junction trees cannot be created *after* the graph is fixed; the restriction on the graph width has to be enforced during the learning process.

10.8 Local and global consistency

Recall that $\mathcal{T}$ denotes the junction tree, the set of cliques which form the nodes of $\mathcal{T}$ is denoted $\mathcal{C}$ and the *intersection of neighbours* in the tree $\mathcal{T}$ are the *separators*, denoted by $\mathcal{S}$. Recall that the potentials associated with $C \in \mathcal{C}$ and $S \in \mathcal{S}$ are denoted by ϕ_C and ϕ_S respectively, and that the *charge* on $\mathcal{T}$, Φ is defined as:

$$\Phi = \{\phi_C : C \in \mathcal{C}, \phi_S : S \in \mathcal{S}\}.$$

Definition 10.10 (Local Consistency) *A junction tree $\mathcal{T}$ is said to be* locally consistent *if whenever $C_1 \in \mathcal{C}$ and $C_2 \in \mathcal{C}$ are two neighbours with separator S, then*

$$\sum_{\mathcal{X}_{C_1 \backslash (C_1 \cap C_2)}} \phi_{C_1} = \phi_S = \sum_{\mathcal{X}_{C_2 \backslash (C_1 \cap C_2)}} \phi_{C_2}.$$

Definition 10.11 (Global Consistency) *A junction tree $\mathcal{T}$ (or its charge) is said to be* globally consistent *if for every $C_1 \in \mathcal{C}$ and $C_2 \in \mathcal{C}$ it holds that*

$$\sum_{\mathcal{X}_{C_1 \backslash (C_1 \cap C_2)}} \phi_{C_1} = \sum_{\mathcal{X}_{C_2 \backslash (C_1 \cap C_2)}} \phi_{C_2}.$$

Global consistency means that the marginalization to $C_1 \cap C_2$ of ϕ_{C_1} and ϕ_{C_2} coincide for every C_1 and C_2 in $\mathcal{C}$. The following results show that, for a junction tree, local consistency implies global consistency.

Proposition 10.7 *After a passage of a fully active schedule of flows, a junction tree $\mathcal{T}$ is locally consistent.*

Proof of Proposition 10.7 The two corollaries of the main result give that for any two neighbouring C_1 and C_2,

$$\sum_{C_1 \backslash S} f_{C_1} = f_S = \sum_{C_2 \backslash S} f_{C_2}.$$

$\square$

An equilibrium, or fixed point has been reached, in the sense that any new flows passed after passage of a fully active schedule do not alter the potentials. The update ratio for another message from C_1 to C_2 becomes

$$\lambda_S = \frac{\sum_{C_1 \backslash S} f_{C_1}}{f_S} = 1.$$

Global consistency of junction trees Here it is shown that for *junction trees*, local consistency implies global consistency.

By definition, a junction tree is a tree such that the intersection $C_1 \cap C_2$ of any pair C_1 and C_2 in $\mathcal{C}$ is contained in *every* node on the *unique* trail in $\mathcal{T}$ between C_1 and C_2. The set $C_1 \cap C_2$ can be empty and, in this case it is therefore (by convention) a subset of every other set.

The following example is instructive.

Example 10.4 Consider the junction tree given in Figure 10.10. Let $C_1 = EK$ and $C_2 = BDE$. Then $C_1 \cap C_2 = E$. There is a unique trail

$$EK \leftrightarrow ELT \leftrightarrow BEL \leftrightarrow BDE$$

from EK to BDE. Clearly $E = C_1 \cap C_2$ is a subset of every separator on the path. The potentials will be denoted (for example) $\phi_{EK}(x_e, x_k) =: \phi_{EK}(ek)$. The following abbreviated notation will be used for a marginalization:

$$\sum_{x_e \in \mathcal{X}_e} \phi_{EK}(x_e, x_k) = \sum_e \phi_{EK}(ek).$$

Now assume that the junction tree is locally consistent. Then, an application of the result that

$$\sum_{\mathcal{X}_{C_1 \backslash (C_2 \cap C_2)}} \phi_{C_1} = \phi_S = \sum_{\mathcal{X}_{C_2 \backslash (C_1 \cap C_2)}} \phi_{C_2}$$

gives

$$\sum_k \phi_{EK}(ek) = \sum_{lt} \phi_{ELT}(elt),$$

$$\sum_t \phi_{ELT}(elt) = \sum_b \phi_{BEL}(bel),$$

$$\sum_l \phi_{BEL}(bel) = \sum_d \phi_{BDE}(bde).$$

Using these, it follows that

$$\sum_k \phi_{EK}(ek) = \sum_l \left(\sum_t \phi_{ELT}(elt) \right) = \sum_l \left(\sum_b \phi_{BEL}(bel) \right)$$

$$= \sum_b \left(\sum_l \phi_{BEL}(bel) \right) = \sum_b \sum_d \phi_{BDE}(bde),$$

so that

$$\sum_k \phi_{EK}(ek) = \sum_b \sum_d \phi_{BDE}(bde).$$

The potential on the right hand side is the marginalization of ϕ_{BDE} to E.

The property of local consistency has therefore been extended to the nodes $C_1 = EK$ and $C_2 = BDE$. The rest of the conditions can be checked; the details are left to the reader.

Proposition 10.8 *A locally consistent junction tree is globally consistent.*

Proof of Proposition 10.8 In a junction tree the intersection $C_1 \cap C_2$ of any pair C_1 and C_2 in $\mathcal{C}$ is contained in *every* node on the unique path in $\mathcal{T}$ between C_1 and C_2. Assume that $C_1 \cap C_2$ is non empty. Consider the unique path from C_1 to C_2. Let the nodes on the path be denoted by $\{C^{(i)}\}_{i=0}^n$ with $C^{(0)} = C_1$ and $C^{(n)} = C_2$, so that $C^{(i)}$ and $C^{(i+1)}$ are neighbours. Denote the separator between $C^{(i)}$ and $C^{(i+1)}$ by

$$S^{(i)} = C^{(i)} \cap C^{(i+1)}.$$

Then, for all i,

$$C_1 \cap C_2 \subseteq S^{(i)}.$$

For a set of variables C, let $\sum_C$ denote $\sum_{X_C}$. The assumption of local consistency means that for any two neighbours

$$\sum_{C^{(i)}\setminus(C^{(i+1)}\cap C^{(i)})} \phi_{C^{(i)}} = \sum_{C^{(i+1)}\setminus(C^{(i)}\cap C^{(i+1)})} \phi_{C^{(i+1)}}.$$

Since $C^{(i)}\setminus(C^{(i+1)} \cap C^{(i)}) = C^{(i)}\setminus S^{(i)}$ and $C^{(i+1)}\setminus(C^{(i+1)} \cap C^{(i)}) = C^{(i+1)}\setminus S^{(i)}$, local consistency may be written equivalently as

$$\sum_{C^{(i)}\setminus S^{(i)}} \phi_{C^{(i)}} = \sum_{C^{(i+1)}\setminus S^{(i)}} \phi_{C^{(i+1)}}.$$

For any two neighbours C_1 and C_2, their associated potentials have to be marginalized to $C_1 \cap C_2$. This is equivalent to computing the sum

$$\sum_{C^{(i)}\setminus(C_1\cap C_2\cap C^{(i)})} \phi_{C^{(i)}} \quad \text{and} \quad \sum_{C^{(i+1)}\setminus(C_1\cap C_2\cap C^{(i+1)})} \phi_{C^{(i+1)}}.$$

Starting with the leftmost marginalization,

$$\sum_{C^{(i)}\setminus(C_1\cap C_2\cap C^{(i)})} \phi_{C^{(i)}} = \sum_{C^{(i)}\setminus S^{(i)}} \sum_{S^{(i)}\setminus(C_1\cap C_2\cap S^{(i)})} \phi_{C^{(i)}},$$

since the nodes (variables) in $C^{(i)}\setminus(C_1 \cap C_2 \cap C^{(i)})$ can be split into two disjoint sets, namely $S^{(i)}\setminus(C_1 \cap C_2 \cap S^{(i)})$ (which can be empty: those outside $C_1 \cap C_2$ but inside the separator $S^{(i)}$) and $C^{(i)}\setminus S^{(i)}$ (those outside the separator).

Next, the order of summation may be exchanged so that

$$\sum_{C^{(i)}\setminus S^{(i)}} \sum_{S^{(i)}\setminus(C_1\cap C_2\cap S^{(i)})} \phi_{C^{(i)}} = \sum_{S^{(i)}\setminus(C_1\cap C_2\cap S^{(i)})} \sum_{C^{(i)}\setminus S^{(i)}} \phi_{C^{(i)}}.$$

Now, using local consistency,

$$\sum_{S^{(i)}\setminus(C_1\cap C_2\cap S^{(i)})} \left(\sum_{C^{(i)}\setminus S^{(i)}} \phi_{C^{(i)}} \right) = \sum_{S^{(i)}\setminus(C_1\cap C_2\cap S^{(i)})} \left(\sum_{C^{(i+1)}\setminus S^{(i)}} \phi_{C^{(i+1)}} \right).$$

This gives

$$\sum_{S^{(i)}\backslash(C_1\cap C_2\cap S^{(i)})}\left(\sum_{C^{(i+1)}\backslash S^{(i)}}\phi_{C^{(i+1)}}\right) = \sum_{C^{(i+1)}\backslash S^{(i)}}\left(\sum_{S^{(i)}\backslash(C_1\cap C_2\cap S^{(i)})}\phi_{C^{(i+1)}}\right)$$

$$= \sum_{C^{(i+1)}\backslash(C_1\cap C_2\cap C^{(i+1)})}\phi_{C^{(i+1)}}.$$

It follows that

$$\sum_{C^{(i)}\backslash(C_1\cap C_2\cap C^{(i)})}\phi_{C^{(i)}} = \sum_{C^{(i+1)}\backslash(C_1\cap C_2\cap C^{(i+1)})}\phi_{C^{(i+1)}}.$$

This operation can be repeated using $C^{(i+1)}$ and $C^{(i+2)}$. Therefore, the first step is to marginalize the potential ϕ_{C_1} to $C_1 \cap C_2$ and the next step is to move over to the next node $C^{(1)}$ on the unique path between C_1 and C_2. The marginalization of ϕ_{C_1} and $\phi_{C^{(1)}}$ coincide. This procedure is continued along the path until the node C_2 is reached. The result is proved. $\qquad\square$

Corollary 10.3 *After the passage of a fully active schedule of flows, a junction tree is globally consistent.*

Proof of Corollary 10.3 This follows from the proposition stating that after passage of a fully active schedule of flows a junction tree $\mathcal{T}$ is *locally* consistent, together with Proposition 10.4. $\qquad\square$

The algorithm for updating considered the cliques of a junction tree, which sent and received messages locally; the global update is performed entirely by a series of local computations. By organizing the variables into cliques and separators on a junction tree and determining a schedule, there is no need for global computations in the inference problem; the global update is achieved entirely by passing messages between neighbours in the tree according to a schedule and the algorithm terminates automatically when the update is completed.

10.9   Message passing for conditional Gaussian distributions

This section uses the junction tree approach for finding a suitable conditional Gaussian approximation for the update of a conditional Gaussian distribution. The problem here is that CG distributions are convenient to use, but while marginalizing a CG distribution over one of its continuous variables gives another CG distribution, marginalizing a CG distribution over one of its discrete variables does not necessarily give a CG distribution. Therefore, in the message passing algorithm, approximating CG distributions are used, which return the true CG distribution after the fully active schedule has been completed.

To ensure that the result is the correct CG distribution, some restrictions have to be made on the variables that are permitted in the cliques and separators. It is therefore convenient to modify the junction tree construction a little, using a *marked graph* to describe the dependence structure. For this section, marked graphs are graphs with two types of nodes, corresponding to the discrete and continuous variables.

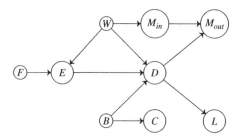

Figure 10.11   Marked graph.

Example 10.5   Consider the following example, taken from [77]. The emissions from a waste incinerator differ because of compositional differences in incoming waste. Another important factor is the way in which the waste is burnt, which can be monitored by measuring the concentration of carbon dioxide in the emissions. The efficiency of the filter depends on its technical state and also on the amount and composition of the waste. The emission of heavy metals depends both on the concentration of metals in the incoming waste and the emission of dust particles in general. The emission of dust is monitored by measuring the penetration of light.

This example may be modelled using the directed acyclic *marked* graph (DAMG) in Figure 10.11. The categorical variables are F: filter state, W: waste type, B method of burning. The continuous variables are M_{in}: metals in the waste, M_{out}: metals emitted, E: filter efficiency, D: Dust emission, C: carbon dioxide concentration in emission and L: light penetration.

Recall the notation introduced for CG distributions in Section 8.9: Δ is the set of *discrete* variables, while Γ is the set of *continuous* variables. For *marked graphs*, the notion of a decomposition has to be extended:

Definition 10.12 (Strong Decomposition) *A triple (A, B, S) of disjoint subsets of the node set V of an undirected marked graph $\mathcal{G}$ is said to form a* strong decomposition *of $\mathcal{G}$ if $V = A \cup B \cup S$ and the following three conditions hold:*

1. *S separates A from B,*

2. *S is a complete subset of V,*

3. *Either $S \subseteq \Delta$, or $B \subseteq \Gamma$, or both.*

When this holds, (A, B, S) is said to decompose $\mathcal{G}$ into the components $\mathcal{G}_{AUS}$ and $\mathcal{G}_{BUS}$.

If only the first two conditions hold, then (A, B, S) is said to form a weak decomposition. *Thus, a weak decomposition ignores the markings of the graph.*

Definition 10.13 (Strongly Decomposable) *An undirected* marked *graph is said to be* strongly decomposable *if it is complete, or if there exists a strong decomposition (A, B, S), where both A and B are non empty, into strongly decomposable sub-graphs $\mathcal{G}_{AUS}$, and $\mathcal{G}_{BUS}$.*

Decomposable unmarked graphs are triangulated; any cycle of length four or more has a chord. Strongly decomposable *marked* graphs are *further* characterized by not having any path between two discrete variables that contains two adjacent continuous variables.

Proposition 10.9 *For an undirected marked graph $\mathcal{G}$, the following are equivalent:*

1. $\mathcal{G}$ *is strongly decomposable.*

2. $\mathcal{G}$ *is triangulated, and for any path $(\delta_1, \alpha_1, \ldots, \alpha_n, \delta_2)$ between two discrete nodes (δ_1, δ_2) where $(\alpha_1, \ldots, \alpha_n)$ are all continuous, δ_1 and δ_2 are neighbours.*

3. *For any α and β in $\mathcal{G}$, every minimal (α, β) separator is complete. If both α and β are discrete, then their minimal separator contains only discrete nodes.*

Proof of Proposition 10.9, 1 $\implies$ 2 The proof, as before for unmarked graphs, is by induction. The inductive hypothesis is: All undirected strongly decomposable graphs with n or fewer nodes are triangulated and satisfy the conditions of Statement 2.

This is clearly true for a graph on one node.

Let $\mathcal{G}$ be a strongly decomposable graph on $n + 1$ nodes.

Either $\mathcal{G}$ is complete, in which case the properties of statement 2 clearly follow,

Or There exist sets A, B, S, where $V = A \cup B \cup S$, where either $B \subseteq \Gamma$ or $S \subseteq \Delta$ or both, and such that $\mathcal{G}_{A \cup S}$ and $\mathcal{G}_{B \cup S}$ are strongly decomposable. Then any cycle of length 4 without a chord must pass through both A and B. By decomposability, S separates A from B. Therefore the cycle must pass through S at least twice. Since S is complete, the cycle will therefore have a chord. Since $\mathcal{G}_{A \cup S}$ and $\mathcal{G}_{B \cup S}$ are triangulated, it follows that $\mathcal{G}$ is also triangulated. If the nodes of S are discrete, it follows that any path between two discrete variable passing through S satisfies the condition of Statement 2. If $B \subseteq \Gamma$, then since all paths in $\mathcal{G}_{A \cup S}$ and all paths in $\mathcal{G}_{B \cup S}$ satisfy the condition of Statement 2, it is clear that all paths passing through C will also satisfy the condition of Statement 2. It follows that $\mathcal{G}$ is strongly decomposable. $\qquad\square$

Proof of Proposition 10.9, 2 $\implies$ 3 Assume that $\mathcal{G}$ is triangulated, with the additional property in Statement 2. Consider two nodes α and β and let S be their minimal separator. Let A denote the set of all nodes that may be connected to α by a trail that does not contain nodes in S and let B denote all nodes that may be connected to β by a trail that does not contain nodes in S. Every node $\gamma \in S$ must be adjacent to some node in A and some node in B, otherwise $\mathcal{G}_{V \setminus (S \setminus \{\gamma\})}$ would not be connected. This would contradict the minimality of S, since $S \setminus \{\gamma\}$ would separate α from β. Suppose that the condition in Statement 2 holds and consider the minimal separator for two discrete nodes α, β, which are not neighbours. The separator is complete. Denote the separator by S. Consider $\hat{S}$, which is S with the continuous nodes removed. Then $\hat{S}$ separates α and β on the sub graph induced by the discrete variables. But the condition of statement 2 implies that α and β are also separated on $\mathcal{G}$. Therefore, $\hat{S}$ separates α and β. It follows that the minimal separator for two discrete nodes contains only discrete nodes. $\qquad\square$

Proof of Proposition 10.9, 3 $\implies$ 1 If $\mathcal{G}$ is complete, it follows that every node is discrete and the result is clear. Let α and β be two discrete nodes that are not contained within their minimal separator. Let S denote their minimal separator. Let A denote the maximal connected component of $V \backslash S$ and let $B = V \backslash (A \cup S)$. Then (A, B, S) provides a decomposition, with $S \subseteq \Delta$. Suppose that two such discrete nodes cannot be found. Let α and β be two nodes that are not contained within their minimal separator, where β is continuous. Let S denote the minimal separator. Let B denote the largest connected component of $V \backslash S$ containing β. Suppose that B contains a discrete node γ. Then S separates γ from α and therefore consists entirely of discrete nodes. Therefore, either $S \subseteq \Delta$, or $B \subseteq \Gamma$, as required.   □

The construction of the junction tree has to be modified. Starting from the directed acyclic graph, the graph is first moralized by adding in the links between all the parents of each variable and then making all the edges undirected, as before. Then, sufficient edges are added in to ensure that the graph is a *decomposable marked graph*.

Next, a junction tree is constructed. As before, this is an organization of a collection of subsets of the variables V into a tree, such that if A and B are two nodes on the junction tree, then the variables in $A \cap B$ appear in each node on the path between A and B.

Definition 10.14 (Strong Root) *A node R on a junction tree is a* strong root *if any pair of neighbours A, B, such that A lies on the path between R and B (so that A is closer to R than B) satisfies*

$$B \backslash A \subseteq \Gamma \qquad or \qquad B \cap A \subseteq \Delta \qquad or\ both.$$

This condition is equivalent to the statement that the triple $(A \backslash (A \cap B), B \backslash (A \cap B), A \cap B)$ forms a strong decomposition of $\mathcal{G}_{A \cup B}$. This means that when a separator between two neighbouring cliques is not purely discrete, the clique furthest away from the root has only continuous nodes beyond the separator.

Theorem 10.1 *The cliques of a strongly decomposable marked graph can be organized into a junction tree with at least one strong root.*

Proof of Theorem 10.1 As before, start with a simplicial discrete node X_1. Then F_{X_1} is a clique. Continue choosing nodes from F_{X_1} that only have neighbours in F_{X_1}. Set i_1 the number of nodes in F_{X_1} that only have neighbours in F_{X_1}. Name the set of nodes in F_{X_1} V_{i_1} and the set of nodes in F_{X_1} that have neighbours *not* in F_{X_1} S_{i_1}.

Now remove the nodes of F_{X_1} that do not have neighbours outside F_{X_1}. Choose a new discrete simplicial node X_2, such that $F_{X_1} \cap F_{X_2} \neq \phi$, the empty set, and repeat the process with index i_2, where i_2 is i_1 plus the number of nodes in F_{X_2} that only have neighbours in F_{X_2}.

Continue this process, by choosing X_j such that $F_{X_j} \cap (\cup_{k=1}^{j-1} F_{X_k}) \neq \phi$, until there are no *discrete* simplicial nodes left with which to continue the process. Let X_n denote the last discrete simplicial node, following this procedure.

Now continue the procedure, for, $j = n + 1, \ldots, N$, choosing X_j such that $F_{X_j} \cap (\cup_{k=1}^{j-1} F_{X_k}) \neq \phi$, until there are no nodes left.

Since the minimal separator of any two discrete nodes contains only discrete nodes, it follows that if there is a *discrete* node in $V \backslash (\cup_{j=1}^{n} V_{i_j})$, then there is a discrete simplicial node among those remaining, such that $F_{X_{n+1}} \cap (\cup_{j=1}^{n} F_{X_n}) \neq \phi$, contradicting the assertion. It follows that all the nodes in $V \backslash (\cup_{j=1}^{n} V_{i_j})$ are continuous.

By construction, the cliques $V_{i_1}, \ldots, V_{i_n}$ may be organised as a junction tree. If cliques V_{i_k} and V_{i_l} are adjacent in the tree, it is clear that their separator is the minimal separator between X_k and X_l. These are both discrete. Because the minimal separator between two discrete nodes contains only discrete variables, it follows that all the separators in this junction tree are discrete.

Now continue the construction of the junction tree as in Theorem 4.3, where the remaining cliques have index greater than i_n. The resulting construction will have the desired properties; any of the cliques $V_{i_1}, \ldots, V_{i_n}$ may be chosen as the strong root. $\square$

To exploit the properties of CG distributions, the following further assumption needs to be made:

Hypothesis 10.1   *No continuous nodes have discrete children.*

This is because, *conditioned on* the discrete variables, the distribution is Gaussian, which makes certain aspects of the computation rather easy.

The prior conditional probability distributions corresponding to the directed acyclic marked graph need to be specified. The assumption is that for a *continuous* variable X, with parents $\Pi(X) = (\Pi_d(X), \Pi_c(X))$, where $\Pi_d(X)$ are the discrete parents and $\Pi_c(X)$ are the continuous parents,

$$X | (\Pi_d(X) = \underline{y}, \Pi_c(X) = \underline{z}) \sim N(\alpha(\underline{y}) + \beta(\underline{y})^t \underline{z}, \gamma(\underline{y})).$$

Following Definition 1.4, the random vectors are taken as *row* vectors when they are several attributes measured on a single run of an experiment. Here, α is a function, β is a (row) vector of the same length as $\underline{z}$ and γ is the conditional variance. The assumption is that the variance is only affected by the discrete parents; the continuous parents only enter linearly through the mean. The conditional density is then a CG potential,

$$\phi_X(\underline{y}, \underline{z}, x) = \frac{1}{(2\pi \gamma(\underline{y}))^{1/2}} \exp \left\{ \frac{(x - \alpha(\underline{y}) - \beta(\underline{y}) \underline{z}^t)^2}{2\gamma(\underline{y})} \right\}.$$

From this, expanding the parentheses, taking logarithms and identifying terms gives the canonical parameters (g_X, h_X, K_X). The log partition function is

$$g_X(\underline{y}) = -\frac{\alpha(\underline{y})^2}{2\gamma(\underline{y})} - \frac{1}{2} \log(2\pi \gamma(\underline{y})),$$

and the other parameters are given by

$$h_X(\underline{y}) = \frac{\alpha(\underline{y})}{\gamma(\underline{y})} \begin{pmatrix} 1 & -\beta(\underline{y}) \end{pmatrix}$$

and

$$K_X(\underline{y}) = \frac{1}{\gamma(\underline{y})} \begin{pmatrix} 1 & -\beta(\underline{y}) \\ -\beta(\underline{y})^t & \beta(\underline{y})^t \beta(\underline{y}) \end{pmatrix}.$$

Marginalization: Continuous variables Suppose $\phi_{\underline{Y},\underline{X}_1,\underline{X}_2}$ is a CG potential, where $\underline{Y}$ are discrete variables and $\underline{X}_1$ and $\underline{X}_2$ are continuous variables. That is, ϕ is given by

$$\phi_{\underline{Y},\underline{X}_1,\underline{X}_2}(\underline{y},\underline{x}_1,\underline{x}_2)$$

$$= \chi(\underline{y})\exp\left\{g(\underline{y}) + h_1(\underline{y})\underline{x}_1^t + h_2(\underline{y})\underline{x}_2^t - \frac{1}{2}(\underline{x}_1,\underline{x}_2)\begin{pmatrix} K_{11} & K_{12} \\ K_{12}^t & K_{22} \end{pmatrix}\begin{pmatrix} \underline{x}_1^t \\ \underline{x}_2^t \end{pmatrix}\right\},$$

where $\chi(\underline{y})$ is a function returning the value 1 if $p_{\underline{Y}}(\underline{y}) > 0$ and 0 if $p_{\underline{Y}}(\underline{y}) = 0$, K is symmetric and the triple (g, h, K) represents the canonical characteristics. Recall the standard result that, taking $\underline{z} \in \mathbf{R}^p$ as a row vector, and K a positive definite $p \times p$ symmetric matrix,

$$\frac{1}{(2\pi)^{p/2}}\int_{\mathbf{R}^p}\exp\left\{-\frac{1}{2}\underline{z}K\underline{z}^t\right\}d\underline{z} = \frac{1}{\sqrt{\det(K)}}$$

and hence that for $\underline{a} \in \mathbf{R}^p$ and K a positive definite $p \times p$ symmetric matrix

$$\int_{\mathbf{R}^p}\exp\left\{(\underline{a},\underline{z}) - \frac{1}{2}\underline{z}^t K\underline{z}\right\}d\underline{z} = \exp\left\{\frac{1}{2}\underline{a}^t K^{-1}\underline{a}\right\}\frac{(2\pi)^{p/2}}{\sqrt{\det(K)}}.$$

From this, it follows, after some routine calculation, that if $\underline{X}_1$ is a random p-vector with positive definite covariance matrix, then

$$\int_{\mathbf{R}^p}\phi_{\underline{Y},\underline{X}_1,\underline{X}_2}(\underline{y},\underline{x}_1,\underline{x}_2)d\underline{x}_1 = \chi(\underline{y})\exp\left\{\tilde{g}(\underline{y}) + \tilde{h}(\underline{y})\underline{x}_2^t - \frac{1}{2}\underline{x}_2\tilde{K}\underline{x}_2^t\right\},$$

where

$$\tilde{g}(\underline{y}) = g(\underline{y}) + \frac{1}{2}\left(p\log(2\pi) - \log\det(K_{11}(\underline{y})) + h_1(\underline{y})K_{11}(\underline{y})^{-1}h_1(\underline{y})^t\right),$$

$$\tilde{h}(\underline{y}) = h_2(\underline{y}), \qquad \tilde{K} = -K_{21}(\underline{y})K_{11}(\underline{y})^{-1}K_{12}(\underline{y}).$$

Marginalization: Discrete variables Consider a CG potential $\phi_{\underline{Y}_1,\underline{Y}_2,\underline{X}}$, where $\underline{Y}_1$ and $\underline{Y}_2$ denote sets of discrete variables and $\underline{X}$ a set of continuous variables. Consider marginalization over $\underline{Y}_2$. Firstly, if $h(\underline{y}_1,\underline{y}_2) = \tilde{h}(\underline{y}_1)$ and $K(\underline{y}_1,\underline{y}_2) = \tilde{K}(\underline{y}_1)$ for some functions $\tilde{h}$ and $\tilde{K}$ (i.e. they do not depend on $\underline{y}_2$), then $\tilde{\phi}$, the marginal of $\phi_{\underline{Y}_1,\underline{Y}_2,\underline{X}}$ is simply

$$\tilde{\phi}(\underline{y}_1,\underline{x}) = \exp\left\{\tilde{h}(\underline{y}_1)^t\underline{x} - \frac{1}{2}\underline{x}^t\tilde{K}(\underline{y}_1)\underline{x}\right\}\sum_{y_2}\chi(\underline{y}_1,\underline{y}_2)\exp\left\{g(\underline{y}_1,\underline{y}_2)\right\}.$$

The potential $\tilde{\phi}$ is therefore CG with canonical characteristics $\tilde{g}(\underline{y}_1) = \log\sum_{\underline{y}_2}\exp\left\{g(\underline{y}_1,\underline{y}_2)\right\}$ and $\tilde{h}$, $\tilde{K}$ as before.

If either h or K depends on $\underline{y}_2$, then a marginalization will not produce a CG distribution, so an *approximation* is used. For this, it is convenient to consider the *mean parameters*, $(p, \mathbf{C}, \mu)$, where $p(\underline{y}_1,\underline{y}_2) = p((\underline{Y}_1,\underline{Y}_2) = (\underline{y}_1,\underline{y}_2))$ and

$$\underline{X}|\{(\underline{Y}_1,\underline{Y}_2) = (\underline{y}_1,\underline{y}_2)\} = N(\mu(\underline{y}_1,\underline{y}_2), \mathbf{C}(\underline{y}_1,\underline{y}_2)).$$

The approximation is as following: $\tilde{\phi}$ is *defined* as the CG potential with mean parameters $(\tilde{p}, \tilde{C}, \tilde{\mu})$ defined as:

$$\tilde{p}(\underline{y}_1) = \sum_{\underline{y}_2} p(\underline{y}_1, \underline{y}_2),$$

$$\tilde{\mu}(\underline{y}_1) = \frac{1}{\tilde{p}(\underline{y}_1)} \sum_{\underline{y}_2} p(\underline{y}_1, \underline{y}_2) \mu(\underline{y}_1, \underline{y}_2),$$

$$\tilde{C}(\underline{y}_1) = \frac{1}{\tilde{p}(\underline{y}_1)} \sum_{\underline{y}_2} p(\underline{y}_1, \underline{y}_2) \left(C(\underline{y}_1, \underline{y}_2) + (\mu(\underline{y}_1, \underline{y}_2) - \tilde{\mu}(\underline{y}_1))^t (\mu(\underline{y}_1, \underline{y}_2) - \tilde{\mu}(\underline{y}_1)) \right).$$

It is relatively straightforward to compute that this approximate marginalization has the correct expected value and second moments.

Marginalizing over both discrete and continuous   When marginalizing over both types of variables, *first* the continuous variables are marginalized, and then the discrete.

The fully active schedule may now be applied. Firstly, the evidence is inserted. This is hard evidence, that certain states of the discrete variables are impossible, or that the continuous variables take certain fixed values. The information then has to be propagated. Start at the leaves, send all messages to a strong root, then propagate back out to the leaves. Since messages are propagated to and from a strong root, all marginalizations are proper marginalizations. This is clear: marginalizing continuous variables gives a proper marginalization. If the separator is purely discrete, then once the continuous variables have been marginalized, the remaining discrete marginalizations are proper marginalizations. The directed acyclic marked graph, and its strong decomposition, have ensured that the propagation is exact.

Having inserted hard evidence and run the schedule, the resulting potentials are not necessarily probability distributions, but after the schedule, the same constant is required to normalize each of them. The updating is finished, therefore, by finding the constant that normalizes potential over one of the cliques or separators to make it a probability distribution.

If the graph is not *strongly decomposable*, then the approximate marginalization may be used, to obtain an approximate update.

10.10   Using a junction tree with virtual evidence and soft evidence

The methods discussed so far in this chapter may be extended to the problem of updating in the light of virtual evidence and soft evidence.

Dealing with virtual evidence is straightforward; for each virtual finding, one adds in a virtual node, as illustrated in Figure 3.3, which will be instantiated according to the virtual finding. This simply adds the virtual finding node to the clique containing the variable for which there is a virtual finding.

Incorporating soft evidence cannot be carried out in such a straightforward manner, because when there is a soft finding on a variable, the DAG is altered by removing the directed arrows from the parent nodes to the variable. One method for incorporating soft evidence is discussed in [134]. The input is a Bayesian network with a collection of soft and hard findings. The method returns a joint probability distribution with two properties:

1. The findings are the marginal distributions for the updated distribution.

2. The updated distribution is the closest to the original distribution (where the Kullback-Leibler divergence is used) that satisfies this constraint (that the findings are the marginals of the updated distribution).

The lazy big Clique algorithm   The method described in chapter 10 is modified to incorporate soft evidence in the following way.

1. Construct a junction tree, *in which all the variables that have soft evidence are in the same clique–the big clique C_1.*

2. Let C_1 (the big clique) be the root node, apply the hard evidence and run the first half of the fully active schedule; that is, propagating from the leaves to the root node.

3. Once the big clique C_1 has been updated with the information from all the other cliques, absorb all the soft evidence into C_1. This is described below.

4. Distribute the evidence according to the method described in Section 10.7 for sending messages from the updated root out to the leaves.

If the big clique is updated to provide a *probability* function (namely a potential that sums to 1), then the distribution of evidence will update the potentials over the cliques and separators to probability distributions over the respective cliques and separators.

Absorbing the soft evidence   Suppose the big clique C_1 has soft evidence on the variables $(Y_1, \ldots, Y_k)$. Suppose soft evidence is received that $Y_1, \ldots, Y_k$ have distributions $q_{Y_1}, \ldots, q_{Y_k}$ respectively. Let q_{C_1} denote the probability function over the variables in C_1 after the soft evidence has been absorbed. Then it is required that, for each $j \in \{1, \ldots, k\}$, $q_{Y_j} = \sum_{X_{C_1 \setminus \{Y_k\}}} q_{C_1}$. That is, the marginal of q_{C_1} over all variables other than Y_k is q_{Y_k}.

The important feature of soft evidence (Definition 3.1) is that after soft evidence has been received, the variable has no parent variables. The *Iterative Proportional Fitting Procedure* (IPFP), therefore, may be employed. It goes in cycles of length k. Firstly, normalize the potential over C_1 (after the hard evidence has been received) so that it is a probability distribution p_{C_1}. Then

$$p_{C_1}^{(0)} = p_{C_1}$$

for $j = 1, \ldots, k$, set $p_{Y_j}^{(mk+j-1)} = \sum_{X_{C_1 \setminus \{Y_j\}}} p_{C_1}^{(mk+j-1)}$, and

$$p_{C_1}^{(mk+j)} = \frac{p_{C_1}^{(mk+j-1)} q_{Y_j}}{p_{Y_j}^{(mk+j-1)}}.$$

This is repeated until the desired accuracy is obtained. It has been well established that, for discrete distributions with finite state space, the IPFP algorithm converges to the distribution that minimises the Kullback-Leibler distance from the original distribution (see [135]).   □

Notes The original paper describing the use of junction trees for updating a Bayesian network is by S.L. Lauritzen and D.J. Spiegelhalter [131]. The terminology *Aalborg formula* is found in [52]. Much of the material is taken from 'Probabilistic Networks and Expert Systems' by R.G. Cowell, A.P. David, S.L. Lauritzen and D.J Spiegelhalter [67]. The proofs or the main results were originally presented in [68]. The reference [136] gives a mathematically rigorous presentation of an alternative message passing scheme known as Schachter's method. This method is also valid for more general influence diagrams. The application of junction tree methods to conditional Gaussian distributions was taken from S.L. Lauritzen [77]. The Iterative Proportion Fitting Procedure dates back to W.E. Deming and F.F. Stephan (1940) [137]; this is the basis for updating a junction tree in the light of soft evidence. The basic technique is taken from [134].

10.11 Exercises: The junction tree and probability updating

1. Let $V = \{X_1, X_2, X_3, X_4, X_5, X_6\}$ be a set of variables, where for $j = 1, \ldots, 6$, variable X_j has state state space $\mathcal{X}_j$. Consider the potentials $\phi_1 : \mathcal{X}_1 \times \mathcal{X}_2 \times \mathcal{X}_3 \to \mathbf{R}_+$, $\phi_2 : \mathcal{X}_2 \times \mathcal{X}_3 \times \mathcal{X}_5 \to \mathbf{R}_+$, $\phi_3 : \mathcal{X}_1 \times \mathcal{X}_3 \times \mathcal{X}_4 \to \mathbf{R}_+$, $\phi_4 : \mathcal{X}_5 \times \mathcal{X}_6 \to \mathbf{R}_+$.

   (a) Determine the domain graph.

   (b) Eliminate X_3 and determine the resulting set of potentials and their domain graph.

   (c) For the original domain graph, determine a perfect elimination sequence ending with X_1.

2. Let $V = \{X_1, X_2, X_3, X_4, X_5, X_6, X_7, X_8\}$ where, for $j = 1, \ldots, 8$, variable X_j has state space $\mathcal{X}_j$. Consider the potentials $\phi_1 : \mathcal{X}_1 \times \mathcal{X}_2 \times \mathcal{X}_3 \to \mathbf{R}_+$, $\phi_2 : \mathcal{X}_2 \times \mathcal{X}_4 \times \mathcal{X}_5 \to \mathbf{R}_+$, $\phi_3 : \mathcal{X}_4 \times \mathcal{X}_6 \times \mathcal{X}_7 \to \mathbf{R}_+$ and $\phi_4 : \mathcal{X}_1 \times \mathcal{X}_6 \times \mathcal{X}_8 \to \mathbf{R}_+$.

   (a) Determine the domain graph.

   (b) Eliminate X_1 and determine the resulting set of potentials and their domain graph.

   (c) For the original domain graph, is there a perfect elimination sequence?

3. Consider the Bayesian network in Figure 10.12.

   (a) Determine the domain graph.

   (b) Does the domain graph have a perfect elimination sequence?

   (c) Triangulate the graph, adding as few fill ins as possible.

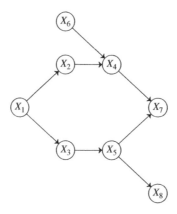

Figure 10.12 Bayesian network for exercises 3 and 4.

4. Consider again the Bayesian network in Figure 10.12.

   (a) Write down the elimination sequence corresponding to your triangulation in the previous exercise.

(b) Suppose each variable has three states and let $X_j = (x_j^{(1)}, x_j^{(2)}, x_j^{(3)})$ for $j = 1, \ldots, 8$. Calculate the total number of entries in all the tables needed to define

$$p_{X_1, X_2, X_3, X_4, X_5, X_6, X_7, X_8}(., ., ., ., ., ., x_7^{(1)}, x_8^{(1)})$$

(c) Consider the calculation of

$$p_{X_1|X_7, X_8}(.|x_7^{(1)}, x_8^{(1)})$$

where the variables are marginalized in the following order: X_2, X_6, X_4, X_5, X_3. Calculate the size of each table to be marginalized in the process.

(d) Try to find an elimination sequence resulting in smaller tables to be marginalized.

5. Let $\mathcal{G} = (V, E)$ be an undirected graph, with node set $V = \{\alpha_1, \ldots, \alpha_d\}$ Recall that an *elimination sequence* of a graph $\mathcal{G}$ is a linear ordering of its nodes. Let σ be an elimination sequence and let Λ denote the fill-ins produced by eliminating the nodes of $\mathcal{G}$ in the order σ. Let $\mathcal{G}^\sigma$ denote the graph $\mathcal{G}$ extended by Λ. Note that in $\mathcal{G}^\sigma$ any node $\alpha \in V$ together with its neighbours of higher elimination order form a *complete* subset in the sense that they are all pairwize linked. Let $N_{\sigma(\alpha)}$ denote the set that contains α and its neighbours of a higher elimination order.

(a) Let $G^{-\alpha_v}$ denote the graph $\mathcal{G}$, after the node α_v has been eliminated. Prove that $G^{-\alpha_v}$ is the domain graph for $\Phi^{-\alpha_v}$, the potential Φ with the variable α_v eliminated.

(b) Prove that the sets $N_{\sigma(\alpha)}$ are the elimination domains corresponding to the elimination sequence σ.

6. Construct a junction tree for the Bayesian network shown in Figure 10.13.

(a) Construct a fully active schedule.

(b) Assume, now, that you have hard evidence that $\{(X_4, X_5, X_8) = (x_4^{(1)}, x_5^{(3)}, x_8^{(2)})\}$. Which communications are necessary to update the network?

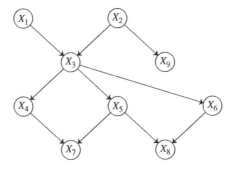

Figure 10.13 Bayesian network.

7. (a) Consider the Bayesian network shown in Figure 10.14, where for each j, the number inside the node j indicates the number of possible states that variable X_j can take. Construct a junction tree using the following 'greedy' algorithm: take a simplicial node, if possible. Otherwise, take the node that requires as few fill-ins as possible. Once the number of fill-ins has been determined, take the node associated with the potential with the smallest table. Once the number of fill-ins and table size has been determined, take the node, satisfying these conditions, with the largest number of states.

   (b) Write a MATLAB code to implement the procedure outlined above, when there are six variables. The input should be the parents of each node. The algorithm should then determine the edges required for the moral graph. The output should be the elimination sequence.

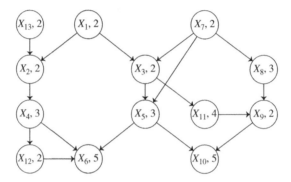

Figure 10.14   Bayesian network for Exercise 7.

8. Consider the Bayesian network in Figure 4.28. Suppose the network receives hard evidence: $\{(X_5, X_6, X_9) = (x_5^{(1)}, x_6^{(3)}, x_9^{(2)})\}$. From the junction tree computed for Exercise 5 in Chapter 4, describe an active schedule to update the distribution of the other variables, conditioned on these instantiations.

9. Let $\mathcal{C}$ denote the set of cliques from a triangulated graph. A *pre-I-tree* is a tree over $\mathcal{C}$ with separators $S = C_1 \cap C_2$ for adjacent cliques C_1 and C_2. The *weight* of a pre-I-tree is the sum of the number of variables in the separators.

   (a) Prove that a junction tree is a pre-I-tree of maximal weight.

   (b) Prove that any pre-I-tree of maximal weight is a junction tree.

10. Consider the DAG given in the Bayesian network in Figure 10.15.

   (a) Determine the minimal set of conditioning variables for the DAG to reduce it to a singly connected DAG.

   (b) The numbers attached to the variables indicate the number of states. Determine a conditioning resulting in a minimal number of singly connected DAGs.

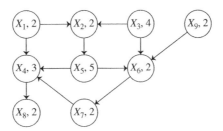

Figure 10.15   Bayesian network for Exercise 10.

11. This exercise is taken from S. Lauritzen [77]. It is a fictitious problem connected with controlling the emission of heavy metals from a waste incinerator. The type of incoming waste W affects the metals in the waste M_{in}, the dust emission D and the filter efficiency E. The quantity of metals in the waste M_{in} affects the metals emission M_{out}. Another important factor is the waste burning regimen B, which is monitored via the carbon dioxide concentration in the emission C. The burning regimen, the waste type and the filter efficiency E affect the dust emission D. The dust emission affects the metals emission and it is monitored by recording the light penetration L. The state of the filter F (whether it is intact or defective) affects E.

The variables F, W, B are qualitative variables with states (the filter is either intact or defective, the waste is either industrial or household, the burning regimen is either stable or unstable). The variables E, C, D, L, M_{in} and M_{out} are continuous. The directed acyclic marked graph is given in Figure 10.11.

- Moralize the graph.

- By adding in as few links as possible, construct a strongly decomposable graph.

- Construct a junction tree. What are the possible strong roots for the junction tree?

11

Factor graphs and the sum product algorithm

The last few chapters treated the algorithmic problem of marginalizing a global potential which is expressed as a product of local potentials, each with a domain that is a subset of the global domain. This chapter addresses the same problem, but introduces a different algorithm is, which does not require (at least explicitly) the construction of junction trees. The algorithm is known as the *sum product algorithm* and it operates on *factor graphs*.

11.1   Factorization and local potentials

As usual, let $\tilde{V} = \{1, \ldots, d\}$, and for each $j \in \tilde{V}$ let $\mathcal{X}_j = (x_j^{(1)}, \ldots, x_j^{(k_j)})$ denote a finite state space. Let $\mathcal{X} = \times_{j=1}^d \mathcal{X}_j$. The space $\mathcal{X}$ is the *configuration space*. Let ϕ denote a potential defined on $\mathcal{X}$.

Let $\underline{x} = (x_1, \ldots, x_d) \in \mathcal{X}$ denote a configuration and, for a subset $D \subseteq \{1, \ldots, d\}$, where $D = \{j_1, \ldots, j_m\}$, let $\underline{x}_D = (x_{j_1}, \ldots, x_{j_m})$ and $\mathcal{X}_D = \times_{v \in D} \mathcal{X}_v$.

A domain $\mathcal{X}_D$ for $D \subset \{1, \ldots, d\}$ (where the subset is strict) is called a *local domain*.

Definition 11.1 (Factorizability) *The potential ϕ is said to be factorizable if it factors into a product of several* local potentials γ_j *each defined on local domains, such that*

$$\phi(\underline{x}) = \prod_{j \in J} \gamma_j(\underline{x}_{S_j}) \tag{11.1}$$

for a collection of local domains $\mathcal{X}_{S_j}$, $j \in J$ where $J = \{1, 2, \ldots, q\}$ and $q \leq d$.

Bayesian Networks: An Introduction   T. Koski, J. Noble
© 2009 John Wiley & Sons, Ltd

For a factorizable potential ϕ, consider the problem of computing the marginal

$$\phi_i(x_i) = \sum_{\underline{z} \in \mathcal{X}_{\hat{V}\backslash\{i\}}} \prod_{j \in J} \gamma_j(\underline{z}, x_i), \tag{11.2}$$

where the domains of the potentials have been extended to $\mathcal{X}$ (Definition 2.25). This is also known as the 'one i (eye) problem'. The aim of this section is to develop an efficient procedure for computing the marginalization, which exploits the way in which the global potential is factorized and uses the current values to update the values assigned to each variable. The method involves a *factor graph*, which is an example of a *bipartite graph*.

Definition 11.2 (Bipartite Graph) *A graph $\mathcal{G}$ is* bipartite *if its node set can be partitioned into two sets W and U in such a way that every edge in $\mathcal{G}$ has one node in W and another in U.*

Bipartite graphs can be (roughly) characterized as graphs that have no cycles of odd length, but this property is not used here. A *factor graph* is a *bipartite graph* that expresses the structure of the factorization given by Equation (11.1). The graph is constructed as follows.

- there is a *variable node* (an element of U) for each variable. A capital letter X will be used to denote the variable node, a small letter the value x in the state space $\mathcal{X}_X$ associated with the variable.

- there is a *function node* (an element of W) for each potential γ_j. γ_j will be used to denote both the potential and the node.

- an undirected edge connecting variable node X_i to factor node γ_j if and only if X_i is in the local domain of γ_j.

In other words, a factor graph is a representation of the relation 'is an argument of'.

11.1.1   Examples of factor graphs

Example 11.1: Error correcting codes   The following example is taken from N. Wiberg [138]. Consider the system of equations:

$$x_1 + x_2 + x_3 = 0$$
$$x_3 + x_4 + x_5 = 0$$
$$x_1 + x_5 + x_6 = 0$$
$$x_2 + x_4 + x_6 = 0 \tag{11.3}$$

The variables are binary (taking values 0 or 1) and addition is binary (modulo 2). This system of equations may be expressed as the factor graph given in Figure 11.1. The boxes correspond to the operation $+$ and the circles correspond to the variables of the system. The notation X_j will be used to denote both the variable and the corresponding variable node in the graph for the variable that takes its values in state space $\mathcal{X}_j$. Similar notation will be used throughout.   □

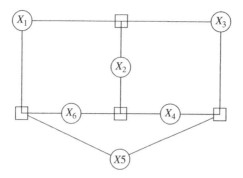

Figure 11.1   A Factor Graph for Equations (11.3).

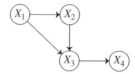

Figure 11.2   A directed acyclic graph.

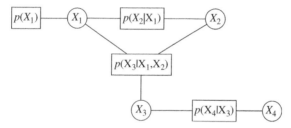

Figure 11.3   The factor graph corresponding to the directed acyclic graph in Figure 11.2.

Example 11.2: Bayesian networks as factor graphs   A Bayesian network has a joint probability distribution that factorizes according to a DAG. This joint distribution can be converted into a factor graph. Each function is the local potential $p_{X_i|\Pi_i}$ and edges are drawn from this node to X_i and to its parents Π_i. The DAG is shown in Figure 11.2 and the corresponding factor graph in Figure 11.3.

Example 11.3: A factor graph of a probability distribution
1. A general joint probability distribution $p(X_1, X_2, X_3, X_4)$ over four variables has the trivial factor graph shown in Figure 11.4.

2. The definition of conditional probability yields

$$p_{X_1,X_2,X_3,X_4} = p_{X_1} p_{X_2|X_1} p_{X_3|X_1,X_2} p_{X_4|X_1,X_2,X_3}.$$

The factor graph corresponding to this chain rule is given in Figure 11.5.

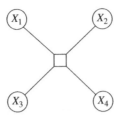

Figure 11.4   Trivial factor graph for a probability distribution.

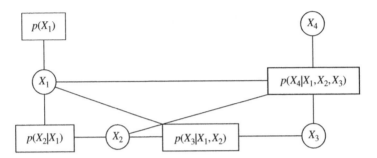

Figure 11.5   The chain rule as a factor graph.

Figure 11.6   A Markov chain as a factor graph.

3. A hidden Markov model (HMM) has

$$p_{X_1,X_2,X_3,Y_1,Y_2,Y_3}$$

$$= p_{X_1} p_{X_2|X_1} p_{X_3|X_2} p_{X_4|X_3} p_{Y_1|X_1} p_{Y_2|X_2} p_{Y_3|X_3}$$

and the corresponding factor graph is given in Figure 11.7. The factor graph corresponding to a Markov chain is illustrated in Figure 11.6.

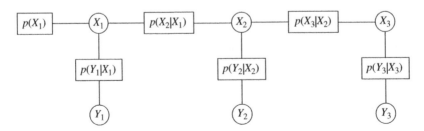

Figure 11.7   A hidden Markov model as a factor graph.

11.2   The sum product algorithm

The following algorithm computes the marginalization for the one eye problem of Equation (11.2) without the introduction of junction trees. Consider Figure 11.8.
The following notation is introduced:

$$\mu_{X_k \to \gamma_j}(x) \qquad x \in \mathcal{X}_k: \qquad \text{Variable to local potential.}$$

This is the message sent from node X to node γ_j in the sum product algorithm and

$$\mu_{\gamma_j \to X_k}(x) \qquad x \in \mathcal{X}_k: \qquad \text{Local potential to variable.}$$

This is the message sent from the function node γ_j to the variable node X.

Recall the definition of *neighbour* (Definition 2.3). N_v will be used to denote the set of neighbours of a node v. A factor graph is undirected. By the definition of a factor graph, all the neighbours of a node will be of the opposite type to the node itself.

The message sent from node v on edge e is the product of the local potential at v (or the *unit* function if v is a variable node) with all messages received at v on edges other than e and then marginalized to the variable associated with e. The messages are defined recursively as follows.

Definition 11.3 (Sum Product Update Rule) *For $x \in \mathcal{X}_k$, and for each $X_k \in N_{\gamma_j}$,*

$$\mu_{X_k \to \gamma_j}(x) = \begin{cases} \prod_{h \in N_{X_k} \setminus \{\gamma_j\}} \mu_{h \to X_k}(x) & \forall x \in \mathcal{X}_k \quad N_{X_k} \neq \phi \\ 1 & N_{X_k} = \phi. \end{cases} \qquad (11.4)$$

and for each $\gamma_j \in N_{X_k}$,

$$\mu_{\gamma_j \to X_k}(x) = \sum_{\underline{y} \in \mathcal{X}_{\hat{V} \setminus \{k\}}} \gamma_j(\underline{y}, x) \prod_{Y \in N_{\gamma_j} \setminus \{X_k\}} \mu_{Y \to \gamma_j}(y_j) \qquad \forall x \in \mathcal{X}_k \qquad (11.5)$$

where ϕ denotes the empty set, and where the domain of γ_j has been extended to $\mathcal{X}$ and variable X_k takes the last position; y_j is the value taken by variable X_j ($j \neq k$).

The flow of computation in a factor graph is illustrated in Figure 11.9.

Definition 11.4 (Initialization) *The initialization is*

$$\mu_{X_k \to \gamma_j}(x) = 1 \qquad \forall x \in \mathcal{X}_j$$

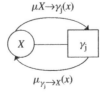

$\mu X {\to} \gamma_j(x)$

$X \qquad \gamma_j$

$\mu_{\gamma_j \to X}(x)$

Figure 11.8   Updates in a factor graph.

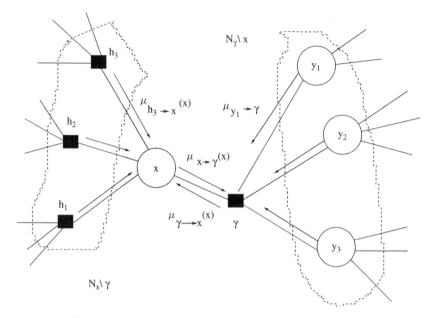

Figure 11.9   Updates in a fragment of a factor graph.

for each $X_k \in N_{\gamma_j}$ and

$$\mu_{\gamma_j \to X_k}(x) = 1 \qquad \forall x \in \mathcal{X}_j$$

for each $\gamma_j \in N_{X_k}$, for each variable node X_k and each function node γ_j.

Definition 11.5 (Termination) *The termination at a node is the product of all messages directed towards that node.*

$$\mu_{X_k}(x) = \prod_{\gamma_j \in N_{X_k}} \mu_{\gamma_j \to X_k}(x), \qquad x \in \mathcal{X}_k \qquad\qquad (11.6)$$

and

$$\mu_{\gamma_j}(\underline{x}_{D_j}) = \prod_{X_k \in N_j} \mu_{X_k \to \gamma_j}(x_k) \qquad \forall \underline{x}_{D_j} \in \mathcal{X}_{D_j}.$$

Note that the function node receives communications from precisely those variables that are in the domain of the function.

After sending sufficiently many messages according to a suitable schedule, the *termination* at the *variable* node yields the *marginalization*, or a suitable approximation to the marginalization, over that variable. That is,

$$\mu_{X_i}(x) = \sum_{\underline{y} \in \mathcal{X}_{\bar{V} \setminus \{i\}}} \phi(\underline{y}, x) \qquad \forall x \in \mathcal{X}_i,$$

where the arguments of ϕ have been rearranged, so that variable X_i appears last.

The schedule   One node is arbitrarily chosen as a root and, for the purposes of constructing a schedule, the edges are *directed* to form a directed acyclic graph, where the root has no parents. If the graph is a tree, then the choice of directed acyclic graph is uniquely defined by the choice of the root node. Computation begins at the leaves of the factor graph.

- Each leaf variable node sends the trivial identity function to its parents.

- Each leaf function node sends a description of γ to its parents.

- Each node waits for the message from all its children before computing the message to be sent to its parents.

- Once the root has received messages from all its children, it sends messages to all its children.

- Each node waits for messages from all its parents before computing the message to be sent to its children.

This is repeated from root to leaves and is iterated a suitable number of times. No iterations are needed if the factor graph is cycle free. This is known as a *generalized forward and backward algorithm*.

The following result was proved by N. Wiberg [138].

Theorem 11.1 (Wiberg).   *Let*

$$\phi(\underline{x}) = \prod_j \gamma_j(\underline{x}_{D_j})$$

*and let $\mathcal{G}$ be a factor graph with **no cycles**, representing ϕ. Then, for any variable node X_k, the marginal of ϕ at $x \in \mathcal{X}_k$ is*

$$\mu_{X_k}(x) = \sum_{\underline{y} \in \mathcal{X}_{\hat{V}\backslash\{k\}}} \phi(\underline{y}, x),$$

where the arguments of ϕ have been rearranged so that the kth variable appears last and $\mu_{X_k}(x)$ is given in Equation (11.6).

Example 11.4   Before giving a proof of Wiberg's theorem, the following example may be instructive. Consider

$$\phi(x_1, x_2, x_3) = \gamma_1(x_1, x_2)\gamma_2(x_2, x_3).$$

The factor graph is then a tree given in Figure 11.10.

Figure 11.10   An example on three variables and two functions.

In this case, the messages are:

$$\mu_{X_1 \to \gamma_1}(x_1) = \mu_{X_3 \to \gamma_2}(x_3) = 1.$$

$$\mu_{\gamma_1 \to X_2}(x_2) = \sum_{x_1 \in \mathcal{X}_1} \gamma_1(x_1, x_2)\mu_{X_1 \to \gamma_1}(x_1) = \sum_{x_1 \in \mathcal{X}_1} \gamma_1(x_1, x_2)$$

$$\mu_{\gamma_2 \to X_2}(x_2) = \sum_{x_3 \in \mathcal{X}_3} \gamma_2(x_2, x_3)\mu_{X_3 \to \gamma_2}(x_3) = \sum_{x_3 \in \mathcal{X}_3} \gamma_2(x_2, x_3)$$

$$\mu_{X_2 \to \gamma_2}(x_2) = \mu_{\gamma_1 \to X_2}(x_2) = \sum_{x_1 \in \mathcal{X}_1} \gamma_1(x_1, x_2)$$

$$\mu_{X_2 \to \gamma_1}(x_2) = \mu_{\gamma_2 \to X_2}(x_2) = \sum_{x_3 \in \mathcal{X}_3} \gamma_2(x_2, x_3)$$

$$\mu_{\gamma_1 \to X_1}(x_1) = \sum_{x_2 \in \mathcal{X}_2} \gamma_1(x_1, x_2)\mu_{X_2 \to \gamma_1}(x_2) = \sum_{(x_2,x_3) \in \mathcal{X}_2 \times \mathcal{X}_3} \gamma_1(x_1, x_2)\gamma_2(x_2, x_3)$$

$$\mu_{\gamma_2 \to X_3}(x_3) = \sum_{(x_1,x_2) \in \mathcal{X}_1 \times \mathcal{X}_2} \gamma_1(x_1, x_2)\gamma_2(x_2, x_3).$$

Note that the variable terminations are

$$\mu_{X_1}(x_1) = \mu_{\gamma_1 \to X_1}(x_1) = \sum_{(x_2,x_3) \in \mathcal{X}_2 \times \mathcal{X}_3} \gamma_1(x_1, x_2)\gamma_2(x_2, x_3)$$

$$\mu_{X_2}(x_2) = \mu_{\gamma_1 \to X_2}(x_2)\mu_{\gamma_2 \to X_2}(x_2) = \sum_{(x_1,x_3) \in \mathcal{X}_1 \times \mathcal{X}_3} \gamma_1(x_1, x_2)\gamma_2(x_2, x_3)$$

$$\mu_{X_3}(x_3) = \mu_{\gamma_2 \to X_3}(x_3) = \sum_{(x_1,x_2) \in \mathcal{X}_1 \times \mathcal{X}_2} \gamma_1(x_1, x_2)\gamma_2(x_2, x_3),$$

which are the required marginalization. The Wiberg's theorem states that if the factor graph is a tree, then after a full schedule, the terminations give the required marginalization.

Proof of Theorem 11.1 Consider Figure 11.11. Suppose that a full schedule has been performed on a tree. The proof proceeds in three steps.   □

Step 1: Decompose the factor graph into n components, $R_1, \ldots, R_n$ Choose a variable X_i and suppose that n edges enter the variable node X_i. Since there are no cycles, the margin $\sum_{\underline{y} \in \mathcal{X}_{\hat{V} \setminus \{i\}}} \phi(\underline{y}, x_i)$ (where the arguments of ϕ have been suitably rearranged) may be written as

$$\sum_{\underline{y} \in \mathcal{X}_{\hat{V} \setminus \{i\}}} \phi(\underline{y}, x_i) = \sum_{\underline{y} \in \mathcal{X} | y_i = x_i} \prod_{j \in R_1} \gamma_j(\underline{y}_{D_j}) \prod_{j \in R_2} \gamma_j(\underline{y}_{D_j}) \cdots \prod_{j \in R_n} \gamma_j(\underline{y}_{D_n})$$

$$= \prod_{k=1}^{n} \sum_{\underline{y}_{R_k} \in \mathcal{X}_{R_k} | y_i = x_i} \prod_{j \in R_k} \gamma_j(\underline{y}_{D_j})$$

$$= \prod_{k=1}^{n} \nu_{R_k}(x_i),$$

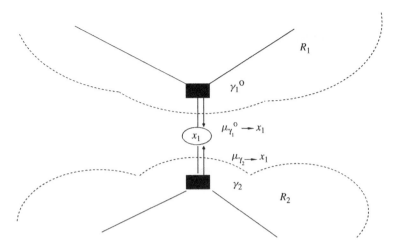

Figure 11.11 Step 1.

where the notation is clear. The last expression has the same form as the *termination formula*. Therefore the assertion is proved if it can be established that

$$v_{R_k}(x_i) = \mu_{\gamma_k^0 \to X_i}(x_i), \quad k = 1, \ldots, n,$$

where $\gamma_1^0, \ldots, \gamma_n^0$ are the n function nodes that are neighbours of X_i. Due to the clear symmetry, it is only necessary to consider one of these.

Step 2: Consider the decomposition of R_1 shown in Figure 11.12, where the function node γ_1^0 has three edges. In the three variable case shown in Figure 11.12, X_1 is the node under consideration and γ_1^0 is outside R_3 and R_4. For this case,

$$
\begin{aligned}
v_{R_1}(x_1) &= \sum_{\underline{y} \in \mathcal{X}_{R_1} | y_1 = x_1} \prod_{j \in R_1} \gamma_j(\underline{y}_{D_j}) \\
&= \sum_{(y_3, y_4) \in \mathcal{X}_3 \times \mathcal{X}_4} \gamma_1^0(x_1, y_3, y_4) \sum_{\underline{z}_{R_3} \in \mathcal{X}_{R_3} | z_3 = y_3} \prod_{j \in R_3} \gamma_j(\underline{z}_{D_j}) \sum_{\underline{z}_{R_4} \in \mathcal{X}_{R_4} | z_4 = y_4} \prod_{j \in R_4} \gamma_j(\underline{z}_{D_j}) \\
&= \sum_{(x_3, x_4) \in \mathcal{X}_3 \times \mathcal{X}_4} \gamma_1^0(x_1, x_3, x_4) \tilde{v}_{R_3}(x_3) \tilde{v}_{R_4}(x_4)
\end{aligned}
$$

where $\mathcal{X}_{R_1}$ denotes all the variable nodes that are neighbours of function nodes with R_1, retaining the same indices as the full set of variables, and similar. It is straightforward to derive a similar expression when the function node has m neighbours.

This expression has the same form as the *update rule* given for $\mu_{\gamma_j \to x}$ in Equation (11.5). In other words, if $\tilde{v}_{R_3}(x_3) = \mu_{X_3 \to \gamma_1^0}(x_3)$ and $\tilde{v}_{R_4}(x_4) = \mu_{X_4 \to \gamma_1^0}(x_4)$, then the result is proved. The algorithm proceeds to the leaf nodes of the factor graph.

Step 3: There are two cases. If the leaf node is a *function* node (as in Step 1, going from a variable to functions), then (clearly from the graph) this is a function (h say) of

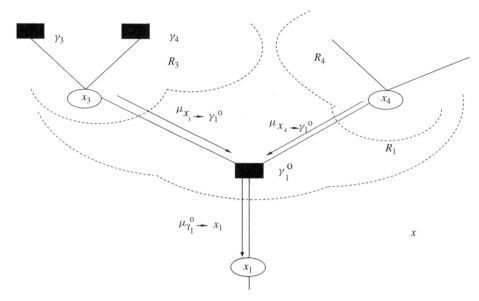

Figure 11.12 Step 2.

a single variable (say Y) and (from Formula (11.5)),

$$v(y) = h(y) = \mu_{h \to Y}(y).$$

If the leaf node is a *variable* node X (as in Step 2, going from functions to variables), then the leaf variable is adjacent to a single function h (or else it is not a leaf), which has neighbours $(Y_1, \ldots, Y_m, X)$, say, then

$$\tilde{v}(x) = 1 = \mu_{X \to h}(x),$$

since if X is a leaf, then h is the only neighbour of X and hence $\mu_{X \to h}(x) \equiv 1$ from Equation (11.4).

By tracing backward from the leaf nodes, it is now clear, by induction, that

$$\sum_{\underline{y} \in \mathcal{X}_{\tilde{v} \backslash \{i\}}} \phi(\underline{y}, x_i) = \prod_{j=1}^{n} \mu_{\gamma_j^0 \to X_i}(x_i),$$

where $(\gamma_1^0, \ldots, \gamma_n^0)$ are the neighbours of node X_i.

Termination   Consider the termination formula

$$\mu_X(x) = \prod_{\gamma_j \in N_X} \mu_{\gamma_j \to X}(x),$$

together with the formula for the message from a variable node to a function node:

$$\mu_{X \to Y_j}(x) = \prod_{h \in N_X \backslash \{Y_j\}} \mu_{h \to X}(x).$$

Suppose the factor graph is a tree. Then, since any variable to function message is the product of all but one of the factors in the termination formula, it is clear that $\mu_X(x)$ may be computed as *the product of the two messages that were passed in opposite directions, a) from the variable X to one of the functions and b) from the function to the variable X.*

11.3   Detailed illustration of the algorithm

The example in this section is taken from taken from Frey [139]. The following figure shows the flow of messages that would be generated by the sum product algorithm applied to the factor graph of the product

$$g(x_1, x_2, x_3, x_4, x_5) = f_A(x_1) f_B(x_2) f_C(x_1, x_2, x_3) f_D(x_3, x_4) f_E(x_3, x_5).$$

shown in Figure 11.13.

The messages are generated in five steps, indicated with circles in the figure. The same flow is detailed below.

Step 1:

$$\mu_{f_A \to x_1}(x_1) = \sum_{\backslash x_1} f_A(x_1) = f_A(x_1).$$

$$\mu_{f_B \to x_2}(x_2) = \sum_{\backslash x_2} f_B(x_2) = f_B(x_2).$$

$$\mu_{x_4 \to f_D}(x_4) = 1$$

$$\mu_{x_5 \to f_E}(x_5) = 1$$

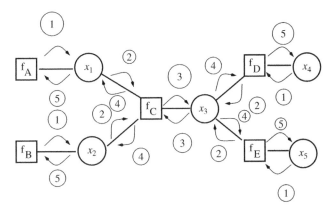

Figure 11.13   Flows of messages.

Step 2:

$$\mu_{x_1 \to f_C}(x_1) = \mu_{f_A \to x_1}(x_1) = f_A(x_1)$$

$$\mu_{x_2 \to f_C}(x_2) = \mu_{f_B \to x_2}(x_2) = f_B(x_2)$$

$$\mu_{f_D \to x_3}(x_3) = \sum_{\backslash x_3} \mu_{x_4 \to f_D}(x_4) f_D(x_3, x_4) = \sum_{x_4} f_D(x_3, x_4)$$

$$\mu_{f_E \to x_3}(x_3) = \sum_{\backslash x_3} \mu_{x_5 \to f_E}(x_5) f_E(x_3, x_5) = \sum_{x_5} f_E(x_3, x_5)$$

Step 3:

$$\mu_{f_C \to x_3}(x_3) = \sum_{\backslash x_3} \mu_{x_1 \to f_C}(x_1) \mu_{x_2 \to f_C}(x_2) f_C(x_1, x_2, x_3)$$

$$= \sum_{x_1, x_2} f_A(x_1) f_B(x_2) f_C(x_1, x_2, x_3)$$

$$\mu_{x_3 \to f_C}(x_3) = \mu_{f_D \to x_3}(x_3) \mu_{f_E \to x_3}(x_3) = \sum_{x_4, x_5} f_D(x_3, x_4) f_E(x_4, x_5)$$

Step 4:

$$\mu_{f_C \to x_1}(x_1) = \sum_{\backslash x_1} \mu_{x_3 \to f_C}(x_3) \mu_{x_2 \to f_C}(x_2) f_C(x_1, x_2, x_3)$$

$$= \sum_{x_2, x_3, x_4, x_5} f_D(x_3, x_4) f_E(x_4, x_5) f_B(x_2) f_C(x_1, x_2, x_3)$$

$$\mu_{f_C \to x_2}(x_2) = \sum_{\backslash x_2} \mu_{x_3 \to f_C}(x_3) \mu_{x_1 \to f_C}(x_1) f_C(x_1, x_2, x_3)$$

$$= \sum_{x_1, x_3, x_4, x_5} f_D(x_3, x_4) f_E(x_4, x_5) f_A(x_1) f_C(x_1, x_2, x_3)$$

$$\mu_{x_3 \to f_D}(x_3) = \mu_{f_C \to x_3}(x_3) \mu_{f_E \to x_3}(x_3)$$

$$= \sum_{x_1, x_2, x_4, x_5} f_A(x_1) f_B(x_2) f_C(x_1, x_2, x_3) f_E(x_3, x_5)$$

$$\mu_{x_3 \to f_E}(x_3) = \mu_{f_C \to x_3}(x_3) \mu_{f_D \to x_3}(x_3)$$

$$= \sum_{x_1, x_2, x_4, x_5} f_A(x_1) f_B(x_2) f_C(x_1, x_2, x_3) f_D(x_3, x_4).$$

Step 5:

$$\mu_{x_1 \to f_A}(x_1) = \mu_{f_C \to x_1}(x_1) = \sum_{x_2, x_3, x_4, x_5} f_B(x_2) f_C(x_1, x_2, x_3) f_D(x_3, x_4) f_E(x_4, x_5)$$

$$\mu_{x_2 \to f_B}(x_2) = \mu_{f_C \to x_2}(x_2) = \sum_{x_1, x_3, x_4, x_5} f_A(x_1) f_C(x_1, x_2, x_3) f_D(x_3, x_4) f_E(x_4, x_5)$$

$$\mu_{f_D \to x_4}(x_4) = \sum_{\backslash x_4} \mu_{x_3 \to f_D}(x_3) f_D(x_3, x_4)$$

$$= \sum_{x_1, x_2, x_3, x_5} f_A(x_1) f_B(x_2) f_C(x_1, x_2, x_3) f_D(x_3, x_4) f_E(x_4, x_5)$$

$$\mu_{f_E \to x_5}(x_5) = \sum_{\backslash x_5} \mu_{x_3 \to f_E}(x_3) f_E(x_3, x_5)$$

$$= \sum_{x_1, x_2, x_3, x_4} f_A(x_1) f_B(x_2) f_C(x_1, x_2, x_3) f_D(x_3, x_4) f_E(x_3, x_5).$$

Termination

$$g_1(x_1) = \mu_{f_A \to x_1}(x_1) \mu_{f_C \to x_1}(x_1) = \sum_{x_2, x_3, x_4, x_5} g(x_1, x_2, x_3, x_4, x_5)$$

$$g_2(x_2) = \mu_{f_B \to x_2}(x_2) \mu_{f_C \to x_2}(x_2) = \sum_{x_1, x_3, x_4, x_5} g(x_1, x_2, x_3, x_4, x_5)$$

$$g_3(x_3) = \mu_{f_C \to x_3}(x_3) \mu_{f_D \to x_3}(x_3) \mu_{f_E \to x_3}(x_3) = \sum_{x_1, x_2, x_4, x_5} g(x_1, x_2, x_3, x_4, x_5)$$

$$g_4(x_4) = \mu_{f_D \to x_4}(x_4) = \sum_{x_1, x_2, x_3, x_5} g(x_1, x_2, x_3, x_4, x_5)$$

$$g_5(x_5) = \mu_{f_D \to x_5}(x_5) = \sum_{x_1, x_2, x_3, x_4} g(x_1, x_2, x_3, x_4, x_5).$$

In the termination step, $g_i(x_i)$ is computed as the product of all messages directed towards x_i. Equivalently, $g_i(x_i)$ may be computed as the product of the two messages that were passed in opposite directions over any single edge incident on x_i, because the message passed on any given edge is equal to the product of all but one of these messages. Therefore (for example), there are three ways to compute $g_3(x_3)$ by multiplying only two messages together:

$$g_3(x_3) = \mu_{f_C \to x_3}(x_3) \mu_{x_3 \to f_C}(x_3) = \mu_{f_D \to x_3}(x_3) \mu_{x_3 \to f_D}(x_3) = \mu_{f_E \to x_3}(x_3) \mu_{x_3 \to f_E}(x_3).$$

Notes   Chapter 11 presents a brief outline of some aspects of S.M. Aji and R.J. McEliece (2000) [140], F.R. Ksischang, B.J. Frey and H.A. Loeliger [141], B.J. Frey [139] and N. Wiberg [138], as related to Bayesian networks. Most of the ideas are originally developed in N. Wiberg [138]. A well written treatise on bipartite graphs is found in [142].

11.4   Exercise: Factor graphs and the sum product algorithm

Consider the directed acyclic graph below.

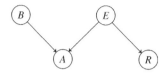

Figure 11.14   Burglary, earthquake and radio.

The variables are B – Burglary, A – Alarm, E – Earthquake and R – news broadcast.

These are random variables with the states (0 – no (false), 1 – yes (true)). The alarm is reliable for detecting burglary, but also responds to minor earthquakes. Radio broadcasts tell about occurrences of such earthquakes, but are not always correct. The conditional probability distributions for this problem are given below.

$$P_{R|E} = \begin{array}{c|cc} R\backslash E & 0 & 1 \\ \hline 0 & 0.99 & 0.05 \\ 1 & 0.01 & 0.95 \end{array}$$

$$P_{A|B,E}(0|.,.) = \begin{array}{c|cc} E\backslash B & 0 & 1 \\ \hline 0 & 0.97 & 0.05 \\ 1 & 0.05 & 0.02 \end{array}$$

$$P_B(1) = 0.01, \quad P_E(1) = 0.999$$

Assume that the joint distribution $P_{A,B,E,R}$ factorizes recursively according to the Bayesian network shown in Figure 11.14. Using the sum-product algorithm, compute

1. The conditional probability $p_{B|A}(1|1)$

2. the conditional probability $p_{B|A,R}(1|1,1)$.

References

1. Pearl, J. (2000) *Causality*, Cambridge University Press, Cambridge.
2. Pearson, K. (1892) *The Grammar of Science*. Elibron Classics, 2006.
3. Hume, D. (1731) *An Enquiry Concerning Human Understanding*, reprinted in Dover Philosophical Classics, 2004.
4. Williamson, J. (2005) *Bayesian Networks and Causality: Philosophical and Computational Foundations*, Oxford University Press, Oxford.
5. Langseth, H. and Portinale, L. (2007) Bayesian networks in reliability. *Reliability Engineering and System Safety*, **92**, 92–108.
6. Woof, D., Goldstein, M. and Coolen, F. (2002) Bayesian graphical models for software testing *IEEE Transactions in Software Engineering*, **28**(5), 510–525.
7. Markowetz, F. and Spang, R. (2007) Inferring cellular networks–a review. *BMC Bioinformatics*, **8** (suppl. 6): S5, available from http://www.biomedcentral.com/1471-2105/8/S6/S5/
8. Gowadia, V. Farkas, C. and Valtorta, M. (2005) PAID: a probabilistic agent-based intrusion detection system. *Computers and Security*, **24**, 529–545.
9. Pearl, J. (1990) *Probabilistic Reasoning in Intelligent Systems*, 2nd edn revised, Morgan and Kaufman Publishers Inc., San Francisco.
10. Halpern, J. (2003) *Reasoning about Uncertainty*, MIT Press, Cambridge, MA, 71–73.
11. Teller, P. and Fine, A. (1975) A Characterisation of conditional probability. *Mathematics Magazine*, **48**(5), 494–498.
12. Diaconis, P. and Zabell, S. L. (1982) Updating subjective probability. *Journal of the American Statistical Association*, **77**(380), 822–830.
13. Cheeseman, P. (1988) In defence of an enquiry into computer understanding. *Computational Intelligence*, **4**, 129–142.
14. Russell, S. and Norvig, P. (1995) *Artificial Intelligence: A Modern Approach*, Prentice Hall, New Jersey.
15. Jeffrey, R. C. (2004) *Subjective Probability: The Real Thing*, Cambridge University Press, Cambridge. UK.

16. Cheeseman, P. (1988) An enquiry into computer understanding. *Computational Intelligence*, **4**, 58–66.

17. Lewis, D. (1986) A subjectivist's guide to objective chance, in *Philosophical Papers*, vol. 2, 83–132.

18. Aumann, R. J. (1976) Agreeing to disagree. *The Annals of Statistics*, **4**, 1236–1239.

19. Suppes, P. (1970) *A Probabilistic Theory of Causation*, Acta Philosophica Fennica, 24, North-Holland Publishing Company, Amsterdam.

20. Freedman, D. and Humphreys, P. (1999) Are there algorithms that discover causal structure? *Synthese*, **121**, 29–54.

21. Woodward, J. (2001) Probabilistic causality, direct causes and counterfactual dependence Stochastic Causality (eds M. Galavotti, P. Suppes and D. Constantini) CSLI Publications, Stanford, California, 39–63.

22. Pearson, K. (1920) The fundamental problem of practical statistics. *Biometrika*, **13**, 1–16.

23. De Finetti, B. (1937) La prévision: ses lois logiques, ses sources subjectives. *Annales de l'Institut Henri Poincaré*, **7**, 1–68.

24. Jeffreys, H. (1939) *Theory of Probability*, Clarendon Press, 3rd edn, 1998.

25. Heckerman, D. (1998) *A tutorial on learning with Bayesian networks*. Report # MSR-TR-95-06, Microsoft Research, Redmont, Washington. http://research.microsoft.com/~heckerman/

26. Robert, C.R. (2001) *The Bayesian Choice From Decision-Theoretic Foundations to Computational Implementation*, Springer.

27. Chung, K.L. (2002) Will the sun rise again? *Mathematical Medley*, **29**, 67–73. Reprinted 2004, in *Chance and Choice: Memorabilia*, World Scientific, New Jersey, 1–8.

28. Lidstone, G.J. (1920) Note on the general case of the Bayes-Laplace formula for the inductive or aposterior probabilities. *Transactions of the Faculty of Actuaries*. Faculty of Actuaries in Scotland, **8**, 182–192.

29. Perks, W. (1947) Some observations on inverse probability including a new indifference rule. *Journal of the Institute of Actuaries*, **72**, 285–334.

30. Good, I.J. (1968) *Estimation of Probabilities*, MIT Press, Cambridge MA.

31. Bernardo, J.M. and Smith, A.F.M. (1994) *Bayesian Theory*. John Wiley & Sons, Ltd.

32. Skyrms, B. (2000) *Choice Chance: An Introduction to Inductive Logic*, 4th edn, Wadsworth, Belmont, California.

33. Hacking, I. (2001) *An Introduction to Probability and Inductive Logic*, Cambridge University Press, Cambridge.

34. Neapolitan, R.E. (2004) *Learning Bayesian Networks*, Pearson Prentice Hall, New Jersey.

35. Korb, K.B. and Nicholson, A.E. (2004) *Bayesian Artificial Intelligence*, Chapman and Hall.

36. Bayes, T., and Price, R. (1763) An essay towards solving a problem in the doctrine of chance. By the late Rev. Mr. Bayes, F.R.S. Communicated by Mr. Price, in a letter to John Canton, A.M.F.R.S. *Philosophical Transactions of the Royal Society of London*, **53**, 370–418. Reprinted in Biometrika (1958) **45**, 296–315.

37. Gillies, D. (1987) Was Bayes a Bayesian? *Historia Mathematica*, **14**, 325–346.

38. Molina, E.C. (1931) Bayes theorem: an expository presentation. *The Annals of Mathematical Statistics*, **2**, 23–37.

39. Stigler, S.M. (1982) Thomas Bayes's Bayesian inference. *Journal of the Royal Statistical Society*, Series A, **145**, 250–258.

40. Stigler, S.M. (1983) Who discovered Bayes's theorem? *American Statistician*, **37**, 290–296.

41. Bellhouse, D.R. (2004) The reverend Thomas Bayes, FRS: a biography to celebrate the tercentenary of his birth. *Statistical Science*, **19**(1), 3–43.

42. Murray, F.H. (1930) Note on the scholium of Bayes. *Bulletin of the American Mathematical Society*, **36**, 129–132.

43. Gillies, D. (2000) *Philosophical Theories of Probability*, Routledge, London.

44. Pearl, J. (1982) Reverend Bayes on inference engines: a distributed hierarchical approach. *AAAI-82 Proceedings*, 133–136.

45. Good, I.J. (1994) Causal tendency, necessitivity, and sufficientivity: an updated review, in P. Humphreys (ed.) *Patrick Suppes: Scientific Philosopher*, Kluwer Academic Publ. Dordrecht, vol. 1, 293–316.

46. Savage, J.L. (1966) *Foundations of Statistics*, John Wiley and Sons, Inc., New York.

47. Dickey, J.M. (1983) Multiple hypergeometric functions: probabilistic interpretations and statistical uses. *Journal of the American Statistical Association*, **78**(383), 628–637.

48. Spirtes, P. Glymour, C. and Scheines, R. (2000) *Causation, Prediction and Search*, 2nd edn, MIT Press.

49. Pẽna, J.M., Nilsson, R., Björkegren, J., and Tegner, J. (2007) Towards scalable and data efficient learning of Markov boundaries *International Journal of Approximate Reasoning*, **45**, 211–232.

50. Schachter, R.D. (1998) Bayes ball: the rational pass time for determining irrelevance and requisite information in belief networks and influence diagrams. *Proceedings of the 14th Annual Conference on Uncertainty in Artificial Intelligence* (eds G.F. Cooper and S. Moral) Morgan Kaufmann, San Francisco. 480–487.

51. Weidl, G. Madsen, A.L. and Israelson, S. (2005) Applications of object-oriented root cause analysis and decision support on operation of complex continuous processes. *Computers and Chemical Engineering*, **29**, 1996–2009.

52. Shafer, G. (1996) *Probabilistic Expert Systems*, SIAM, Philadelphia.

53. Vorobev, N. (1963) Markov measures and Markov extensions. *Theory of Probability and Applications*, **8**, 420–429.

54. Langseth, H. and Bangsø, O. (2001) Parameter learning in object oriented Bayesian networks. *Annals of Mathematics and Artificial Intelligence*, **32**, 221–243.

55. Andersson, S.A., Madigan, D., Perlman, M.D. and Triggs, C.M. (1997) A graphical characterisation of lattice conditional independence models. *Annals of Mathematics and Artificial Intelligence*, **21**, 27–50.

56. Pearl, J., Geiger, D. and Verma, T. (1989) Conditional independence and its representations. *Kybernetica*, **25**(2), pp. 33–44.

57. Wright, S. (1921) Correlation and causation, *Journal of Agricultural Research*, **20**, 557–585.

58. Shipley, B. (2000) *Cause and Correlation in Biology: A User's Guide to Path Analysis, Structural Equations and Causal Inference*, Cambridge University Press.

59. Kiiveri, H., Speed, T.P., and Carlin, J.B. (1984) Recursive causal models. *Journal of the Australian Mathematical Society* (Series A), **36**, 30–52.

60. Pearl, J. and Verma, T. (1987) The logic of representing dependencies by directed acyclic graphs. *Proceedings of the AAAI*, Seattle, 374–379.

61. Geiger, D. Verma, T. and Pearl, J. (1990) Identifying independence in Bayesian networks. *Networks*, **20**, 507–534.

62. Verma, P. and Pearl, J. (1992) *An Algorithm for Deciding if a Set of Observed Independencies has a Causal Explanation* in Uncertainty in Artificial Intelligence, Proceedings of the Eighth Conference (eds D. Dubois, M.P. Welman, B. D'Ambrosio and P. Smets), Morgan Kaufman, San Francisco, 323–330.

63. Murphy, K. (1998) *A brief introduction to graphical models and Bayesian Networks*. http://www.cs.ubc.ca/~murphyk/Bayes/bnintro.html

64. Garcia, L.D., Stillman, M. and Sturmfels, B. (2005) Algebraic geometry of Bayesian networks *Journal of Symbolic Computation*, **39**, 331–355.

65. Jeffrey, R.C. (1965) *The Logic of Decision*, McGraw-Hill, New York; University of Chicago Press, Chicago, 2nd edn revised, 1990.

66. Chan, H. and Darwiche, A. (2005) A distance measure for bounding probabilistic belief change. *International Journal of Approximate Reasoning*, **38**, 149–174.

67. Cowell, R.G., Dawid, A.P., Lauritzen, S.L. and Spiegelhalter, D.J. (1999) *Probabilistic Networks and Expert Systems*, Springer, New York.

68. Dawid, A.P. (1992) Applications of a general propagation algorithm for probabilistic expert systems. *Statistics and Computing* **2**, 25–36.

69. Dechter, R. (1999) Bucket elimination: a unifying framework for reasoning. *Artificial Intelligence* **113**, 41–85.

70. Dawid, A.P. (1997) Conditional independence for statistics and AI. Tutorial at AISTATS–97. citeseer.ist.psu.edu/article/dawid97conditional.html.

71. Dawid, A.P. (1997) Conditional independence, in *Encyclopedia of Statistical Sciences*. Update Volume 2, eds S. Kotz, C.B. Read and D.L. Banks, Wiley-Interscience, pp. 146–155.

72. Heckerman, D. Geiger, D. and Chickering, D.M. (1995) learning Bayesian networks: the combination of knowledge and statistical data. *Machine Learning*, **20**, 197–243.

73. Murphy, K. (2002) Dynamic Bayesian networks. http://www.cs.ubc.ca/~murphyk/Papers/dbnchapter.pdf

74. Chickering, D.M., Heckerman, D. and Meek, C. (2004) Large sample learning of Bayesian networks is NP-Hard. *Journal of Machine Learning Research*, **5**, 1287–1330.

75. Häggström, O. (2002) *Finite Markov Chains and Algorithmic Applications*, Cambridge University Press.

76. Hastings, W.K. (1970) Monte Carlo sampling methods using Markov chains and their applications, *Biometrika*, **57**, 97–109.

77. Lauritzen, S. (1992) Propagation of probabilities, means and variances in mixed graphical association models. *Journal of the American Statistical Association*, **87**(420), 1098–1108.

78. Lunn, D.J., Thomas, A., Best, N. and Spiegelhalter, D. (2000) WinBUGS–A Bayesian modelling framework: concepts, structure, and extensibility. *Statistics and Computing*, **10**, 325–337.

79. Corander, J., Koski, T. and Gyllenberg, M. (2006) Bayesian model learning based on a parallel MCMC Strategy. *Statistics and Computing*, **16**, 355–362.

80. Moore, A. and Soon Lee, M. (1998) Cached sufficient statistics for efficient machine learning with large data sets. *Journal of Artificial Intelligence Research*, **8**, 67–91.

81. Chan, H. and Darwiche, A. (2005) On the revision of probabilistic beliefs using uncertain evidence. *Artificial Intelligence*, **163**, 67–90.

82. Cooper, G.F. (1990) The computational complexity of probabilistic inference using Bayesian belief Networks. *Artificial Intelligence*, **42**, 393–405.

83. Pearl, J. (1987) Evidential reasoning using stochastic simulation of causal models. *Artifical Intelligence*, **32**, 245–257.

84. Arnborg, S. (1985) Efficient algorithms for combinatorial problems with bounded decomposability: a survey. *BIT*, **25**, 2–23.

85. Arnborg, S. (1993) Graph decompositions and tree automata in reasoning with uncertainty *Journal of Experimental and Theoretical Artificial Intelligence*, **5**, 335–357.

86. Frydenberg, M. (1990) The chain graph Markov property. *Scandinavian Journal of Statistics*, **17**, 333–353.

87. Madigan, D., Andersson, S A., Perlman, M.D. and Volinsky, C.T. (1996) Bayesian model averaging and model selection for markov equivalence classes of acyclic digraphs. *Communications in Statistics: Theory and Methods*, **25**(11), 2493–2519.

88. Golumbic, M.C. (2004) *Algorithmic Graph Theory and Perfect Graphs*, Elsevier.

89. Studený, M. (2005) *Probabilistic Conditional Independence Structures*, Springer.

90. Geiger, D. and Heckerman, D. (1997) A characterization of the Dirichlet distribution through global and local parameter independence. *The Annals of Statistics*, **25** (3), 1344–1369.

91. Ramoni, M. and Sebastiani, P. (1997) Parameter estimation in Bayesian networks from incomplete databases, KMI-TR-97-22, *Knowledge Media Institute*, The Open University.

92. Jensen, F.V. (2001) *Bayesian Networks and Decision Graphs*, Springer

93. Antal, P., Fannes, G., Timmerman, D., *et al.* (2004) Using literature and data to learn Bayesian networks as clinical models of ovarian tumors. *Artificial Intelligence in Medicine*, **30**, 257–281.

94. Castelo, R. and Siebes, A. (2000) Priors on network structures. Biasing the search for Bayesian networks. *International Journal of Approximate Reasoning*, **24**, 39–57.

95. Cooper, G.F. and Herskovitz, E. (1992) A Bayesian method for the induction of probabilistic networks from data. *Machine Learning*, **9**, 309–347.

96. Robinson, R.W. (1977) *Counting unlabelled acyclic digraphs*, in Springer Lecture Notes in Mathematics: Combinatorial Mathematics *V*, C.H.C. Little (ed.) 28–43.

97. Koivisto, M., and Sood, K. (2004) Exact Bayesian structure discovery in Bayesian networks *Journal of Machine Learning Research*, **5**, 549–573.

98. Kontkanen, P. Myllymäki, P. *et al.* (2000) On predictive distributions and Bayesian networks. *Statistics and Computing*, **10**, 39–54.

99. Chow, C.K. and Liu, C.N. (1968) Approximating discrete probability distributions with dependence trees. *IEEE Transactions on Information Theory*, **IT-14**, (3), 462–467.

100. Gyllenberg, M. and Koski, T. (2002) Tree augmented classification of binary data minimizing stochastic complexity. Technical Report, Matematiska institutionen, University of Linköping, LiTH-MAT-R-2002-4.

101. Suzuki, J. (1999) Learning Bayesian belief networks based on the minimum description length Principle: Basic Properties. *IEICE Transactions on Fundamentals*, **E82**, (10) 2237–2245.

102. Lazkano, E. Sierra, B. *et al.* (2007) On the use of Bayesian networks to develop behaviours for mobile robots. *Robots and Autonomous Systems*, **55**, 253–265.

103. Tsamardinos, I., Brown, L.E. and Aliferis, C.F. (2006) The max-min hill-Climbing Bayesian network structure learning algorithm *Machine Learning*, **65**, 31–78.

104. Pēna, J.M. (2007) Approximate counting of graphical models via MCMC. *Proceedings of the 11th Conference in Artificial Intelligence*, 352–359.

105. Fowlkes, E.B., Freeny, A.E. and Ladwehr, J.M. (1988) Evaluating logistic models for large contingency tables. *Journal of the American Statistical Association*, **83**(403), 611–622.

106. Lacampagne, C.B. (1979) An evaluation of the Women and Mathematics (WAM) program and associated sex-related differences in the teaching, learning and counseling of mathematics. Unpublished Ed. D. thesis, Columbia University Teacher's College.

107. Corander, J., Ekdahl, M. and Koski, T. (2008) Parallel interacting MCMC for learning of Bayesian network topologies. *Data Mining and Knowledge Discovery*, **17** (3), 431–456.

108. Castillo, E., Gutiérrez, J.M. and Hadi, A.S. (1996) A new method for efficient symbolic propagation in discrete Bayesian networks *Networks*, **28** (1), 31–43.

109. Castillo, E., Gutiérrez, J.M. and Hadi, A.S. (1997) Sensitivity analysis in discrete Bayesian networks. *IEEE Transactions on Systems, Man and Cybernetics*. Part A: Systems and Humans, **27**(4).

110. Coupé, V.M. and van der Gaag, L.C. (1998) Practicable sensitivity analysis of Bayesian belief networks. *Prague Stochastics '98: Proceedings of the Joint Session of the 6th Prague Symposium of Asymptotic Statistics and the 13th Prague Conference on Information Theory, Statistical Decision Functions and Random Processes*, 81–86.

111. Chan, H. and Darwiche, A. (2002) When do numbers really matter? *Journal of Artificial Intelligence Research*, **17**, 265–287.

112. Neal, R. (1992) Correctionist learning of belief networks. *Artificial Intelligence*, **56**, 71–113.

113. Barndorff-Nielsen, O. (1978) *Information and Exponential Families in Statistical Theory*, John Wiley & Sons, Ltd.

114. Jordan, M.I., Ghahramani, Z., Jaakkola, T.S. and Saul, L.K. (1999) An Introduction to Variational Methods for Graphical Models. *Machine Learning*, **37**, 183–233.

115. Wainright, M.J. and Jordan, M.I. (2003) *Graphical models, exponential families and variational Inference*. Technical report 649, Department of Statistics, University of California, Berkeley.

116. Humphreys, K. and Titterington, D.M. (2000) Improving the mean-field approximation in belief networks using Bahadur's reparameterisation of the multivariate binary distribution *Neural Processing Letters*, **12**, 183–197.

117. Strotz, R.H. and Wold, H.O.A. (1960) Recursive versus nonrecursive systems: an attempt at synthesis. *Econometrica* **28**, 417–427.

118. Pearl, J. (1995) Causal diagrams for empirical research. *Biometrika*, **82**, 669–710.

119. Eaton, D., and Murphy, K. (2007) Exact Bayesian structure learning from uncertain interventions AI & Statistics. http://www.cs.ubc.ca/~murphyk/my papers.html

120. Wold, H.O.A. (1963) *Orientering i det statistiska arbetsfältet*, biblioteksförlaget, Stockholm.

121. Edwards, D. (2000) *Introduction to Graphical Modelling*, Chapter 9: Causal inference. Springer.

122. Freedman, D. (1999) From association to causation: some remarks on the history of statistics. *Statistical Science*, **14**(3), 243–258.

123. Lindley, D.V. (2002) Seeing and doing: the concept of causation. *International Statistical Review/Revenue International de Statistique*, **70**, 191–197.

124. Meek, C. and Glymour, C. (1994) Conditioning and intervening. *British Journal of Philosophy of Science*, **45**, 1001–1021.

125. Lauritzen, S. (2001) *Causal Inference from Graphical Models*, in *Complex Stochastic Systems*, Chapman and Hall, 63–108.

126. Tian, J. and Pearl, J. (2002) A General Identification Condition for Causal Effects. *Proceedings of the Eighteenth National Conference on Artifical Intelligence*, AAAI Press, Menlo Park, California 567–573.

127. Huang, Y. Valtorta, M. (2006) Pearl's calculus of intervention is complete. *Proceedings of the 22nd Conference on Uncertainty in Artifical Intelligence*, UAI Press, 217–224.

128. Pearl, J. (1995) Causal inference from indirect experiments. *Artificial Intelligence in Medicine*, **7**, 561–582.

129. Spohn, W. (2001) Bayesian nets are all there is to causal dependence. *Stochastic Causality* (eds Galavotti, M., Suppes, P. and Constantini, D.), CLSI Publications, Stanford, California, 157–172.

130. Becker, A., and Geiger, D. (2001) A sufficiently fast algorithm for finding close to optimal clique trees. *Artificial Intelligence*, **125**, 3–17.

131. Lauritzen, S. and Spiegelhalter, D. (1988) Local computations with probabilities on graphical structures and their application to expert systems. *Journal of the Royal Statistical Society*, Series B, **50**, 157–224.

132. Lauritzen, S. (1996) *Graphical Models*, Clarendon Press, Oxford.

133. Bach, F.R. and Jordan, M.I. (2002) Thin junction trees. *Advances in Neural Information Processing Systems (NIPS)* **14**.

134. Valtorta, M. Kim, Y.G. and Vomlel, J. (2002) Soft evidential update for probabilistic multi-agent systems. *International Journal of Approximate Reasoning*, **29**(1), 71–106.

135. Brown, D.T. (1959) A note on approximations to discrete probability distributions. *Information and Control*, **2**, 386–392.

136. Hájek, P. Havranek, T. and Jirousek, R. (2000) *Uncertain Information Processing in Expert Systems*, CRC Press, Boca Raton, Florida.

137. Deming, W.E., and Stephan, F.F. (1940) On a least squares adjustment of a sampled frequency table when the expected marginal totals are known. *Annals of Mathematical Statistics*, **11**, 427–444.

138. Wiberg, N. (1996) *Codes and decoding on general graphs*. Linköping Studies in Science and Technology. Dissertation 440, Linköpings Universitet, Linköping.

139. Frey, B.J. (1998) *Graphical Models for Machine Learning and Digital Communication*, MIT Press, Cambridge MA.

140. Aji, S.M., and McEliece, R.J. (2000) The generalised distributive law. *IEEE Transactions on Information Theory*, **46**, 325–343.
141. Ksischang, F.R., Frey, B.J., Loeliger, H.A. (2001) Factor graphs and the sum product algorithm *IEEE Transactions on Information Theory*, **47**, 498–519.
142. Asratian, A.S., Denley, T.M.J. and Häggkvist, R. (1998) *Bipartite Graphs and their Applications*, Cambridge University Press, Cambridge.

Index

WILEY SERIES IN PROBABILITY AND STATISTICS

Established by WALTER A. SHEWHART AND SAMUEL S. WILKS

Editors: *David J. Balding, Noel A. C. Cressie, Garrett M. Fitzmaurice, Harvey Goldstein, Geert Molenberghs, David W. Scott, Adrian F.M. Smith, Ruey S. Tsay, Sanford Weisberg*

Editors Emeriti: *Vic Barnett, J. Stuart Hunter, David G. Kendall, Jozef L. Teugels*

The Wiley Series in Probability and Statistics is well established and authoritative. It covers many topics of current research interest in both pure and applied statistics and probability theory. Written by leading statisticians and institutions, the titles span both state-of-the-art developments in the field and classical methods.

Reflecting the wide range of current research in statistics, the series encompasses applied, methodological and theoretical statistics, ranging from applications and new techniques made possible by advances in computerized practice to rigorous treatment of theoretical approaches.

This series provides essential and invaluable reading for all statisticians, whether in academia, industry, government, or research.

*Now available in a lower priced paperback edition in the Wiley Classics Library.

BENDAT and PIERSOL · Random Data: Analysis and Measurement Procedures, Third Edition

BERNARDO and SMITH · Bayesian Theory

BERRY, CHALONER and GEWEKE · Bayesian Analysis in Statistics and Econometrics: Essays in Honor of Arnold Zellner

BHAT and MILLER · Elements of Applied Stochastic Processes, Third Edition

BHATTACHARYA and JOHNSON · Statistical Concepts and Methods

BHATTACHARYA and WAYMIRE · Stochastic Processes with Applications

BIEMER, GROVES, LYBERG, MATHIOWETZ and SUDMAN · Measurement Errors in Surveys

BILLINGSLEY · Convergence of Probability Measures, Second Edition

BILLINGSLEY · Probability and Measure, Third Edition

BIRKES and DODGE · Alternative Methods of Regression

BISWAS, DATTA, FINE and SEGAL · Statistical Advances in the Biomedical Sciences: Clinical Trials, Epidemiology, Survival Analysis, and Bioinformatics

BLISCHKE and MURTHY (editors) · Case Studies in Reliability and Maintenance

BLISCHKE and MURTHY · Reliability: Modeling, Prediction and Optimization

BLOOMFIELD · Fourier Analysis of Time Series: An Introduction, Second Edition

BOLLEN · Structural Equations with Latent Variables

BOLLEN and CURRAN · Latent Curve Models: A Structural Equation Perspective

BOROVKOV · Ergodicity and Stability of Stochastic Processes

BOSQ and BLANKE · Inference and Prediction in Large Dimensions

BOULEAU · Numerical Methods for Stochastic Processes

BOX · Bayesian Inference in Statistical Analysis

BOX · R. A. Fisher, the Life of a Scientist

BOX and DRAPER · Empirical Model-Building and Response Surfaces

* BOX and DRAPER · Evolutionary Operation: A Statistical Method for Process Improvement

BOX · Improving Almost Anything *Revised Edition*

BOX, HUNTER and HUNTER · Statistics for Experimenters: An Introduction to Design, Data Analysis and Model Building

BOX, HUNTER and HUNTER · Statistics for Experimenters: Design, Innovation and Discovery, Second Edition

BOX and LUCEÑO · Statistical Control by Monitoring and Feedback Adjustment

BRANDIMARTE · Numerical Methods in Finance: A MATLAB-Based Introduction

BROWN and HOLLANDER · Statistics: A Biomedical Introduction

BRUNNER, DOMHOF and LANGER · Nonparametric Analysis of Longitudinal Data in Factorial Experiments

BUCKLEW · Large Deviation Techniques in Decision, Simulation and Estimation

CAIROLI and DALANG · Sequential Stochastic Optimization

CASTILLO, HADI, BALAKRISHNAN and SARABIA · Extreme Value and Related Models with Applications in Engineering and Science

CHAN · Time Series: Applications to Finance

CHARALAMBIDES · Combinatorial Methods in Discrete Distributions

CHATTERJEE and HADI · Regression Analysis by Example, Fourth Edition

CHATTERJEE and HADI · Sensitivity Analysis in Linear Regression

CHERNICK · Bootstrap Methods: A Practitioner's Guide

CHERNICK and FRIIS · Introductory Biostatistics for the Health Sciences

CHILÉS and DELFINER · Geostatistics: Modeling Spatial Uncertainty

CHOW and LIU · Design and Analysis of Clinical Trials: Concepts and Methodologies, Second Edition

CLARKE · Linear Models: The Theory and Application of Analysis of Variance

CLARKE and DISNEY · Probability and Random Processes: A First Course with Applications, Second Edition

* COCHRAN and COX · Experimental Designs, Second Edition

CONGDON · Applied Bayesian Modelling

*Now available in a lower priced paperback edition in the Wiley Classics Library.

*Now available in a lower priced paperback edition in the Wiley Classics Library.

*Now available in a lower priced paperback edition in the Wiley Classics Library.

*Now available in a lower priced paperback edition in the Wiley – Interscience Paperback Series.

MONTGOMERY, PECK and VINING · Introduction to Linear Regression Analysis, Fourth Edition

MORGENTHALER and TUKEY · Configural Polysampling: A Route to Practical Robustness

MUIRHEAD · Aspects of Multivariate Statistical Theory

MULLER and STEWART · Linear Model Theory: Univariate, Multivariate and Mixed Models

MURRAY · X-STAT 2.0 Statistical Experimentation, Design Data Analysis and Nonlinear Optimization

MURTHY, XIE and JIANG · Weibull Models

MYERS and MONTGOMERY · Response Surface Methodology: Process and Product Optimization Using Designed Experiments, Second Edition

MYERS, MONTGOMERY and VINING · Generalized Linear Models. With Applications in Engineering and the Sciences

† NELSON · Accelerated Testing, Statistical Models, Test Plans and Data Analysis

† NELSON · Applied Life Data Analysis

NEWMAN · Biostatistical Methods in Epidemiology

OCHI · Applied Probability and Stochastic Processes in Engineering and Physical Sciences

OKABE, BOOTS, SUGIHARA and CHIU · Spatial Tesselations: Concepts and Applications of Voronoi Diagrams, Second Edition

OLIVER and SMITH · Influence Diagrams, Belief Nets and Decision Analysis

PALTA · Quantitative Methods in Population Health: Extentions of Ordinary Regression

PANJER · Operational Risks: Modeling Analytics

PANKRATZ · Forecasting with Dynamic Regression Models

PANKRATZ · Forecasting with Univariate Box-Jenkins Models: Concepts and Cases

PARDOUX · Markov Processes and Applications: Algorithms, Networks, Genome and Finance

PARMIGIANI and INOUE · Decision Theory: Principles and Approaches

* PARZEN · Modern Probability Theory and Its Applications

PEÑA, TIAO and TSAY · A Course in Time Series Analysis

PIANTADOSI · Clinical Trials: A Methodologic Perspective

PORT · Theoretical Probability for Applications

POURAHMADI · Foundations of Time Series Analysis and Prediction Theory

POWELL · Approximate Dynamic Programming: Solving the Curses of Dimensionality

PRESS · Bayesian Statistics: Principles, Models and Applications

PRESS · Subjective and Objective Bayesian Statistics, Second Edition

PRESS and TANUR · The Subjectivity of Scientists and the Bayesian Approach

PUKELSHEIM · Optimal Experimental Design

PURI, VILAPLANA and WERTZ · New Perspectives in Theoretical and Applied Statistics

PUTERMAN · Markov Decision Processes: Discrete Stochastic Dynamic Programming

QIU · Image Processing and Jump Regression Analysis

RAO · Linear Statistical Inference and its Applications, Second Edition

RAUSAND and HØYLAND · System Reliability Theory: Models, Statistical Methods and Applications, Second Edition

RENCHER · Linear Models in Statistics

RENCHER · Methods of Multivariate Analysis, Second Edition

RENCHER · Multivariate Statistical Inference with Applications

RIPLEY · Spatial Statistics

RIPLEY · Stochastic Simulation

ROBINSON · Practical Strategies for Experimenting

ROHATGI and SALEH · An Introduction to Probability and Statistics, Second Edition

ROLSKI, SCHMIDLI, SCHMIDT and TEUGELS · Stochastic Processes for Insurance and Finance

ROSENBERGER and LACHIN · Randomization in Clinical Trials: Theory and Practice

†Now available in a lower priced paperback edition in the Wiley – Interscience Paperback Series.
*Now available in a lower priced paperback edition in the Wiley Classics Library.

*Now available in a lower priced paperback edition in the Wiley Classics Library.

TIERNEY · LISP-STAT: An Object-Oriented Environment for Statistical Computing and Dynamic Graphics

TSAY · Analysis of Financial Time Series

UPTON and FINGLETON · Spatial Data Analysis by Example, Volume II: Categorical and Directional Data

VAN BELLE · Statistical Rules of Thumb

VAN BELLE, FISHER, HEAGERTY and LUMLEY · Biostatistics: A Methodology for the Health Sciences, Second Edition

VESTRUP · The Theory of Measures and Integration

VIDAKOVIC · Statistical Modeling by Wavelets

VINOD and REAGLE · Preparing for the Worst: Incorporating Downside Risk in Stock Market Investments

WALLER and GOTWAY · Applied Spatial Statistics for Public Health Data

WEERAHANDI · Generalized Inference in Repeated Measures: Exact Methods in MANOVA and Mixed Models

WEISBERG · Applied Linear Regression, Second Edition

WELSH · Aspects of Statistical Inference

WESTFALL and YOUNG · Resampling-Based Multiple Testing: Examples and Methods for p-Value Adjustment

WHITTAKER · Graphical Models in Applied Multivariate Statistics

WINKER · Optimization Heuristics in Economics: Applications of Threshold Accepting

WONNACOTT and WONNACOTT · Econometrics, Second Edition

WOODING · Planning Pharmaceutical Clinical Trials: Basic Statistical Principles

WOODWORTH · Biostatistics: A Bayesian Introduction

WOOLSON and CLARKE · Statistical Methods for the Analysis of Biomedical Data, Second Edition

WU and HAMADA · Experiments: Planning, Analysis and Parameter Design Optimization

WU and ZHANG · Nonparametric Regression Methods for Longitudinal Data Analysis: Mixed-Effects Modeling Approaches

YANG · The Construction Theory of Denumerable Markov Processes

YOUNG, VALERO-MORA and FRIENDLY · Visual Statistics: Seeing Data with Dynamic Interactive Graphics

ZACKS · Stage-Wise Adaptive Designs

* ZELLNER · An Introduction to Bayesian Inference in Econometrics

ZELTERMAN · Discrete Distributions: Applications in the Health Sciences

ZHOU, OBUCHOWSKI and McCLISH · Statistical Methods in Diagnostic Medicine

Printed and bound by CPI Group (UK) Ltd, Croydon, CR0 4YY

27/10/2024

14580216-0004